Springer Series in Plasma Science and Technology

Series editors

Michael Bonitz, Kiel, Germany
Liu Chen, Hangzhou, China
Rudolf Neu, Garching, Germany
Tomohiro Nozaki, Tokyo, Japan
Jozef Ongena, Brussel, Belgium
Hideaki Takabe, Dresden, Germany

Plasma Science and Technology covers all fundamental and applied aspects of what is referred to as the "fourth state of matter." Bringing together contributions from physics, the space sciences, engineering and the applied sciences, the topics covered range from the fundamental properties of plasma to its broad spectrum of applications in industry, energy technologies and healthcare.

Contributions to the book series on all aspects of plasma research and technology development are welcome. Particular emphasis in applications will be on high-temperature plasma phenomena, which are relevant to energy generation, and on low-temperature plasmas, which are used as a tool for industrial applications. This cross-disciplinary approach offers graduate-level readers as well as researchers and professionals in academia and industry vital new ideas and techniques for plasma applications.

More information about this series at http://www.springer.com/series/15614

Hideaki Takabe

The Physics of Laser Plasmas and Applications - Volume 1

Physics of Laser Matter Interaction

Hideaki Takabe
Institute of Radiation Physics
Helmholtz Zentrum Dresden Rosendorf
Dresden, Germany

ISSN 2511-2007   ISSN 2511-2015 (electronic)
Springer Series in Plasma Science and Technology
ISBN 978-3-030-49615-9   ISBN 978-3-030-49613-5 (eBook)
https://doi.org/10.1007/978-3-030-49613-5

© Springer Nature Switzerland AG 2020
This work is subject to copyright. All rights are reserved by the Publisher, whether the whole or part of the material is concerned, specifically the rights of translation, reprinting, reuse of illustrations, recitation, broadcasting, reproduction on microfilms or in any other physical way, and transmission or information storage and retrieval, electronic adaptation, computer software, or by similar or dissimilar methodology now known or hereafter developed.
The use of general descriptive names, registered names, trademarks, service marks, etc. in this publication does not imply, even in the absence of a specific statement, that such names are exempt from the relevant protective laws and regulations and therefore free for general use.
The publisher, the authors, and the editors are safe to assume that the advice and information in this book are believed to be true and accurate at the date of publication. Neither the publisher nor the authors or the editors give a warranty, expressed or implied, with respect to the material contained herein or for any errors or omissions that may have been made. The publisher remains neutral with regard to jurisdictional claims in published maps and institutional affiliations.

This Springer imprint is published by the registered company Springer Nature Switzerland AG
The registered company address is: Gewerbestrasse 11, 6330 Cham, Switzerland

Snap shot of fusion target irradiated by Gekko XII laser at Osaka Univ.

Preface

This year, 2020, is the 60th anniversary of the birth of the laser. When the laser was invented by Maiman in 1960, the news spread as a shocking technology of the future. Soon after, the laser was used in a scene from the film "Goldfinger" in the "007 series" in 1964, and this film made people worldwide realize how powerful the laser is. Soon after, Maiman's small laser had scaled up the purpose of fusion energy after the publication of laser fusion concept in 1972. I was very much interesetd in reseraching fusion energy when I was a student at Osaka University and when the oil crisis of 1973 had crippled the global economy.

Along with research of the laser fusion as graduate student and young faculty member, I faced numerous challenges in integrated physical phenomena to research on laser fusion and analyze experimental data from many angles with large-scale lasers. Such intergrated physics is the coupled phenomenon of compressible fluid, radiation, dense matter state, atomic physics, kinetic physics, and so on in mm scale fusion target materials. These physical phenomena are also key to research in astrophysics, and there are several connections in the non-dimensional scale of time and space.

Toward the end of the 1980s, chirped pulse amplification (CPA) technique made it possible to increase the laser intensity more than 10^4 times in the intense lasers used for laser fusion and related plasma physics. The electron motion in such an ultra-intense laser is highly relativistic and the Lorentz factor becomes roughly $\gamma = 10^4$. These new lasers have opened the window for relativistic plasma physics and quantum electrodynamics in a small-scale laboratory.

In the last 60 years since Maiman's invention, the physics of laser plasma has grown into a mature science and a challenging subject with the help of precise diagnostics with ultra-short and small-scale resolutions. Honestly speaking, however, greater effort is required to understand the physics in real experiments and to control the laser plasma for many applications from fundamental to industrial ones.

The author has worked for laser plasma and related physics for 40 years as a computational theorist. I was fortunate because my institute in Osaka is a place where the majority of research is experimental in nature, and big laser Gekko-XII for fusion research and LFEX laser for ultra-intense physics are installed there. I have benefited from discussions with experimentalists from all over the world.

I have tried to write these three books so that the description is always with intuitive understanding as I explain to experimentalists. I tried not to use complicated equations in order to keep the books easy to read for non-specialists. Many color figures are used to help non-experts obtain concrete images about the physics. Since the idea of laboratory astrophysics was conceived from the similarity of the physics of turbulence in laser implosion and supernova explosion, some ideas and examples of experimental laboratory astrophysics are also described.

While writing this book, I realized that one book is not enough to transfer my knowledge to the readers clearly; therefore, I decided to write three volumes.

Volume 1 discusses the physics of how laser energy is absorbed by plasma and how electrons obtain laser energy in non-thermodynamic equilibrium process. Importance of non-linear and chaos physics are described. Assuming that readers have preliminary knowledge about this field, I explain the following subjects in Volume 1: Electromagnetism, Mechanics, Analytical mechanics, Quantum mechanics, and Relativity.

Volume 2 is devoted to the physics of plasma energized by laser absorption and related physics. Abrupt heating by laser generates extremely high pressure, driving strong shock waves to compress matter to a high density like the core of planets and stars. The physics of compressible hydrodynamics becomes important, and the subject of turbulence driven by the hydrodynamic motion is a challenging subject. Volume 2 discusses the following topics of physics to the readers: Fluid dynamics, Thermodynamics, Statistical physics, Quantum statistical physics, and Atomic physics.

In Volume 2, it is assumed that the plasma is collisional and is in thermodynamic equilibrium locally. On the other hand, the physics of plasma without collision is discussed in Volume 3. In relatively low-density region, high-energy electrons generated by laser are freely running without collision, and they do not collide with each other, the so-called collisionless plasma. The electrons interact with electric and magnetic field fluctuations to generate a variety of plasma instabilities. The field and charged particle interaction provides particle acceleration, anomalous transport, anti-matter generation, and so on. Volume 3 provides insights into the following areas of physics: Electromagnetism, Physical kinetics, Statistical physics, Relativity, and Quantum electrodynamics.

As you already know, the laser-plasma physics is not a single subject discipline and the expression of "integrated physics" would be nice expression. As a metaphor, the laser plasma is a kind of "decathlon", not 100m dash. Note that it requires a constant study and effort.

Since I am a theoretical plasma physicist, most topics discuss about theoretical physics of laser plasmas. Although I have worked with many experimentalists, I have not written about the methods of diagnostics for laser plasma. Several experimental data are shown in this book, but the accuracy of diagnostics is not discussed. Note that I selectively cited a limited number of papers so that those papers are also very useful for senior researchers to know more about each topic.

Preface

I would like to express my special acknowledgment to Prof. C. Yamanaka who guided me in the research of laser plasma and its applications, especially in laser fusion, and Prof. K. Mima who navigated me to the research of theoretical plasma physics. I thank sincerely Dr. R. Sauerbrey and Dr. T. Cowan for providing me the best opportunities and constant encouragement to write the current books. Topics on relativistic lasers in Vol. 1 and on high-density plasma in Vol. 2 are reflection of the subjects of HiBEF (*Helmholtz* International Beamline for *Extreme* Fields) project in Dresden where I work. I would like to thank Prof. P. Mulser and Prof. R. L. Morse for their supervision of my research in Munich and Tucson, respectively.

I also thank the following individuals for their fruitful discussions: R. Kodama, Y. Sakawa, Y. Kuramitsu, S. Yamada, T. Morita, T. N. Kato, R. M. More, D. Saltzmann, F-L Wang, S. Blinnikov, A. Titov, S. Atzeni, B. Remington, H-S. Park, M. E. Campbell, D. Kraus, K. Falk, J. Zhang, Z-M. Sheng, L-G Huang, M. Murakami, H. Nagatomo, T. Yabe, N. Ohnishi, A. Mizuta, J-F Ong, T. Kato, J. Myer-ter-Vehn, S. Witkovski, L. Montieth, and P. McKenty.

I would like to express my thanks to the people involved in HiBEF project at HZDR for their valuable comments on this topic. I also thank H. Niko who suggested and encouraged me to write these books.

Relating to the proposal of laboratory astrophysics, I would like to express sincere thanks to K. Nomoto who guided me to the world of supernova explosion physics after the explosion of SN1987a. I also thank my friends in astrophysics, I. Hachisu, K. Shibata, R. Yamazaki, and M. Hoshino, who introduced me to the concept of laboratory astrophysics and shared their ideas to find topics on astrophysics to be possibly modeled by laser plasma.

Finally, I would like to thank my family, Yoko, Yugo, Ayana, and Ayako, for their unwavering support to my research for a long time and for understanding my decision to go to Germany by taking early retirement from Osaka University.

Dresden, Germany
March 10, 2020

Hideaki Takabe

About the Book

After 60 years of the invention of laser, it has grown as a new light source indispensable for our life. The application to industrial and research has spread over wide fields. The book discusses how the power lasers are absorbed by matter and how high temperature and high density plasmas are produced. It is noted that such plasmas are mimics of the plasmas in the Universe, namely, plasma inside of stars, exploding plasma like supernovae, accelerating plasma like cosmic ray. The book (Vol. 1) gives the physics of the intense laser and material interaction to produce such plasmas. The physics is explained intuitively so that non-specialists can obtain clear images about how the laser energy is converted to the non-thermal plasma energy; simply saying, how laser heats matter. For intense laser of 10^{13-16} W/cm^2, the laser energy is mainly absorbed via collisional process, where the oscillation energy is converted to thermal energy by non-adiabatic Coulomb collision with the ions. Collisionless interactions with the collective modes in plasma are also described. The main topics are the interaction of ultra-intense laser and plasma interaction for the intensity near and over 10^{18} W/cm^2. In such regime, relativistic dynamics become essential. A new physics appears due to the relativistic effects, such as mass correction, relativistic nonlinear force, chaos physics of particle motions, and so on. The book provides clearly the theoretical base for challenging the laser-plasma interaction physics in the wide range of power lasers.

Contents

1 **Introduction** 1
   1.1   Brief History of Intense Lasers 3
        1.1.1   Invention of Laser 3
        1.1.2   High-Power Lasers for Nuclear Fusion 5
        1.1.3   Ultra-Intense and Ultra-Short Lasers 8
   1.2   What Is Plasma? 9
        1.2.1   Ionization 9
        1.2.2   High-Density Plasmas 10
        1.2.3   Magnetic Fusion 12
   1.3   Basic Equations 13
        1.3.1   Maxwell Equations 13
        1.3.2   Electron Motion in Laser Field 14
        1.3.3   Normalized Laser Strength 15
   1.4   Non-relativistic Laser-Plasma Interaction 16
        1.4.1   Collisional Absorption 16
        1.4.2   Collisionless Absorption and Ponderomotive Force 18
   1.5   Relativistic Laser-Plasma Interaction 20
        1.5.1   Relativistic Electron Oscillation 20
        1.5.2   Relativistic Laser and Solid Interaction 21
        1.5.3   Theory of Chaotic and Stochastic Heating 21
   1.6   PIC Simulations 23
   References 25

Part I   Non-relativistic Lasers

2 **Laser Absorption by Coulomb Collision** 29
   2.1   Plasma Generation by Lasers 29
        2.1.1   Field-Induced Electron Emission 29
        2.1.2   Ionization by Multiphoton Absorption 32
        2.1.3   Tunneling and Over-Threshold Ionizations 34
        2.1.4   Experimental Data 37

	2.2	Laser as Electromagnetic Waves	39
		2.2.1 Maxwell Equations	40
		2.2.2 Electromagnetic Waves in Vacuum	43
		2.2.3 Lasers as Coherent Electromagnetic Waves	44
	2.3	Electron Current Induced by Laser Fields	45
		2.3.1 Antenna and Thomson Scattering	48
		2.3.2 Electron Current in Matters	50
	2.4	Electron Coulomb Collision by Ions in Plasma	53
		2.4.1 Debye Shielding	56
		2.4.2 Yukawa Potential	57
		2.4.3 Coulomb Logarithm (Log)	58
		2.4.4 Collision Frequency and Electrical Resistivity	61
		2.4.5 Relaxation Time to Thermal Equilibrium	62
	2.5	Lasers in Plasmas	65
		2.5.1 Classical Absorption of Laser Energy	69
	2.6	Laser Absorption in Plasma	72
		2.6.1 Physical Image of Classical Absorption	72
		2.6.2 Simple Diffusion Model	73
		2.6.3 Kinetic Derivation by Dawson and Oberman	75
		2.6.4 Quasi-Linear Model of Absorption	78
	2.7	Bremsstrahlung and Collisional Absorption	79
	References		80
3	**Ultra-Short Pulse and Collisionless Absorption**		81
	3.1	Ultra-Short Pulse in Non-relativistic Intensity	81
	3.2	Self-Consistent Analysis of Short Pulse Absorption	88
	3.3	Quantum Theory of Electric Conductivity in Dense Plasmas	95
	3.4	Nonlinear Inverse Bremsstrahlung (IB) Absorption ($v_e < v_{os}$)	99
	3.5	Electron Plasma Waves and Collisionless Absorption	104
	3.6	Resonance Absorption	107
		3.6.1 Collisional Absorption Model	110
		3.6.2 Singular Point Integral Model	111
	3.7	Resonance in Pendulum	112
	3.8	Linear Mode Conversion in Resonance Absorption	115
		3.8.1 Absorption Rate and Pump Depletion	117
	3.9	Large Amplitude Electron Plasma Waves	118
		3.9.1 Wave-Breaking	119
	3.10	Vacuum Heating	123
		3.10.1 Physical Image and Absorption Rate	126
		3.10.2 Skin Depth at Sharp Boundary	129
	References		130
4	**Nonlinear Physics of Laser-Plasma Interaction**		131
	4.1	Ponderomotive (PM) Force	131
	4.2	Nonlinear Schrodinger Equation	133

4.3	Filament Instability of Lasers		136
4.4	Ion Fluid and Ion Acoustic Waves		139
	4.4.1	Charge Neutral Plasma Fluids	140
	4.4.2	Ion Sound Waves	141
4.5	Density Profile Modification		144
4.6	Principle of Parametric Instabilities		146
	4.6.1	Coupled Oscillator Model-1	149
	4.6.2	Thermal Noise of Electrostatic Fluctuations	150
	4.6.3	Coupled Oscillator Model-2	151
4.7	Stimulated Raman and Brillouin Scattering		152
4.8	Decay-Type Parametric Instabilities		154
4.9	Experimental Data for SRS		156
4.10	Physics of Saturation of SRS Instability		157
	4.10.1	Effect of Inhomogeneity	159
	4.10.2	Nonlinear Saturation of SRS	160
4.11	Broadband Effect		161
References			163

Part II Relativistic Lasers

5 Relativistic Laser-Electron Interactions 167

5.1	Introduction		167
5.2	Special Relativity for Electron Motion		167
	5.2.1	Equation of Relativistic Electron Motion	169
	5.2.2	Lorentz Transformation of Time and Space	172
	5.2.3	Lorentz Transformation of Velocities	174
	5.2.4	Lorentz Transformation of Fields	175
	5.2.5	Plane Electromagnetic Waves in Vacuum	177
5.3	Electron Motion in a Relativistic Strong Field		179
	5.3.1	Constant of Motion in Vacuum	180
	5.3.2	Normalizations	181
	5.3.3	Electron Motion in Vacuum	182
	5.3.4	Free Electron Orbit	183
	5.3.5	Electrons in Plasmas	186
	5.3.6	Linear Polarization	187
	5.3.7	Circular Polarization	188
5.4	Nonlinear Radiation Scattering		188
	5.4.1	Linear Thomson Scatterings	189
	5.4.2	Compton and Inverse Compton Scatterings	191
	5.4.3	Nonlinear Thomson Scattering	194
	5.4.4	Relativistic Beaming Effect and Doppler Shift	197
	5.4.5	Nonlinear Compton Scattering	199
References			201

6 Relativistic Laser Plasma Interactions 203
- 6.1 Charge Separation in Low-Density Plasma 203
  - 6.1.1 Charge Separation by Photon Force 203
  - 6.1.2 Wake Field Generation and Energy Deposition 205
- 6.2 Laser Propagation in Plasmas 207
  - 6.2.1 Relativistic Transparency 207
  - 6.2.2 Higher Harmonic Generation (HHG) 209
  - 6.2.3 Electrostatic Field Excitation by vxB Force 212
  - 6.2.4 Density Bunching Current 213
  - 6.2.5 Simulation for $a_0 = 10$ in Low Density 214
- 6.3 Ponderomotive Force in Relativistic Field 217
  - 6.3.1 PM Force by Electrostatic Wave 219
  - 6.3.2 Validity of Electrons in Plasmas Assumption 219
- 6.4 Relativistic Raman Scattering 220
- 6.5 Relativistic Self-Focusing 221
  - 6.5.1 Self-Focusing Condition 223
  - 6.5.2 Strong Laser Limit 224
  - 6.5.3 Difference of 2D and 3D Focusing 225
  - 6.5.4 Frequency Shifts 226
  - 6.5.5 Filamentation Instability 226
- 6.6 Relativistic Skin Depth 229
- 6.7 JxB Force and Heating 231
- 6.8 Moving Mirror Model and Higher Harmonic Generation from Solid Surface 234
- References 238

7 Relativistic Laser and Solid Target Interactions 239
- 7.1 Pre-formed Plasma in Laser-Solid Interaction 239
  - 7.1.1 Pedestal of Laser Pulse 239
  - 7.1.2 Model Experiments with Controlled Pre-formed Plasmas 241
- 7.2 Laser Absorption at Solid Targets 244
- 7.3 Absorption Enhancement by Hot Electron Re-circulation 248
- 7.4 Hole Boring by Ponderomotive Force 252
  - 7.4.1 Density Dependence of Hole Boring 254
- 7.5 Laser Interaction in Long Pre-formed Plasmas 255
- 7.6 Absorption Efficiency Based on Conservation Laws 260
- 7.7 Enhanced Coupling with Foam Layered Targets 263
  - 7.7.1 Experimental Result 271
- 7.8 Efficient Absorption in Structured Targets 273
  - 7.8.1 Micro-pillar Array Targets 273
  - 7.8.2 Nano-pillar Array Target 275
  - 7.8.3 Micro-tube Plasma Lenses 277
  - 7.8.4 Common Rule of Better Coupling 279

	7.9	Magnetic Field Generation	279
	7.10	Multi-dimensional Physics in Pre-formed Plasmas	282
	References		285

8 Chaos due to Relativistic Effect ... 287
- 8.1 Basic Relation of an Electron in Relativistic Laser Field ... 287
- 8.2 Laser Direct Acceleration ... 292
  - 8.2.1 Acceleration (1) ... 293
  - 8.2.2 Acceleration (2) ... 294
  - 8.2.3 PIC Simulations ... 294
- 8.3 Direct Acceleration after Interaction with Longitudinal Field ... 297
  - 8.3.1 PIC Simulation ... 301
- 8.4 Chaotic Motion due to External Force ... 305
  - 8.4.1 Simple Example [1] (Random Force) ... 305
  - 8.4.2 Simple Example [2] (Periodic Force) ... 307
  - 8.4.3 Chaos in Propagating Relativistic Wave ... 309
- 8.5 Electron Heating by Laser Field and Induced Plasma Waves ... 313
  - 8.5.1 Modeling PIC Simulation ... 314
  - 8.5.2 Quasi-linear Diffusion ... 315
- 8.6 Hot Electron Generation ... 316
  - 8.6.1 Hot Electrons by Femtosecond Lasers ... 317
  - 8.6.2 Hot Electrons by Picosecond Lasers ... 318
  - 8.6.3 Sheath Potential Effect ... 319
  - 8.6.4 Multi-dimensional Effects ... 320
- 8.7 Electron Motion in Two Counter-Propagating Relativistic Lasers ... 322
  - 8.7.1 Counter-Propagating Two Laser Systems ... 324
  - 8.7.2 One-Electron Orbit ... 325
  - 8.7.3 Lyapunov Exponent ... 327
  - 8.7.4 Hot Electron Temperature Scaling ... 329
- References ... 330

9 Theory of Stochasticity and Chaos of Electrons in Relativistic Lasers ... 331
- 9.1 Vlasov-Fokker-Planck Equations ... 331
  - 9.1.1 Stochastic Diffusion Equation ... 332
  - 9.1.2 Vlasov-Fokker-Planck Equation ... 334
- 9.2 Stochastic Heating by Laser Filamentation ... 334
  - 9.2.1 Numerical Calculation of Test Particles ... 337
  - 9.2.2 PIC Simulation of Stochastic Diffusion ... 339
- 9.3 Nonlocal Jump in Energy Space ... 340
  - 9.3.1 Levy's Flights ... 340
  - 9.3.2 Integral Form of Random Walk ... 342
  - 9.3.3 Fokker-Planck Diffusion Model ... 342
  - 9.3.4 Fractional Fokker-Planck Model ... 343

	9.4	Time Evolution of Distribution and Fractal Index α	345
		9.4.1 Application to Hot Electron Scaling	347
		9.4.2 Local and Nonlocal: Gaussian and Lorentzian	349
		9.4.3 Evaluation of Fractional Index α from Experimental Data	351
	9.5	Model Experiment of Cosmic Ray Physics in the Universe	354
		9.5.1 Relativistic EM Wave Generation by Relativistic Shocks	355
		9.5.2 Model Experiments with Relativistic Lasers	357
	9.6	Fractal Index $\alpha = 0.8$ in Big Tokamak Transports	358
	9.7	Chaos in Standard Map Model	360
	9.8	Analytical Mechanics of Electron Motions	362
		9.8.1 One Laser Relation	362
		9.8.2 Adiabatic Constant	363
		9.8.3 Numerical Solution for Counter Beam Interaction	365
		9.8.4 Perturbation Method in Hamilton Equation	367
		9.8.5 Adiabatic Approximation	369
		9.8.6 Further Discussion	370
References			371

Appendices 373
   Appendix-1: Rutherford Scatterings 373
   Appendix-2: Adiabatic and Nonadiabatic 375
      Periodic Motion 375
      Derivation of Adiabatic Constants 377
      Nonadiabatic Case 378
   Appendix-3: PIC Simulation 379
      Particles in Computer 379
      Maxwell Equations in Computer 380
      Surprising Progress of Computing 381
   References 382
   Appendix 4: Tsallis Statistics at Maximum Entropy 383
   Reference 384

Index 385

About the Author

Hideaki Takabe is a Professor at the Institute of Radiation Physics, Helmholtz Zentrum Dresden Rossendorf (HZDR), Dresden, Germany and Professor-Emeritus at Osaka University, Japan. He received a Ph.D. degree from Osaka University and continued research at Max-Planck Institute for Plasma Physics, Germany, and the University of Arizona, USA. He continued his research and education activity at the Institute of Laser Engineering, Osaka University until early retirement in 2015. His research fields are plasma physics, laser fusion, laboratory astrophysics, and HED physics. He is known for Takabe formula in laser fusion and also as a pioneer of laboratory astrophysics concept. He has taught Plasma Physics and Computational Physics at the School of Engineering and School of Science for more than 20 years. He committed the globalization of Osaka University. He was also a council member of the Physical Societies of Japan (JPS) and the Association of Asia-Pacific Physical Society (AAPPS), where he founded the Division of Plasma Physics (DPP). He received the Edward Teller Medal in 2003 and John Dawson Award on Excellence of Plasma Physics Research in 2020. He is a fellow of the American Physical Society (2000~).

The original version of this book was revised. The book was inadvertently published without index. Index is now updated in the book.

Chapter 1
Introduction

The laser invented in 1960 is now widely used for our life, in science researches, etc. Its technology is still progressing, and its applications have spread over very wide fields from industrial applications to fundamental research. The topics of this book are focused on the physics of the plasma interacting with intense and ultra-intense lasers. It is studied mainly on how laser energy is converted to the energy of matter in the state of plasma. It is the topic of plasma heating by laser irradiation and/or laser absorption due to the interaction with matters. The laser intensity discussed in the book is extremely intense in the range of 10^{13} W/cm^2 – 10^{22} W/cm^2. The pulse durations of such intense lasers are from nanoseconds to sub-picoseconds. The total energy of such laser pulses is about $1 \sim 10^6$ joule. Such lasers are roughly called high and ultra-high power lasers; the brief history of technology progress of which is given, for example, in [1].

It is intuitively clear how it is intense by comparing the laser power to the energy flux by the sun which is of the order of W/cm^2. When such high-power lasers irradiate on any matters, they are soon evaporated and ionized to become plasma before the main part of the laser photons irradiates the matters. Most of the laser photons interact with the laser-produced plasmas. This is the reason for the name of physics as the laser-plasma interaction.

There are typically two types of such high-power lasers. One is high-power laser delivering sub-kilo joule to a few Mega Joule of photon energy with the pulse duration of nanosecond (ns: 10^{-9} s) rage. Such a laser is focused on solid materials with its focused intensity of 10^{13-16} W/cm^2. The focusing diameter is of ~100 μm to several mm, roughly speaking. The intensity and pulse width are designed so that most of the laser energy is converted to the thermal energy of plasma. The physics of laser energy absorption is discussed in this text.

Since the plasma expands with the ion sound velocity, extremely high pressure is generated in the plasma near the surface by the confinement of absorbed energy due

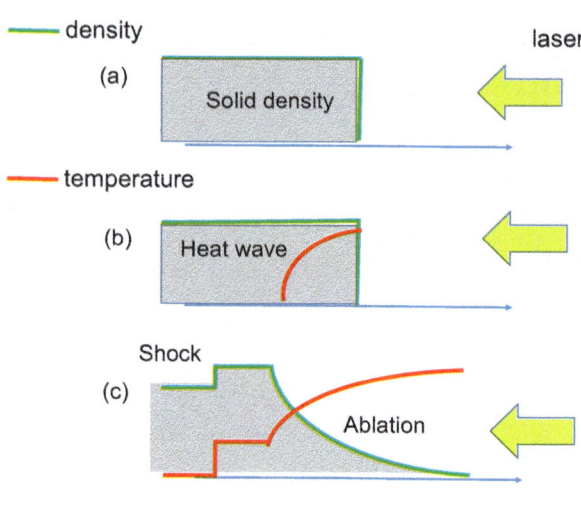

Fig. 1.1 Schematics of the density and temperature evolution when an intense laser is irradiated on the solid. (**a**) Laser is irradiated on the surface of solid target. (**b**) The electrons near the solid surface are heated abruptly, and the heat is transferred by the electrons inside the solid before the hydrodynamic expansion is induced by the ion motion. (**c**) Due to extremely high pressure near the surface, strong shock waves propagate into the solid target to compress and heat the solid. At the same time, the materials being plasma near the surface ablate and expand like rocket exhaust finally to accelerate the matter to the opposite direction via the shock waves and subsequent compression

to relatively slow expansion velocity. The generated pressure is around 100 Mbar (10^8 atmospheric pressure), and the plasma with temperature of about keV (10^7 K) expands to the vacuum, and a strong shock wave is produced to the inside of the target solid. This is schematically shown in Fig. 1.1. Producing such shock wave and high-density plasmas is one of the most important applications of intense laser for fundamental science study.

Such plasmas produced by the shock waves are characterized by density higher than solids and high temperature. Such state of matter is called **high-energy density (HED) plasma**, and the physics of HED state is called HED physics (**HEDP**) [2]. The intense laser is almost a unique tool to study HEDP in laboratory. Recently, the physics of the matter relatively low temperature (T < 100 eV) at high-density state is widely studied relating to bridge the physics of condensed matter to the HED plasma. It is called **warm dense matter (WDM)** [3], and this study with intense lasers is very important to study the planetary physics such as inside of Earth, Jupiter, Saturn, and many extra-galaxy planets recently founded. Such plasma is also called laser-produced plasma or laser-driven matter, and the physics of such plasma is treated in Volume 2 in this series.

Around the mid-1980s, the so-called **chirped pulse amplification (CPA)** technique was developed to compress the intense laser pulse by 10^{3-4} [4]. The invention and success of the CPA technique have made it possible to study the physics of laser-plasma interaction at ultra-high intensity up to 10^{22} W/cm², and pulse duration is 10 fs (fs: 10^{-15} s) – 1 ps (ps: 10^{-12} s), where laser pulse is 10 cycle for 30 fs pulse for 1 μm laser. This accomplishment made it possible to study the

Fig. 1.2 Gérard Mourou and Donna Strickland. The inventors of CPA technique for ultra-intense ultra-short lasers. They are awarded Nobel Prize for Physics in 2018. The prize citation is *for their method of generating high-intensity, ultra-short optical pulses*. (From Nobel Prize Foundation)

physics related to relativistic plasma, nonlinear physics of wide range of photon energy, and so on. Since the construction of the CPA laser is not so expensive and laser is also compact compared to the high-power laser, many groups in the world now study the physics at extreme condition with the use of such lasers. The 2018 Nobel Prize in Physics was awarded jointly to Gérard Mourou and Donna Strickland (Fig. 1.2). The citation is *for their method of generating high-intensity, ultra-short optical pulses*. It is noted that this is a PhD thesis research by Strickland under supervisor Mourou. In the present book, the relativistic physics of the electrons being irradiated by lasers is shown after the review of recent experimental results.

1.1 Brief History of Intense Lasers

1.1.1 Invention of Laser

The history of laser began when Einstein invented the physics of photoelectric effect in 1905. This theory proposed that light also delivers its energy in a form of discrete quantum particles now called photons. Einstein proposed the physical process of stimulated emission that makes lasers possible. Einstein showed theoretically the possibility that besides light being absorbed and emitted spontaneously, an electron could be stimulated to emit light of a particular energy. After his theoretical prediction, it took almost 40 years to demonstrate laser emission technically.

Schawlow and Towns proposed the **laser** in 1958 as continuous extension in frequency from the **maser**. The maser is an abbreviation of microwave amplification by stimulated emission of radiation (MASER), while laser is an abbreviation of replacement of microwave to light (LASER). Towns was awarded Nobel Prize for physics in 1964 for the invention of the concept of maser and laser with two Russian

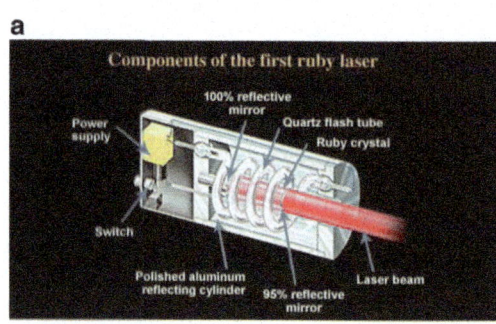
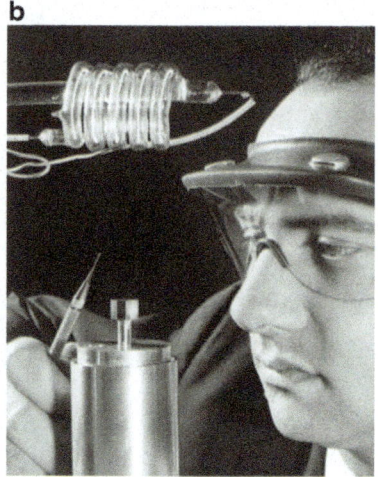

Fig. 1.3 (**a**) Maiman and his ruby laser. (**b**) The structure of the inside of cylindrical case

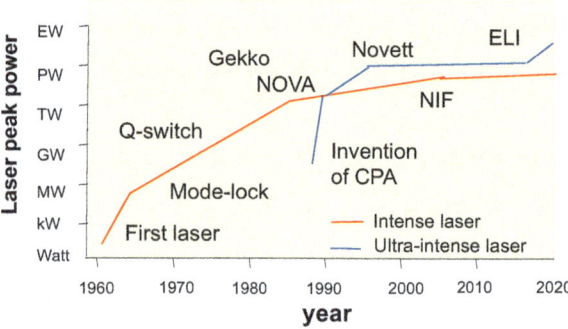

Fig. 1.4 Historical evolution of output power of high-power laser for laser fusion research. The NOVA laser is converted to ultra-intense laser after the usage of NOVA for laser fusion research. NIF is the world's biggest laser constructed for laser fusion ignition

physicists, Vasov and Prokorov. Towns wrote his autobiography from the invention of the laser to finding of maser from the universe. The observed maser by radio telescope is the emission from the cloud of water molecules far in space, and the molecules are continuously excited by the photons from an ambient star [5]. The early work that finally led to the invention of maser and laser was the research to develop the radiation source for the radar during WWII. Although they thought that laser would be possible from excited gas, Maiman has realized the first laser using ruby rod pumped by flash lump as shown in Fig. 1.3 [6].

The peak power of lasers in the unit of watt (W) has increased as shown in Fig. 1.4. Maiman's first laser was only 100 W; however, it increased exponentially by subsequent technology innovations, namely, Q-switching and mode locking. Metal-doped glass laser has been developed as better amplifier for high-power laser in 1961, and a single pulse extraction was demonstrated in 1968 with 50 GW at 1 ns pulse (50 J). Since the breakdown of the glass amplifiers limits the laser intensity (W/cm^2) in the amplification process, the further increase of the power has

1.1 Brief History of Intense Lasers

been achieved by the use of large-sized glass amplifiers. As a result, the large energy intense laser has been developed as a large-scale facility.

1.1.2 High-Power Lasers for Nuclear Fusion

The construction of large-scale laser facilities was greatly accelerated by a paper published in 1972 by Nuckolls et al. [7] on **the inertial confinement fusion (ICF) driven by laser implosion**. The fusion scenario is schematically shown in Fig. 1.5. A spherical plastic capsule with frozen deuterium-tritium (DT) layer is irradiated by intense laser beams uniformly on the surface to generate about 100 Mbar of pressure. Extremely high pressure exhausts the capsule material to drive strong shock wave toward the center. Such hydrodynamics is called implosion. The imploding kinetic energy is converted to thermal energy at the center to ignite DT fusion reaction. The alpha particles produced by the fusion heat the DT fuel to generate large amount of fusion energy.

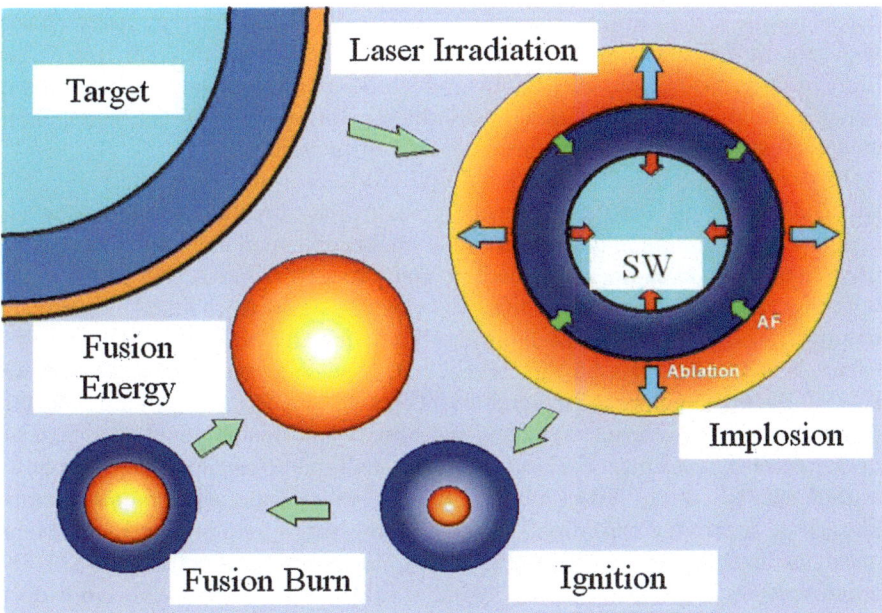

Fig. 1.5 Laser fusion scenario in spherically symmetric assumption. The DT fuel capsuled by plastic shell with its diameter of about 1 cm is irradiated by many laser beams to drive ablation. The shock waves (SW) accelerate the fuel to the center of the target by implosion. The kinetic energy of implosion is converted to heat at the center to initiate fusion ignition; consequently, the fusion burn starts. About one-third of the DT is expected to nuclear burn before the expansion of the fuel, and about 100 times larger energy is produced compared to the incident laser energy

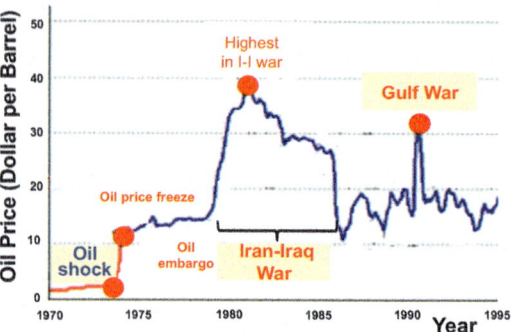

Fig. 1.6 The year change of the price of crude oil per barrel from before the oil shock in 1972. The surprising increase of the price over 10 years after the oil shock has pushed the progress of fusion energy research especially in the 1980s

The paper is based on theoretical and computational study on the nuclear fusion energy production by spherical capsule implosion driven by the surface pressure induced by laser heating. It concluded that 10 kJ of laser energy is enough to achieve ICF by laser implosion. Then, Janus, Siva, and Nova lasers were constructing at Lawrence Livermore National Laboratory (**LLNL**) as shown in Fig. 1.4.

It should be noted that the great events of world politics have boosted fusion energy research. It is the **oil shock** in 1973 and the subsequent second oil shock 1980 as shown in Fig. 1.6 [8]. The price of oil increased over ten times after the first shock, and all people in the world expected some new energy. **Fusion energy research** was then supported financially by each government, and much progress of research has been accomplished. Soon after the first oil shock, intensive review on laser-driven fusion was published by Brueckner and Jorna [9]. This review was also an important element for spreading the laser fusion research worldwide.

In Japan, for example, the Gekko XII (GXII) laser facility with 30 kJ (Fig. 1.7) was constructed in 1983, and the author was heavily involved in the theory and computation of design and analysis of the implosion experiment with GXII. After many implosion experiments, it became clear that 10 kJ prediction in [7] is too optimistic. Hydrodynamic instabilities and energy transports are found to inhibit the optimistic scenario. It became clear that more fundamental research to study a variety of related plasma physics is required. Review on experimental result and analysis over the first decade research with GXII are given in Sec. 3 of Ref [10]. It is demonstrated that the turbulent mixing in the final implosion phase is critical physics to degrade the performance of implosion. Around the same time, it was found in astrophysics that the turbulent mixing and neutrino transport are also critical in the physics of supernova explosion. The X-ray observation required such physics in understanding the explosion of SN1987A identified in February 1987. The interdisciplinary collaboration has come soon to the proposal of **laboratory astrophysics** as fundamental research with large laser facilities as described in Sec. 4 of Ref [10]. Laser fusion, HED physics, and laboratory astrophysics are mainly discussed in Vols. 2 and 3 in this series.

In order to demonstrate nuclear fusion ignition and fusion burn in the imploded deuterium-tritium core plasma, LLNL has constructed National Ignition Facility (NIF, Fig. 1.8) and started fusion experiment from 2009. The NIF is the biggest

1.1 Brief History of Intense Lasers

Fig. 1.7 Gekko XII laser system at Osaka University. Picture shows the amplifier room, and four beam amplifications chain out of 12 beams. The final sized of each laser beam is 30 cm in diameter. Implosion experiment with Gekko XII started in 1983

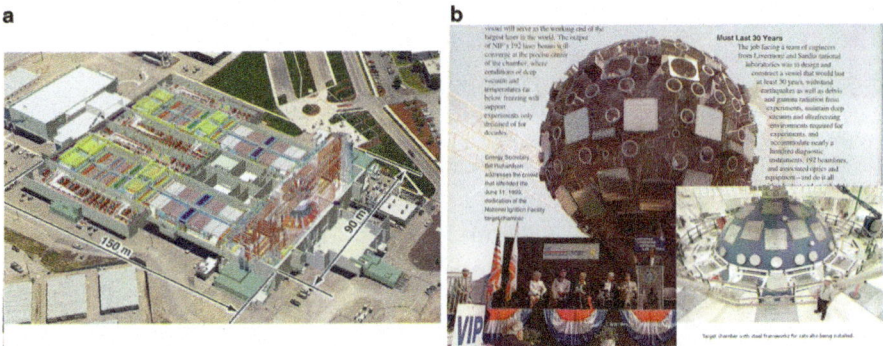

Fig. 1.8 (a) The bird view of the National Ignition Facility. The left side is laser amplifier chains. The 192 beams of laser are bundled to two directions, up and down, to irradiate to a target at the center of vacuum chamber on the right with blue color. (b) Snapshot of ceremony for installing the target chamber. The right bottom is the chamber after install. The size of building and chamber is imaged with the diameter of chamber 10 m

laser facility with 192 laser beams with each beam size 40cm × 40cm. The big size of laser is to deliver a large amount of laser energy 1.8 megajoule. However, nuclear fusion ignition and burn have not demonstrated yet with NIF fusion implosion experiments. It is proofed through the NIF experiment that the physics of implosion is not simple, and plasma is compressed with turbulent motion; the physics of which is still under investigation with the supercomputing. Since the electricity production design of the laser fusion reactor limits the maximum energy of laser pulse less than 5 MJ as explained in Vol. 2, it is required to demonstrate a medium gain with enough ignition with the scale of the NIF class energy. If the laser energy required for fusion reactor is more than 2 ~ 3 times the NIF, it may be difficult to control the fusion output for continuous electricity production with about 10 Hz operation, 10 laser shots per second. The laser fusion is still the stage of science and not engineering even at the present time after a half century from Nuckolls paper. In addition, thermodynamic properties of WDM and HED states are not well understood as well as the physics of transports in such extreme states. All of such physics are required to know the physical phenomena in the universe, and it is suggested that the physics of laser plasma is still a challenging subject.

It should be noted that the biggest laser NIF is funded to promote the science-based stockpile stewardship (SBSS) [11]. A variety of experiment has been carried out with NIF laser to study the main topics of SBSS. They are the equation of states, material opacities, radiation hydrodynamics, and shock physics of high-energy density state of matters [12]. At the same time, the development of computer codes to simulate the physics has been promoted intensively by developing super computers. The verification and validation of developed codes are one of the most important activities in comparing the related experimental data. The project aims at the prediction capability of HEDP dynamics and phenomena.

1.1.3 Ultra-Intense and Ultra-Short Lasers

Laser scientists have worked for new method to increase the laser power to TW and higher with compact laser system. Application of the concept of CPA used for radar to optical range had been carried out. As shown in Fig. 1.9, using the technique with two grating plates, a laser pulse is expanded in time, amplified, and finally compressed. This CPA technique allows the construction of a compact TW laser system in the mid-1980s. The CPA technique was also used for a large facility NOVA, and two beams of NOVA were converted to PW laser with 1 ps and 1 kJ in the mid of 1990s. Such short pulse and ultra-intense laser is used to study the laser wakefield acceleration, photonuclear interaction physics, QED positron production, etc. Such topics are discussed in Volume 3 in this series. The science of extreme state of matter and vacuum is now studied in many laboratories with CPA lasers. The facilities PW lasers are reviewed by Danson et al. at the stage of 2019 [13].

ICF experiment continues to be carried out with NIF, and 50 kJ of fusion energy production has been reported [14]. In such experiment, it is found that the fusion product alpha particles provide additional heating in the compressed core to enhance the fusion yield by three times over due to heating by implosion. In addition, model experiments of radiation hydrodynamics, equation of state, nuclear physics, and the astrophysics in laboratory have been studied.

The biggest three facilities for science research with CPA ultra-intense lasers are almost ready to start experiments at three countries with several lasers of 1–10 PW

1.2 What Is Plasma?

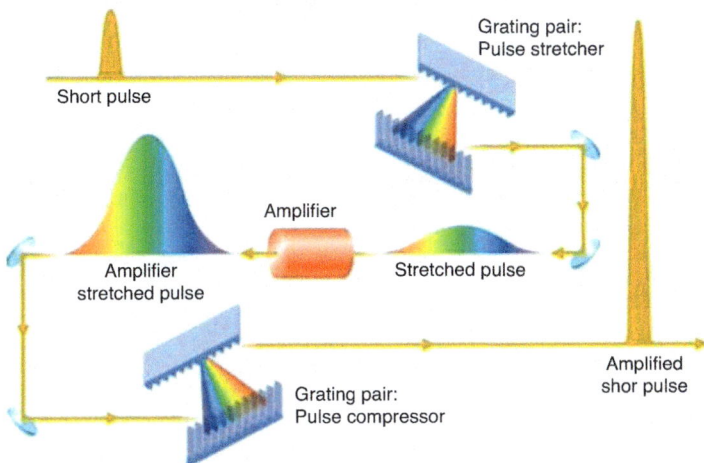

Fig. 1.9 A schematics of CPA method. Generated short pulse is stretched by a small grating to be amplified. The amplified long pulse laser is compressed by a large grating by about $10^{3\sim4}$ times. As the result, the laser power increases orders of magnitude compared to the conventional high-power lasers

output [15]. ELI-project is funded by the EU. Particle acceleration is mainly studied at ELI-Beams in Czech, atto-second photon science will be developed by ELI-ALPS in Hungary, and photonuclear physics is the main subject in ELI-NP in Romania. Present status is reported in [16].

Most of the high- and ultra-high-power lasers are used for research as users' facilities for the international community and most of recent experiments have been carried out by international teams. The style of research is similar to that of the high-energy experiment with a large-scale facility, e.g., Large Hadron Collider (LHC). The characteristics of high-power laser experiment are that a variety of experiments from condensed matter physics to astrophysics are possible by changing laser and target conditions. Precise nanotechnology has been used to fabricate complex structured targets.

1.2 What Is Plasma?

1.2.1 Ionization

Any substance becomes plasma with the increase of its temperature to roughly more than 1 eV, because any kind of neutral atom binds an outer electron with a binding energy of order of eV. Note that in this book, the temperature T is given in the unit of energy without showing Boltzmann constant k_B. The T in eV has the relation:

$$1[\text{eV}] = 1.16 \times 10^4 [\text{K}] \qquad (1.2.1)$$

$$1\text{eV} = 1.6 \times 10^{-19} \text{ [Joule]}$$

1 eV is the energy that an electron obtains by the electric field with a potential difference of 1 volt. Note that most of molecule dissociates around T ~ 0.1 eV, since the molecular binding energy is of the order of 0.1 eV. The fraction of dissociated and slightly ionized air is a function of temperature in the atmosphere. When the air is heated to T = 0.2 eV, oxygen and nitrogen are still molecule at T = 0.6 eV, most of oxygen is dissociated to neutral atom at T = 1 eV, most of nitrogen dissociates and ionization starts, and at T = 1.5 eV both are fully ionized.

The ionization of condensed matter is also a function of density. Since the fraction of electron n(ε) with energy ε has the following general form in thermodynamic equilibrium condition:

$$n(\varepsilon) \propto g(\varepsilon) f(\varepsilon), \qquad (1.2.2)$$

where g(ε) is the density of state and f(ε) is Fermi-Dirac distribution and a function of T such as exp(ε/T). The increase of free electron at high temperature is due to the latter contribution, while the density effect is due to the former term. Consider two extreme situations. In low-density limit like the interstellar media in space, density is about one particle per cm^3, for example. In low-density case, there are many states for free electrons escaped from bound state, and g(ε) is large for free state, while small number for bound state.

Consider a hydrogen atom. It is reasonable to assume the principal quantum number n* as the maximum bound state, since the orbit radius rapidly increases with the principal number n from the core and the electron with n > n* is easily de-trapped. Then, the total number of the bound state is 2(n* + 1)^2. Even if the temperature is nearly zero, the probability of free electron is much larger than the bound electron at low density. This is the reason why space is assumed to be filled by plasma, although T ~ 0.

1.2.2 High-Density Plasmas

The second extreme condition is the case of very high density. It is well-known that hydrogen becomes metal at high pressure limit even at T = 0. For the giant planet Jupiter, it has been assumed that the core region is of metallic hydrogen, and strong current is generated by rotation to keep strong magnetic field on the surface. This is a metallic hydrogen problem which is still an open question in experiment as studied in Volume 2.

It is predicted that any material changes to metal by strong shock compression like the shock wave for laser fusion at relatively low temperature. Such a phase transition is called plasma phase transition in condense matter physics. Metallic

1.2 What Is Plasma?

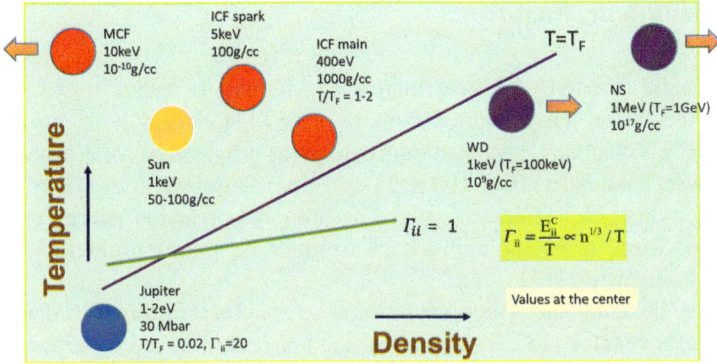

Fig. 1.10 The rough diagram in density and temperature of plasmas both in logarithmic scale. The fusion plasmas in MCF and ICF are shown with the central parameters of high-density astrophysics objects. At high density, quantum effect like Fermi degeneracy and strongly coupling effect play important role

transition means rapid increase of free electrons. In high density, the electron wave functions of the upper energy levels of bound state overlap with those of nearby electrons; consequently such electrons become free electrons, and $g(\varepsilon)$ of the bound state consists of only the ground state and a few lower states. The number of bound state becomes smaller than the number of electrons per an atom even at $T = 0$, and the other electrons become free. Such ionization at high pressure is called pressure ionization as discussed in Volume 2.

In analyzing the plasma state, the **local thermodynamic equilibrium (LTE)** assumption is usually employed as above. In LTE assumption, the thermodynamic and transport properties of plasma are assumed to be only the function of temperature and density. The advantage of LTE assumption is that it is appropriate to describe simple plasma, and theoretical analysis is simple compared to the case of **non-LTE** plasma. The readers are always careful about these differences in considering the laser-produced plasmas. Non-LTE assumption becomes essential for the laser plasma in low-density region such as expanding ablation plasma.

In Fig. 1.10, high-density and extremely high-density plasma parameters are shown in temperature and density diagram. It is clear that even the same fusion plasma for ICF and MCF, the density is different more than ten orders of magnitudes. MCF plasma is classical one, while the laser-imploded core is expected to be as low temperature as possible to increase fusion energy product. The main fuel of the compressed core is partially degenerated, and the temperature is about a few hundred eV near corresponding **Fermi temperature** at high density. Even ICF core plasma is high density, it is much lower than the centers of **white dwarfs** and **neutron starts**. They are highly degenerate extreme plasmas. The center of giant planets like Jupiter is in WDM state, and electrons are in quantum state. In addition, plasma is nonideal where ion-ion Coulomb interaction energy is larger than the thermal energy.

1.2.3 Magnetic Fusion

Research scale of MCF is bigger than ICF. MCF is to realize fusion energy by confining the fusion temperature plasma (~10 keV) stationary with magnetic field. The three big Tokamaks have accomplished great progress of MCF research in the 1980s. These Tokamaks (JET, TFTR, JT-60) were shutdown after DT or DD fusion reaction demonstration independently. The hot days of fusion energy research are vividly described, for example, in a book from the beginning of the fusion research including laser fusion [17].

The big Tokamak activities are unified to the **international thermonuclear experimental reactor** (**ITER**) project, which is initiated by the superpower summit at Geneva in 1985 as international project for peaceful purpose proposed by Gorbachev to Regan as a symbol of the end of Cold War [18]. Tokamak ITER has a complicated structure as shown in Fig. 1.11. The plasma is confined in the vacuum of D-shaped donut by the wall shown with orange color. The vacuum donut is designed to be encircled by superconducting current to generate strong magnetic field.

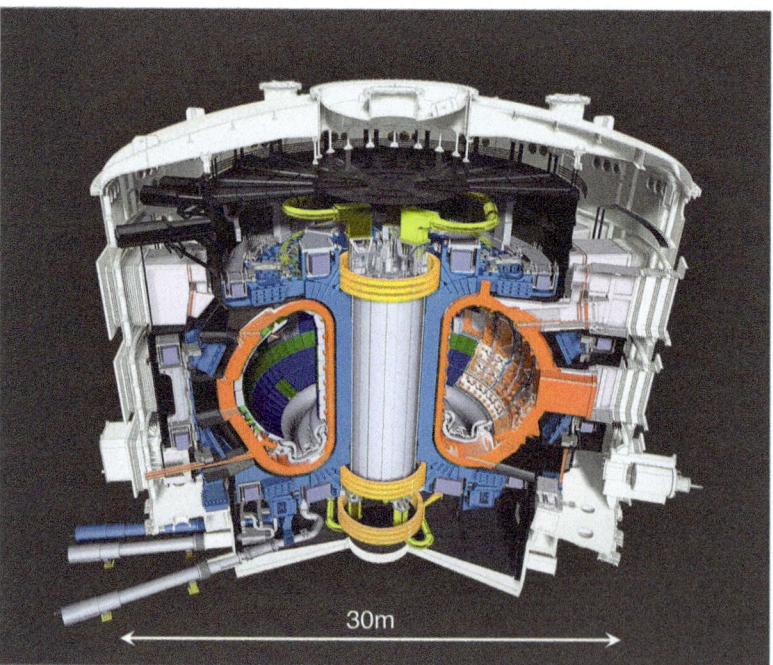

Fig. 1.11 Artistic view of the structure of ITER Tokamak machine. The central torus-colored orange is the vacuum vessel for confining fusion plasma by magnetic field

1.3 Basic Equations

The intense lasers irradiate a variety of solid materials located in a vacuum chamber. The molecules or atoms are ionized by multiphoton absorption during the time scale of femtosecond (fs) or less, where fs $= 10^{-15}$ s. The dynamics of such nonlinear atom-laser interaction cannot be solved by the standard perturbation method. The physics is well studied by solving **time dependent Schrodinger equation (TDSE)** in multidimensional space. Such computational result and experimental results with intense lasers have verified the model equations of ionization rate by field ionization via tunneling effect. Theoretical work has been developed for the case where the ionization time is shorted than the laser oscillation period. These topics are briefly discussed in Chap. 2.

1.3.1 Maxwell Equations

The subsequent intense laser photons, therefore, interact with the free electrons produced after the ionization. For the case of ns pulse, laser interacts with the expanding plasma with the density lower than the **cut-off density** ($n_e \sim 10^{21}$ cm^{-3}), about 100 times less density than the solid. Electric and magnetic fields of laser are governed by Maxwell equation. In the present book, MKS unit is employed. The Maxwell equations are given in the SI unit as the following coupled partial differential equations to the electric field **E** and magnetic flux density **B** in the plasma with charge density ρ and current density **j**.

$$\textbf{Faraday's Law} \quad \nabla \times \mathbf{E} = -\frac{\partial \mathbf{B}}{\partial t} \quad (1.3.1)$$

$$\textbf{Ampere's Law} \quad \frac{1}{\mu_0} \nabla \times \mathbf{B} = \mathbf{j} + \varepsilon_0 \frac{\partial \mathbf{E}}{\partial t} \quad (1.3.2)$$

$$\textbf{Poisson Equation} \quad \varepsilon_0 \nabla \cdot \mathbf{E} = \rho \quad (1.3.3)$$

$$\textbf{Absence of Magnetic Monopole} \quad \nabla \cdot \mathbf{B} = 0, \quad (1.3.4)$$

where ε_0 and μ_0 are the permittivity and permeability of vacuum, respectively.

From two Eqs. (1.3.1) and (1.3.2), the following relations are exactly derived.

$$\frac{\partial}{\partial t} W + \nabla \cdot \mathbf{S} = -\mathbf{j} \cdot \mathbf{E} \quad (1.3.5)$$

In (1.3.5), W is the energy density, and **S** is the energy flux of the laser in plasma. Eq. (1.3.5) represents the relation that the time change of electromagnetic wave energy density W balances the divergence of its energy flux, the **Poynting vector S**, and the energy source on RHS. It is important to note that no net **<jE>** heating is expected for the condition that oscillating laser field **E** and **j** are in 90 degree phase deference. In such a case, the plasma is called **reactive** (lossless). If the current induced by **E** has some phase difference from 90 degree, the plasma is **resistive** (lossy) or **active** (source) resulting net energy transfer between the field and matter.

In the present book, the **laser intensity** which is equal to the absolute value of the Poynting vector **S** is expressed as I_L in the unit of [W/cm^2]. This unit is mixture of SI and cgs, while we use "cm" for the unit of length according to the custom in laser plasma physics. The field strengths E and $B/\mu_0 = H$ are given as the function of laser intensity I_L [W/cm^2].

$$E = 20\sqrt{I_L} \quad [\text{V/cm}]$$
$$H = 3.3 \times 10^{-5}\sqrt{I_L} \quad [\text{Tesla}] \tag{1.3.6}$$
$$= 33\sqrt{I_L} \quad [\text{Gauss}]$$

For the laser intensity 10^{14-16} W/cm^2 to be mainly treated in Chaps. 2, 3, and 4, the electric field of laser is $2 \times 10^{8-9}$ V/cm. This value corresponds to the electric field to bind an electron in a hydrogen atom (13.6 V/0.5A ~ 10^9 V/cm).

The highest laser intensity ever achieved experimentally is about $I_L = 10^{22}$ [W/cm^2], where (1.3.6) indicates an electric field of 2×10^{12} [V/cm] and a magnetic field of 3×10^{12} [G]. It is important to know that in such ultra-strong laser field, the laser field is stronger than the atomic-binding electric field. The electrons oscillate by laser field, and the force by nuclei is regarded to be perturbation. The usual magnetic field strength of a neutron star (pulsar) is about 10^{12}[G]. It is interesting to note that ultra-intense lasers are approaching the value of the strongest magnetic field in the universe. It is noted that the electron oscillation in such extreme field is relativistic and nonlinear. Interaction physics changes dramatically from non-relativistic case. The laser-matter interaction in such ultra-high intensity will be discussed in Chaps. 5, 6, 7, and 8.

1.3.2 Electron Motion in Laser Field

The electron motion in laser field is given by its equation of motion in the laser field. Since the electron motion induces charge and current densities in plasma, self-consistent analysis is required in solving the field and particle quantities. Although the background ion is more than 10^3 times heavier than the electron and less mobile, the electron expansion from the target surface heated by laser should accompany the ions to keep charge neutrality. In Fig. 1.12, the electron and ion energy distributions near the solid surface just after the laser front heats the electrons are schematically plotted.

The electrons are heated to expand into the vacuum, while the produced electrostatic potential $-e\phi$ shown in Fig. 1.12 confined the expanding electron. This is called a sheath field, and its field accelerates the ions into the vacuum.

1.3 Basic Equations

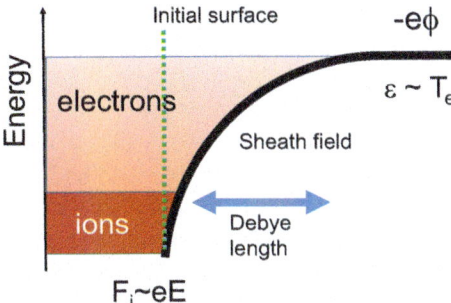

Fig. 1.12 Schematics of energy distribution near the surface of solid target at the beginning of laser irradiation. Heated electrons tend to expand to vacuum to form the sheath electric field over the distance of Debye length, which soon after pull out the heavy ions to the vacuum with electrons to produce ablating plasma as stationary state

Since the electrons are low mass and its expansion velocity is of the order of thermal velocity $v_s = (T_e/m_e)^{1/2}$, the ions are heavy and can move with the ion sound velocity via the sheath field $C_s = (T_e/m_i)^{1/2}$ even the ions are cold. The charge separation distance is roughly Debye length λ_{De}:

$$\lambda_{De} = \left(\frac{\varepsilon_0 T_e}{e^2 n_e}\right)^{1/2} \qquad (1.3.7)$$

It is better to assume that after the initial short time, the electrons interact with laser field in the background ions with same charge density of plus sign. Therefore, it is required to solve the equation of motion for electrons in the background ions.

The electrons are accelerated by electric and magnetic fields, and their motions are governed by the equation of motion with Lorentz force. The basic equation is given in the form:

$$\frac{d}{dt}\mathbf{p} = -e(\mathbf{E} + \mathbf{v} \times \mathbf{B}) \qquad (1.3.8)$$

$$\mathbf{p} = m\gamma\mathbf{v}, \qquad (1.3.9)$$

where **p** is the momentum of an electron, **v** is the velocity, m is the electron mass, and γ is the relativistic Lorentz factor. It is not simple to calculate the local electron charge and current density for many electrons randomly moving with thermal motion. Usually a fluid assumption is used to obtain these densities.

1.3.3 Normalized Laser Strength

The amplitude of the electron oscillation motion in non-relativistic regime (v/c << 1) is easily calculated to be:

$$\frac{|\mathbf{v}|}{c} = a_0 \qquad (1.3.10)$$

$$a_0 \equiv \frac{eE_0}{m\omega c} = \frac{eA_0}{mc}, \qquad (1.3.11)$$

where E_0 and A_0 are amplitude of laser electric field and its Poynting vector. It is clear that the condition of non-relativistic motion is $a_0 \ll 1$, while the situation with a_0 increasing to unity requires relativistic analysis of electron motion. In addition, $\mathbf{v} \times \mathbf{B}$ term in (1.3.8) is small enough to be neglected in non-relativistic case, while it plays important role in relativistic regime. The a_0 in (1.3.11) is called laser "strength parameter." Note that a_0 is Lorentz invariant, and E_0 changes in the moving frame to compensate the relativistic Doppler shift of laser field to keep E_0/ω constant. It is calculated as a function of laser intensity I_L [W/cm^2] and its wavelength λ_L [μm].

$$a_0 = 8.5 \times 10^{-10} \lambda_L \sqrt{I_L} \qquad (1.3.12)$$

Since the conventional high-power laser has the wavelength around μm, the critical laser intensity giving $a_0 = 1$ is about 10^{18} W/cm^2.

The equation to the time evolution of electron energy can be derived from (1.3.8) to be:

$$\frac{d}{dt}\left(mc^2 \gamma\right) = -e\mathbf{v} \cdot \mathbf{E} \qquad (1.3.13)$$

Even in highly relativistic case, the electron energy increases only through the interaction with the laser electric field. The $\mathbf{v} \times \mathbf{B}$ force changes only the direction of the momentum, but no contribution to energy change. If summing up the contribution of all electrons at a local unit volume in (1.3.13) and adding it to (1.3.5), both RHS terms are cancelled, and the total energy of field and electrons conserves.

1.4 Non-relativistic Laser-Plasma Interaction

1.4.1 Collisional Absorption

In the case without absorption, lasers propagate in plasma. This is called **adiabatic interaction** of laser with electrons in plasma. The laser fields are modified in plasma and reflected near the critical density before the solid region. In the wave theory, the adiabatic propagation of laser can be described with well-known **WKB** method. For the laser to be absorbed in such plasma, **nonadiabatic interaction** with electrons should be taken place. The nonadiabatic interaction represents a rapid change of electron motion by another force during a time interval Δt shorter than the electron quivering motion.

1.4 Non-relativistic Laser-Plasma Interaction

Namely, for the oscillation period $\tau_L = 2\pi/\omega_L$, where τ_L is laser quivering period (3 fs for $\lambda_L = 1$ μm) and ω_L is laser frequency, the condition is given as:

$$\omega_L \Delta t < 1 \qquad (1.4.1)$$

In the case where nonadiabatic force is due to random collisions with background ions, it is called **collisional absorption**. It is also called **inverse Bremsstrahlung absorption** or **classical absorption**. This is very basic physics in laser-plasma interaction, and the absorption rate is derived after obtaining the property of Coulomb collision in Chap. 2.

The classical absorption for the case of ultra-short pulse is an important topic in the case where intense laser with pulse width of ~100 fs is irradiated on the solid surface. Such laser is widely used for laser welding and many applications. In this case, the ions have no time to expand, and the laser electric field penetrating into the solid interacts with the free electrons in the condense matter. The quantum effects of free electrons become important. The quantum statistics is necessary in calculating the absorption rate. The quantum statistical physics is also required to study the WDM and HEDP as explained in Vol. 2. The absorption of ultra-short pulse is discussed in Chap. 3 for the case of non-relativistic intensity.

When the plasma becomes very high temperature due to such a classical absorption, the laser-plasma interaction becomes almost collisionless. The most subsequent topics in the book are considered under the condition that plasma can be assumed collisionless. To identify the plasma is collisional or collisionless, electron collision **mean free path** with velocity v by the ion Coulomb field l_{ei} derived in Chap. 2 is used. For fully ionized plasma with ion charge Z and electron density n_e, the electron mean free path by ion scattering is:

$$\ell_{ei} \sim 3 \times 10^{13} \frac{\varepsilon_{eV}^2}{Zn_e} \quad [\text{cm}], \qquad (1.4.2)$$

where $\varepsilon_{eV} = 1/2mv^2$ is the kinetic energy of electron in unit of eV.

Inserting a typical cut-off density $n_e = 10^{21}$ cm^{-3}, $Z = 1$, and $\varepsilon = 1$ keV, the collision mean free path is about 300 μm. This length is shorter than the expanding plasma size for long pulse lasers so that substantial absorption is obtained by collisional process. For long pulse of ~ns, it is reasonable to assume that the collisional absorption is most dominant. For ultra-short pulse, the laser interacts with almost solid density, and the mean free path would be less than the laser penetration length (skin depth) so that enough collisional absorption is expected near the solid surface.

As shown in Fig. 1.12, the electron cloud near the surface starts to expand to the vacuum, once the electrons absorb laser energy and the temperature increases. The heavy ions are then pulled by the ambipolar field to slowly expand to the vacuum with the electron cloud. Then, the energy is confined in the layer determined by the ion expansion as shown in Fig. 1.1 with expanding plasma. Its length is roughly equal to the ion sound velocity C_s per unit time. Assuming all laser energy is absorbed in this region, we can roughly obtain the following relation:

$$C_s P_A \sim I_L, \quad (1.4.3)$$

where the plasma energy density is approximated with its pressure P_A, which is called **ablation pressure** generated by the expanding plasma. With the use of the relation that the photon pressure P_L satisfies the relation, $P_L = I/c$, a new relation is obtained:

$$\frac{P_A}{P_L} \sim \frac{c}{C_s} \quad (1.4.4)$$

It is clear that the ablation pressure is 3 ~ 4 order of magnitude higher than the laser pressure. This is the principle why the laser heating is used to study high-density plasma by the ablation pressure instead of direct laser pressure. Thanks to the ablation pressure, the laser implosion fusion is possible as reviewed in a book [19]. The physical principle of ICF and the present status of laser fusion research are given in Vol. 2. The present accomplishment of ICF by NIF experiment and its difficulty compared to the optimistic scenario are reviewed by mainly focusing on hydrodynamic instability and turbulent mixing.

1.4.2 Collisionless Absorption and Ponderomotive Force

Plasma as a continuous medium has a resonance frequency ω_{pe}.

$$\omega_{pe} = \left(\frac{e^2 n_e}{\varepsilon_0 m}\right)^{1/2} \quad (1.4.5)$$

It is called **electron plasma frequency** or **plasma frequency** in short. The plasma frequency is only a function of electron density, and the oscillation is kept by the balance between induced electric field and electron inertia in the fixed ion background. Since the induced electric field by charge separation is large for higher density, the plasma frequency is proportional to (electron density)$^{1/2}$. When the electron oscillating current is evaluated from (1.3.8) and inserting it to (1.3.2), it is found that the RHS of (1.3.2) cancels each other when the plasma electron density satisfies the relation

$$\omega_{pe} = \omega_L \quad (1.4.6)$$

This determines the cut-off density already mentioned. Physical meaning is that the induced current generate coherent electromagnetic wave at the cut-off density so that

1.4 Non-relativistic Laser-Plasma Interaction

this wave is 180 degree dephasing to the laser field to prevent the penetration of laser into the over region. In addition, the generated wave also propagates to the vacuum as reflected wave from the cut-off density layer.

The fact that the plasma is a kind of resonator indicates that a resonant interaction may happen at the point of the cut-off density in an inhomogeneous plasma. Then, it is natural to expect that the laser energy is converted to the energy of plasma oscillation due to resonant coupling. This is called **resonant absorption** and discussed in Chap. 3. Resonantly excited plasma wave becomes large amplitude, and the so-called **wave-breaking** happens to generate **hot electrons**.

In the laser-plasma interaction, nonlinear laser force should be taken into account for the case where the laser intensity is high, and its energy density is comparable to the plasma energy density. This effect is derived to be a form of pressure or force. It is called **ponderomotive force** and given as the force by ponderomotive (**PM**) potential Up.

$$\mathbf{f}_{PM} = -\nabla U_p$$
$$U_p = \langle \gamma_L - 1 \rangle mc^2, \qquad (1.4.7)$$

where γ_L is Lorentz factor of electron motion in laser field, and $\langle \ \rangle$ means a time average. This can be reduced as follows in non-relativistic regime:

$$U_p^{NR} = \left\langle \frac{m}{2} v_{os}^2 \right\rangle = \frac{\varepsilon_0}{2} \left\langle |\mathbf{E}_L|^2 \right\rangle, \qquad (1.4.8)$$

where v_{os} is the electron quivering velocity by laser electric field $\mathbf{E}_L$. It is easy to understand this force that the electron oscillation motion also transfers the electron momentum flux like the thermal pressure, which is proportional to the velocity average of mv^2 of thermal motion.

It is noted that the PM force is always dominant near the front of the expanding plasma into the vacuum. The plasma dynamics is governed by PM force, and the density filament and laser intensity filament are easily induced. This is called **filamentation instability**. In the expanding plasma, the PM force also induces the **parametric instability** of **stimulated Brillouin scattering (SBS)** and **stimulated Raman scattering (SRS)**. SBS induces exponential growth of reflected light and ion acoustic waves in plasma, contributing anomalous laser reflection. SRS induces exponential growth of reflected light and plasma wave in plasma. SRS also causes anomalous laser reflection. In addition, hot electrons are produced, since the plasma wave has phase velocity higher than the thermal electrons. The subject is discussed in Chap. 4.

Historically, the book by Kruer [20] has obtained a high reputation as a well-written book on the laser-plasma interaction physics. In the book, detail mathematics is shown to derive many important relations for the case of a non-relativistic laser intensity. In order to avoid the overlapping of derivation, etc. most of physical description is developed in an intuitive way. The same way is also adopted in studying the relativistic laser-plasma interaction to avoid complicated mathematical derivation.

1.5 Relativistic Laser-Plasma Interaction

1.5.1 Relativistic Electron Oscillation

The physics of laser-plasma interaction in relativistic laser intensity is rather new topics and most of studies have been focused on applications of such laser for particle accelerations, gamma-ray source, and so on. The physics of laser plasma coupling and laser absorption has left as still ongoing research. Well-written review and book are not available yet to understand the basic physics systematically, although the author referred to two books already published [21]. It is noted that most of the experimental data have been published in the last decade, and PIC simulations have been done to analyze the data. The author tries to review systematically the experimental and simulation result to compare to the theoretical understanding.

From Chap. 5, the laser-plasma interaction for ultra-short ultra-intense lasers is explained. When the electron oscillation velocity approaches unity, relativistic effect such as electron mass increase, strong $\mathbf{v} \times \mathbf{B}$ force, finiteness of laser propagation velocity, etc. modify the non-relativistic interaction, and new physics such as chaos or stochastic electron dynamics becomes essential. Brief review of relativistic theory directly related to the interaction physics is given in Chap. 5. Exact analytic solution of an electron motion in relativistic laser field is obtained. Using this solution, nonlinear scatterings of laser by a single electron are studied. **Higher harmonic generation (HHG)** is derived in Chap. 5.

The relativistic ponderomotive force in (1.4.7) is clear to become very important at extremely high intensity. It induces the oscillation of the solid surface with a period of 10s of femtosecond, before electron expansion to the vacuum makes the surface blurred. The HHG is observed and can be explained theoretically with the use of **relativistic oscillating mirror (ROM)** model. Then, the propagation of relativistic laser in plasma with density below the cut-off density is studied. In the relativistic case, the **parametric instabilities** grow very fast because of the stronger PM force. The laser spectrum becomes broad, and energy cascade is seen in frequency space. In the relativistic case, the relativistic mass correction causes filamentation instability and others. The physics induced by such new nonlinearity in relativistic regime is studied in Chap. 6.

1.5.2 Relativistic Laser and Solid Interaction

For the purpose of applications, many experiments have been done in the last two decades by irradiating a variety of solid targets with relativistic lasers. Many papers have been published by varying mainly the laser intensity and target conditions. Most of the data have been analyzed with use of **Particle-in-Cell (PIC)** simulations. One-, two-, and three-dimensional simulations have been carried out to explain the corresponding experimental results. It is surprising that in most of cases, good agreement with the experimental data is obtained. However, there are limited cases where the effort aims to clarify the physics and propose some model equation. In the early time of research, the experimental results were dependent on the condition of each laser. One of the big issues giving the laser dependence is the so-called the **pedestal** accompanying the main ultra-short pulse. The duration of the pedestal pulse is in the range of nanosecond.

Even if the intensity ratio of the pedestal to the main pulse is about 10 orders of magnitude different, the solid surface is melted, and **pre-formed plasma** is formed. The density of pre-formed plasma is usually lower than the cut-off density, and the relativistic laser dominantly interacts with the pre-formed plasma. For example, the intensity dependence of laser absorption fraction strongly depends on the energy ratio of the pedestal to the main pulse. Such a fact has been clarified as phenomena after comparison of experiments among different facilities. Review of the result of such interaction experiments is given in Chap. 7. A brief survey is also given about the **structured targets** widely used recently.

1.5.3 Theory of Chaotic and Stochastic Heating

Final Chaps. 8 and 9 are devoted to theoretical study of electron acceleration by relativistic Lorentz force in low-density plasma. In the case where laser linearly polarized in the y-direction is propagating to the x-direction, it is shown that an electron momentum normalized by mc is obtained analytically after solving (1.3.8).

$$p_y = a_0 \sin(kx - \omega t) + \beta \quad (1.5.1)$$

$$\begin{aligned} p_x &= \gamma - \alpha \\ &= \frac{1}{2\alpha}\left(p_y^2 + 1 - \alpha^2\right), \end{aligned} \quad (1.5.2)$$

where α and β are integration constants, and the first term on RHS in (1.5.1) is the vector potential of laser with the normalized amplitude defined in (1.3.11).

The constants are given from the initial condition, and $\beta = 0$ is assumed without loss of generality. The value of the constant α (>0) is very important for efficient acceleration of electrons. High-energy acceleration is expected if some way can

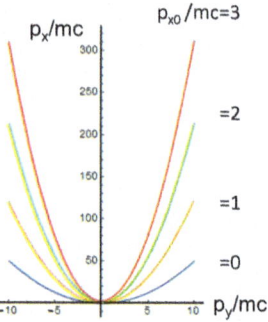

Fig. 1.13 Nonlinear oscillation of electron in relativistic laser field for the case of $a_0 = 10$ and four different initial momentum in x-direction. The initial momentum is 0, 1, 2, and 3 in unit of mc. With its increase, electron obtains more energy. This nonlinear motion causes the electron motion in chaos under nonadiabatic random force. Especially, the acceleration in x-direction gives electrons higher energy

change the value of α much smaller than unity as clear from (1.5.2). Note that $\alpha = 1$ is the case of adiabatic interaction allowing the initial condition, $p_x(0) = 0$. In Fig. 1.13, electron trajectories in the momentum space are plotted for the case of $a_0 = 10$ for four different initial momentums in x-direction, $p_x(0)/mc = 0, 1, 2, 3$, respectively. The relativistic oscillations are characterized by the periodic motions in two dimensions, x and y. The oscillation frequencies are ω_L in y-direction, while it is $2\omega_L$ in x-direction. This is a dramatic change from non-relativistic case where the oscillation is one-dimension only in the y-direction with frequency ω_L.

It is shown that nonadiabatic interaction with the other force, such as electrostatic wave field, sheath field, reflecting laser field, or laser dephasing, can change an effective adiabatic constant α. The nonlinearity in the relativistic oscillation results quite different dynamics of an electron compared to the non-relativistic case. If α becomes smaller than unity, the electron obtains larger energy though the nonadiabatic interaction. One of the standard problems of such interaction is the case where less intense but relativistic 2-nd laser is added to the system. It has been clarified theoretically that above a certain threshold of the intensity of the 2-nd laser, the simple Eq. (1.3.8) can have solution of **chaos**. Then, the electron motion becomes stochastic. Therefore, there are lucky electrons so that they continuously gain energy by each nonadiabatic interaction with the 2-nd laser, and high-energy electron component appears in energy space. This heating of the electrons in plasma and better coupling between laser and electrons are expected. Several numerical results are shown in Chap. 8 as the diffuse of the orbits from the lines in Fig. 1.13.

Note that the chaos is possible only in the case of the electron motion in relativistic intensity. Studying the physics of chaos is to know how the laser energy is absorbed in plasma and what type of electron energy distribution is produced. It is not enough to study with computation about the statistical dynamics of the electron distribution. In Chap. 9, theoretical approaches to such **stochasticity** are introduced from simple diffusion-type **Fokker-Planck equation** to the so-called **fractional**

Fokker-Planck equation (FFPE) in the momentum space. Such study is most important in using the relativistic laser for any kind of applications. **Levy's flights** or jumps in the momentum space become important to allow anomalous and **nonlocal transports** of electrons in the energy space. Comparing with computational and experimental data, it is shown that the FFPE is a simple but strong tool to investigate the collisionless heating of electrons in relativistic laser. Model experiment of **cosmic ray** in universe is also discussed based on the FFPE by strong electromagnetic waves generated by relativistic collisionless shock waves.

1.6 PIC Simulations

There have been many challenges to find the theory to understand the laser-plasma coupling in the case of non-relativistic laser with relatively long pulse. Theoretical model of classical absorption in non-relativistic laser-plasma interaction is relatively easy to derive. However, it is a rather new topic to develop the theory for understanding and predicting the laser-plasma coupling in the case of relativistic laser irradiation. This is partially because relativistic nonlinearity makes the problem complicated.

The laser-plasma interaction and resultant hydrodynamics have been studied by different simulations. The former has been studied mainly by Particle-in-Cell (PIC) simulations, while the latter has been studied with hydrodynamic one. The general method is that some model equations from theory and/or PIC simulations are installed in hydrodynamic code to see a long time evolution of plasma. Roughly speaking, hydrodynamics simulation can run about thousand times linger time compared to the PIC code. This is due to the fact that electron motion is thousand times faster than the ion motion.

Since in this book, the laser-electron interaction is discussed frequently, and many PIC simulation results are compared to theories and experiments. It is useful to briefly understand the principle of PIC simulation. In PIC scheme, (1.3.8) is regarded as the equation of the average momentum of N_G electrons ($N_G \gg 1$). Such a particle is called a giant particle with the mass and charge:

$$M_G = N_G m \\ Q_G = N_G e \quad (1.6.1)$$

Then, regard **p** and **r** of (1.3.8) are average momentum and position of N_G electrons. In the codes, the partial differential Maxwell Eqs. (1.3.1) and (1.3.2) are solved with finite difference method at each grid point in space for a time step Δt determined by numerical stability condition. However, the equations of motion (1.3.8) for all particles are solved in Lagrangian frame (particle frame). Time integration of (1.3.8) is carried out for each Δt by finite difference method or other methods. The field quantities and particle quantities are defined at different points, and interpolation method is used to couple both equations.

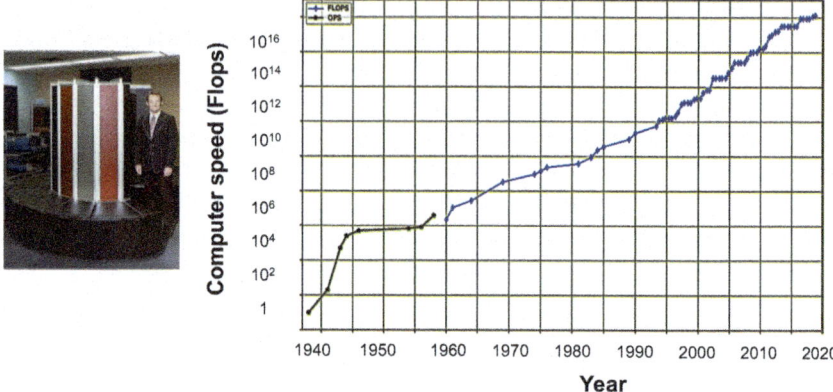

Fig. 1.14 History of the increase of computational speed by supercomputers. The first supercomputer CRAY-1 is in use in the 1980. The era of supercomputer began. The speed has constantly increased 10 times every 5 years. The rapid growth of the speed has made the research method of computer simulation popular. The picture is CRAY-1 and Dr. S. Cray. It is called "most expensive chair"

The PIC codes have been developed soon after the invention of computer in the 1950s by Dawson et al. One of the first papers on laser plasma with PIC code was published soon after the paper by Nuckolls in 1972. The PIC simulation has been done with the code developed to study electron Weibel instability in two-dimensional system. Due to the restriction of computer speed and memory at that time, the space grid in x and y was 50×50 and 10^5 giant particles, and the calculation is 2000 times Δt. Even under such restriction of computer speed, the physics are well obtained and compared to the corresponding theoretical results.

In order to study the physical phenomena with available computation time, giant particles are used to represent N particles, each of which has a charge Nq and mass Nm. For example, a solid foil with thickness 10 μm irradiated by laser with a focusing diameter 10 μm contains 10^{14} electrons and ions. This number is still a big number for the top supercomputer today and a very huge number in the beginning of PIC research in 1970–1980.

With a huge progress of computer speed as shown in Fig. 1.14, PIC codes have been widely used as tool to compare experimental results, especially in the case of ultra-short pulse relativistic laser-plasma interaction. Since the pulse duration is very short less than 1 ps, the full-time simulation is now possible with PIC code even in two or three dimensions. The number of giant particles used now in PIC code reached about 10^{10} or more. A brief introduction on the numerical scheme of PIC code is given in Appendix-3.

References

1. T. Ditmire, Am. Sci. **98**, 394 (2010). M. Rose, H. Hogana, History of the Laser: 1960–2019. https://www.photonics.com/Article.aspx?AID=42279
2. E.M. Campbell et al., Laser Part. Beams **15**, 607 (1997). P. R. Drake, *High-Energy-Density Physics; Foundation of Inertial Fusion and Experimental Astrophysics* (Springer, 2018)
3. M. Koenig et al., Plasma Phys. Controll. Fusion **47**, B441 (2005). F. Graziani et al (edt.), *Frontiers and Challenges in Warm Dense Matter*, edited by (Springer, 2014)
4. A.D. Strickland, G. Mourou, Opt. Commun. **56**, 212 (1985)
5. C.H. Towns, *How the Laser Happened: Adventures of a Scientist* (Oxford University Press, 1999)
6. T.H. Maiman, Nature **187**, 493–494 (1960). C. Townes, *Theodore H. Maiman (1927–2007)*. Nature **447**, 654 (2007)
7. J. Nuckolls et al., Nature **239**, 139 (1972)
8. H. Root et al., SSRN Electronic Journal https://doi.org/10.2139/ssrn.2716757 January (2016)
9. K.A. Bruckner, S. Jorna, Rev. Mod. Phys. **46**, 325 (1974)
10. H. Takabe, Nucl. Fusion **44**, S149–S170 (2004)
11. V.H. Reis et al., Phys. Today **69**, 8 (2016)
12. R. Jeanloz, Phys. Today **53**, 12–44 (2000). K. O'Nions et al., Nature 415, 853 (2002)
13. C.N. Danson et al., High Power Laser Sci. Eng. **7**, 03000e54 (2019)
14. J.L. Kline et al., Nucl. Fusion **59**, 112016 (2019)
15. S. Wills, Optics and Photonics News, January (2020), p. 30
16. Web-site: https://eli-laser.eu/
17. T.A. Heppenheimer, *The Man-Made Sun* (Omni Press, 1984)
18. Web-site, https://www.iter.org/
19. S. Atzeni, J. Meyer-ter-Vehn, *The Physics of Inertial Fusion* (Clarendon Press, 2004)
20. W.L. Kruer, *The Physics of Laser Plasma Interactions* (Addision_Wesley, 1988)
21. A. Macchi, *A Superintense Laser-Plasma Interaction Theory Primer* (Springer, 2013). P. Gibbon, *Short Pulse Laser Interaction with Matter* (Imperial College Press, 2007)

Part I
Non-relativistic Lasers

Chapter 2
Laser Absorption by Coulomb Collision

2.1 Plasma Generation by Lasers

2.1.1 Field-Induced Electron Emission

Plasma production in laboratory has started with gas discharge in a low-pressure gas in a glass tube by applying high voltage between the minus and pulse electrodes. The material of the minus side, cathode, is metal in general. The electric potential structure of binding energy to an electron is schematically shown in Fig. 2.1. Metal confines many free electrons, and their maximum energy at zero temperature is called **Fermi energy**, E_F, which is only a function of electron density. The electric potential near the metal surface is in general of a profile given by black in Fig. 2.1. The potential difference from the inside of metal to the vacuum is called **work function** W. For example, usual metals have the work function of, for example, 4.4 eV [Cu], 4.25 eV [Al], 4.5 eV [W], and so on.

When a high voltage is imposed to the surface of such metals, the electric field by the high voltage alters the electric potential structure as shown by green in Fig. 2.1. In the case where the external electric field is high enough, it is possible for the electrons in the metal to come out through the metal surface. This is due to the **tunneling effect** well-known in quantum mechanics. Such ionization is called **field-induced emission**.

The emission rate of the electrons from the metal surface is approximately calculated by the use of **WKB method**. Given the external electric field E in z direction perpendicular to the metal surface, the effective potential to the electron is in the form:

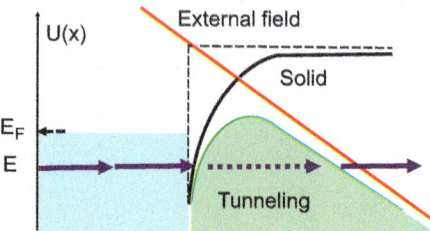

Fig. 2.1 Schematics of the potential structure near solid surface of metal. The electrons are confined in metal by bonding potential ($-e\phi$) given by black line. When a strong external electric field is applied to the metal, the surface potential to electrons is altered as green line. Then, electrons have a probability to escape from the metal due to the tunneling effect in quantum mechanics. This is called field-induced electron emission

$$U(z) = -f_s \frac{e^2}{4\pi\varepsilon_0 z} - eEz \qquad (2.1.1)$$

where f_s is a surface factor and is taken to be 1/4 [1]. Note that this can be used for the case of field-induced ionization of atoms as seen later soon. It is easy to calculate the potential maximum U_{max} and its position z_m. For $fs = 1/4$, they are:

$$U_{max} = -\frac{e}{2}\sqrt{\frac{eE}{\pi\varepsilon_0}} \qquad (2.1.2)$$

$$z_m = \sqrt{\frac{e}{16\pi\varepsilon_0 E}} \qquad (2.1.3)$$

If $|U_{max}| \approx W$ (= work function) is required as view of classical mechanics, it is almost the same electric field of atomic binding energy, which is roughly $E \sim E_0/(er_B)$, where E_0 and r_B are the binding energy of hydrogen at the grand state and Bohr radius, respectively. Inserting the constants, the critical electric field becomes $E \sim 10^9$ [V/cm]. This is unrealistically higher value technically. It is clear that with the help of the quantum tunneling effect, the electrons are extracted from the metal by the field-induced emission mechanism.

The calculation of the tunneling probability is easily understood by modifying the potential structure of green in Fig. 2.1 to the potential structure made of three steps as shown in Fig. 2.2. This is a simple problem, and the electron wave function confined in the metal is partially reflected from the surface and partially penetrated in the barrier to go out to the right direction. Given potential, we can solve to obtain the fraction of C_2 and C_3. Then, $|C_3|^2$ is the tunneling probability. It is straightforward to calculate the electron probability fluxes of red and yellow arrows in Fig. 2.2 to obtain the number of electrons ionizing per unit time.

Without showing the detail calculation, the electron emission current J from the metal surface is given in the relation:

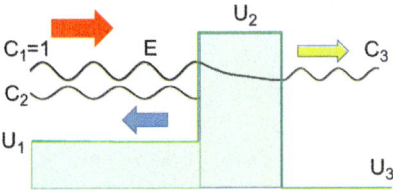

Fig. 2.2 Modeling the green curve potential in Fig. 2.1 to three steps, it is very easy to obtain the electron flux escaping to the right with the use of undergraduate level quantum mechanics. The flux of C_3 decays exponentially in the region of U_2 potential

$$J = a\frac{E^2}{W} \exp\left(-b\frac{W^{3/2}}{E}\right) \qquad (2.1.4)$$

where a and b are constants given in [1]. It is noted that with the increase of the strength of electric field, the factor of the exponent in (2.1.4) increases very fast.

The work function is about several eV, and an electric field of $E = 10^6$ V/cm is required to extract large current. Such discharge is called cold cathode discharge or **electric field induced discharge**. In addition, if there are fine protrusions on the surface, the electric field in the vicinity thereof becomes strong. Then, the electron emission from the protruding point occurs in a relatively low electric field. Like the lightning rod, the electric field becomes stronger at the protruding point such as the apex of the cone. Therefore, with the progress of nanotechnology, a method using a cathode having many protruding structures came to be used.

Before finishing here, electron emission from the metal due to heating of the metal surface is briefly explained. This phenomenon is called **thermionic emission**. It is clear that the electron energy distribution in the metal expands to higher energy levels suggested by Fermi-Dirac distribution, when the metal surface is heated by a certain method. In order to obtain efficient emission of electrons, however, metal temperature should be kept lower than the melting temperature that is lower than the work function W.

A material having a high melting temperature and a low work function is suitable. Tungsten is often used as a thermal electron source. The melting temperature is 3695 K degrees (0.32 eV), and the work function is 4.5 eV. Even the value of exp $(-4.5/0.32) = 10^{-6}$, many thermionic electrons appear. By integrating the Fermi-Dirac distribution for given temperature T, the thermal current density from the metal surface with the positive energy is found to have the relation [1]:

$$J \propto T^2 \exp\left(-\frac{W}{T}\right) \qquad (2.1.5)$$

where T is in energy unit and Boltzmann constant is omitted.

2.1.2 Ionization by Multiphoton Absorption

When the laser intensity increases very high, ionization by nonlinear photo-absorption appears to be important. This is called **multiphoton ionization (MPI)**. The picture of the physics is very clear as shown in Fig. 2.3, where electron at the grand state is excited to each virtual quantum state and finally obtain enough energy to be in the free state after pumped up like ladder-like virtual states. MPI is nonlinear process and observed over a certain threshold of laser intensity, while the photoionization shown in Fig. 2.3 is linear process and happens when the photon energy is lesser than the ionization potential. It is, however, not so simple to calculate, for example, the threshold laser intensity for MPI even for a simplest atom hydrogen. From the first principle, we have to solve the **time-dependent Schrodinger equation (TDSE)**:

$$i\hbar \frac{\partial}{\partial t} \Psi(\mathbf{r}, t) = \left[-\frac{\hbar^2}{2m} \nabla^2 - \frac{e^2}{4\pi\varepsilon_0 r} - i\hbar \mathbf{A} \cdot \nabla + \frac{e^2}{2m} \mathbf{A}^2 \right] \Psi(\mathbf{r}, t) \quad (2.1.6)$$

Detail discussion of MPI has been given in [2] by solving (2.1.6) numerically. It is out of the scope of the present book, and therefore let us challenge this MPI problem based on an intuitive way as shown below. In (2.1.6), the equation to the wave function of the virtual state is found to satisfy approximately the following relation:

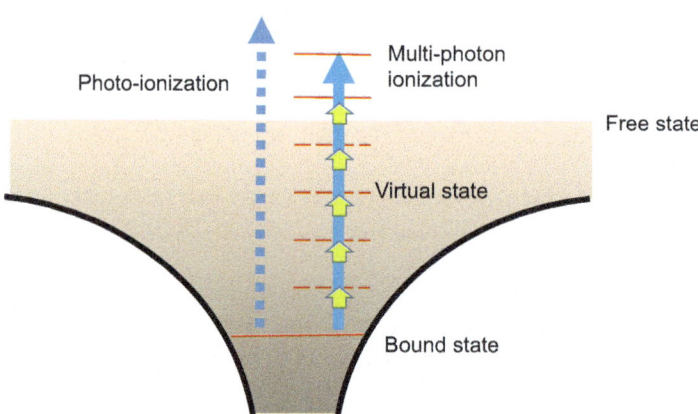

Fig. 2.3 Schematics of multiphoton ionization process. The black lines are potential structure by the nucleus, and the electron grand state is shown by solid red line. The vertical states are shown by red dashed lines so that energy gap is equal to laser photon energy. In the multiphoton ionization, the electron at the grand state climbs up like a ladder to the ionized state by absorbing multiphotons. If the photon energy from out is larger than the ionization energy, the electron is easily ionized by Einstein's photoelectric effect

2.1 Plasma Generation by Lasers

$$\Delta E \cdot \Delta t \geq \frac{\hbar}{2} \qquad (2.1.7)$$

This is the **uncertain principle,** and it takes time Δt for the electron at the ground state to excite to the virtual state with its energy ΔE above the ground state.

The size of the electron cloud in a hydrogen atom is approximately 1 Å ($= 10^{-8}$ cm). The ionization energy of the ground state electron is 13.6 eV. Consider the interaction with laser photons whose wavelength 1.06 μm of glass laser. Then, each photon energy is about 1 eV. A simplest image of the MPI is that if an electron can simultaneously absorb 14 or more photons, the electron in the ground state becomes a free electron.

Let us evaluate coarsely using only uncertainty principle, how much laser intensity can be expected from multiphoton absorption. Calculate for case of laser intensity $I = 10^{15}$ W/cm^2. The number of photons that pass through the cross-sectional area of hydrogen atoms is:

$$\frac{10^{15}}{1.6 \times 10^{-19}} \times \left(10^{-8}\right)^2 = 6.3 \times 10^{17} \quad [1/\text{s}] \qquad (2.1.8)$$

If many photons interact with the wave function of hydrogen atom, as can be seen from the uncertainty principle, there is a possibility of absorbing laser photons before virtual energy levels disappear.

Assuming $\Delta E = 1$eV in (2.1.7) as a virtual state that absorbs 1 eV photon, then the life time of the virtual state is:

$$\Delta t = \frac{1.05 \times 10^{-34}}{2 \times 1.6 \times 10^{-19}} = 3.3 \times 10^{-16} \quad [\text{s}] \qquad (2.1.9)$$

The time required for the electron to climb up the all virtual levels as shown in Fig. 4.12 is the time τ_{ion} that about 14 photons are absorbed by the electron until it is ionized:

$$\tau_{\text{ion}} = 14 \times 3.3 \times 10^{-16} \approx 5 \times 10^{-15} \quad [\text{s}] \qquad (2.1.10)$$

Therefore, if 14 or more photons collide with the wave function of an electron bound to atom in a time shorter than 5 fs, the bound electron can absorb the energy of the laser photons from the ground state by the multiphoton absorption. The phenomenon of absorbing energy of laser via jumping of electron through such virtual levels is called **multiphoton absorption.**

The above intuitive evaluation shows that when the laser intensity is 10^{15} W/cm^2, in the life time of the virtual levels, the following number of photons interact with each electron:

$$Np = 6.3 \times 10^{17} \times 5 \times 10^{-15} = 3000 \tag{2.1.11}$$

The threshold intensity at which multiphoton ionization begins can be roughly evaluated:

$$I_L > \frac{14}{3000} \times 10^{15} \approx 5 \times 10^{12} \quad [\text{W/cm}^2] \tag{2.1.12}$$

Regarding almost any molecular gas, the molecular dissociation by multiphoton absorption occurs first, and ionization of dissociated atoms occurs as above.

2.1.3 Tunneling and Over-Threshold Ionizations

As seen above, the field-induced emission is due to the tunneling effect of the quantum mechanics. As long as the ionization time is much smaller than the laser oscillation period, the stationary condition can be assumed as in the DC field. In Landau-Lifshitz's textbook of quantum mechanics (section 77 in [3]), the eigenvalue problem when the electrostatic field of the external field $U = -zE$ is applied to hydrogen atoms is solved using the parabolic coordinates, since atom is 3 dimension than 1-dimension in Fig. 1.1. Then, they asked the readers to calculate the probability of tunnel ionization for hydrogen atom case. In Fig. 2.4, the cut view of the atomic potential with strong external field is shown. When the bound state is lower than the barrier of the potential (a), the tunneling ionization occurs. Its rate is given in, for example, in [2, 3] as follows:

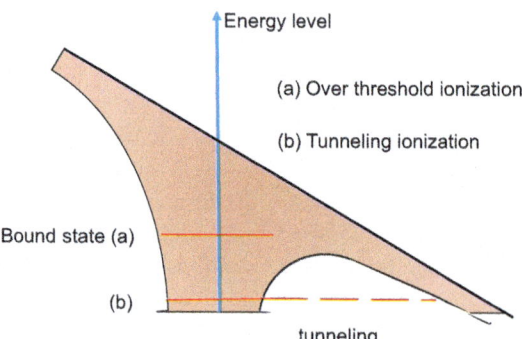

Fig. 2.4 Schematics of the atomic potential structure to electron when a static strong electric field is applied to the atom. In the case where the ground state without the external field shown as (*a*) becomes higher than the maximum potential energy on the right, the bounded electron can be free, and such ionization is called over-threshold ionization (OTI). On the other hand, although the field strength is not so strong to induce OTI and the electron grand state is like (*b*), the bounded electrons can escape from the atoms via quantum tunneling effect. This is called tunneling ionization (TI)

2.1 Plasma Generation by Lasers

The physical quantities E_H and ω_H are defined as:

$$\omega_H = \frac{me^4}{(4\pi\varepsilon_0)^2 \hbar^3} \Rightarrow \hbar\omega_H = \frac{e^2}{4\pi\varepsilon_0 r_B} \quad (2.1.13)$$

$$E_H = \frac{m^2 e^5}{(4\pi\varepsilon_0)^3 \hbar^4} = \frac{e}{4\pi\varepsilon_0 r_B^2} \quad (2.1.14)$$

where r_B is the Bohr radius. Note that ω_H is the classical rotation frequency of an electron in the hydrogen atom potential, and E_H is the strength of electric field at the Bohr radius in hydrogen atom. They have obtained the ionization probability W by the tunneling effect in the form:

$$\widehat{W} = \frac{4}{\widehat{E}} \exp\left(-\frac{2}{3\widehat{E}}\right) \quad (2.1.15)$$

$$\widehat{W} = \frac{W}{\omega_H}, \quad \widehat{E} = \frac{E}{E_H} \quad (2.1.16)$$

where E is the external electric field by laser. The W is the **tunneling ionization (TI) rate**.

Landau-Lifshitz calculated the probability of tunnel ionization with perturbation theory and **WKB approximation** (Appendix-2) for the condition that the normalized electric field is much smaller than unity $\widehat{E} \ll 1$. When the laser intensity approaches the critical value 10^{15} W/cm^2, the laser electric field becomes comparable to the atomic field, $\widehat{E} = 1$. In such case, the ground state is located above the maximum of the modified potential as seen in case (b) in Fig. 2.4. It is said approximately that the tunneling ionization occurs with an intensity weaker than this intensity.

Then, what happens when approaching this intensity? As shown in Fig. 2.4, the bound state of atom disappears, and the bound electron becomes free to instantaneously ionize. In other words, only the bound state orbit radius is enough for an electron accelerated by the laser electric field to get ionization energy. This is called **over-threshold ionization (OTI)**.

The condition of applicability of the result (2.1.15) by the perturbation analysis is not clear. Keldysh, however, made exact calculations for the case of the electrostatic field of arbitrary intensity. M. V. Keldysh published a famous paper on the field ionization in 1965 [2, 4]. He considered the effect of strong field modifying the Coulomb potential in atom distorted by external force as shown in Fig. 2.4 and studied the physics of bound electrons ionized by the tunnel effect. He obtained the transition probability from the bound electron to the **Volkov solution** of a relativistic electron, which is the exact solution of the Dirac equation in the electromagnetic field of arbitrary amplitude.

He introduced the reciprocal of the time that an electron leave the constraint of the nucleus and introduced the tunnel ionization period ω_t as follows:

$$\omega_t \equiv \frac{eE}{\sqrt{2mE_i}} \qquad (2.1.17)$$

This is obtained under the condition that electron is accelerated by the electric field E in the time (ω_t^{-1}) to obtain the ionization energy E_i.

The ratio of this frequency to the laser frequency ω_0 is introduced.

$$\gamma = \frac{\omega_0}{\omega_t} \qquad (2.1.18)$$

This γ is called the **Keldysh parameter**. In the case of large γ (weak laser intensity), the static field assumption is not appropriate, and the physics becomes MPI, while smaller γ (strong laser intensity), intuitively fits to the case of TI and OTI:

$$\gamma \gg 1 \text{ (MPI)} \qquad (2.1.19)$$

$$\gamma \leq 1 \text{ (TI or OTI)} \qquad (2.1.20)$$

Writing the laser wavelength in μm unit and ionization energy in eV unit, the laser intensity at which $\gamma = 1$ is:

$$I = 10^{13} \times \frac{E_{i,eV}}{\lambda_{\mu m}^2} \quad [W/cm^2] \qquad (2.1.21)$$

It is very troublesome to reproduce Keldysh's calculations here, so skipping mathematics for derivation, the ionization probability of TI and OTI is given to be [2]:

$$W \propto E^{5/2} \exp\left(-\frac{4}{3} \frac{\sqrt{2m}E_i^{3/2}}{e\hbar E}\right) \qquad (2.1.22)$$

After Keldysh, more precise calculations have been carried out. The readers interested in more precise discussion are recommended to refer the book of [2].

Finally, consider the case of field ionization by ultra-intense and relativistic lasers. The electron momentum distribution after the field ionization is possibly dominant in the direction of electric field of laser, namely, laser polarization direction. If the free electron energy after the ionization is small enough, the distribution function will become isotropic due to Coulomb collision process to be discussed soon later. It is, however, very different in case of field ionization by relativistic lasers. Since the electron energy after the ionization is MeV range, and they are collisionless. So, the electron distribution function has large value in the direction of ionization force. In the case of non-relativistic intensity, the force is mainly due to the laser electric field. In the case of relativistic intensity, however, the dominant force is due to **v** × **B** force, and the electrons are ionized in the laser propagation direction dominantly. Such an isotropy of the distribution function with

2.1 Plasma Generation by Lasers

MeV energy plays an important role in the interaction with relativistic laser after the ionization.

The higher harmonic generation (HHG) from an electron oscillating in the binding force of atom is also interesting. If (2.1.6) is solved in Coulomb potential by the atomic nucleus, the electron wave function oscillates with laser frequency, but it is not the harmonic oscillation. The time evolution of the current by the electron in (1.3.2) has many higher harmonic Laplace components, as imagined with the classical electron motion. With the increase of laser intensity, the bound electron starts to emit HH light, and it can be used for many applications. Even the electron is abruptly ionized by the field, some electrons recombine to the atom emit the excess energy as photons in X-ray range. Since this phenomenon happens in a very short time like laser oscillation period, the generated X-ray pulse will be of the pulse duration of atto-second. Atto-second X-ray pulse is attractive photon source to study the fundamental science in many fields. Details are given in [2], and this book stops this topics here.

2.1.4 Experimental Data

Intense laser is irradiated into the rarefied argon gas to observe the fraction of ionization of argon atom from Ar^+ to Ar^{7+} [5]. The first ionization energy of argon atom is 15.76 eV. In order to get precise data, short pulse laser of 600 fs pulse length is used, and laser intensity is varied from 10^{14} to 10^{17} W/cm^2 as seen in Fig. 2.5, where each mark represents a single laser short result. The laser wavelength is 1.053 mm, and it almost corresponds to the photon energy of 1 eV. It is clear that above the intensity of 10^{14} W/cm^2. The sharp increase of the number of ions shows that the phenomena are nonlinear, and each ionization has a threshold laser intensity above which the ionization starts. The threshold intensity of the first ionization is about 2×10^{14} W/cm^2 in Fig. 2.5. The rough evaluation based on the uncertain principle in (2.1.12) is lower estimate of this value, but it is useful for rough evaluation of the threshold.

There are several models to predict such fraction of ionization over one pulse with typically Gaussian pulse shape. In Ref. [5], many data for different kinds of gases are also shown, and all data are compared to six different theoretical works. After the comparison of the experimental data and the theoretical models, the authors concluded that a much more primitive model based on Coulomb-barrier suppression has best fit to the data. This theory ignores the tunneling and other quantum mechanical effects. This is called **barrier-suppression ionization (BSI)** mechanism, and it is given only by the threshold intensity in the form [5]:

Fig. 2.5 Argon ion production rate as a function of peak laser intensity. The solid lines are calculated from BSI model for each ionization potential

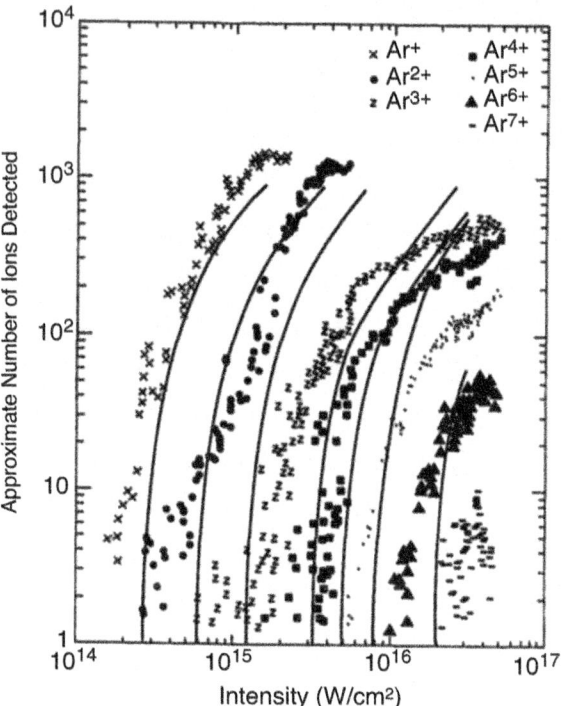

$$I_{th} = 4 \times 10^9 \frac{E_i^4 (eV)}{Z^2} \quad [W/cm^2] \tag{2.1.23}$$

This is derived from the relation that the maximum barrier potential given by (2.1.2) is equal to the ground state binding energy.

With use of the temporal profile of the intensity and (2.1.23), the fraction of ionization is given as function of average laser intensity and the ionization potential of each ionization stage. The solid lines in Fig. 2.5 are the results obtained by such calculations. It is surprising to know that such simple theory gives good agreement with the experimental data. The compared theories all assume the ionization of one electron without interaction with the other bound electrons. It is pointed out that non-sequential (NS) double ionization becomes important before the single electron ionization threshold. Here, "NS double ionization" refers to the simultaneous removal of two electrons rather than a sequential process of removing one electron then removing another short time later.

In Fig. 2.6, the newly taken data are plotted [6]. The theoretical lines are shown with dotted, dashed, and solid lines. In this case, theoretical model for one electron ionization is ADK model from Keldysh plus electron scattering model. The solid lines showing good agreement are the case with the addition of NS double and triple ionization process. The dotted lines are without NS ionization. The dashed line is the

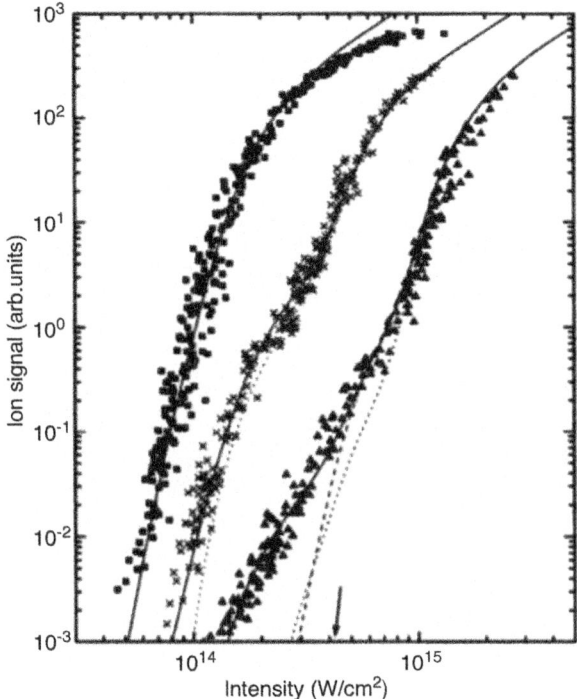

Fig. 2.6 Same as Fig. 2.5 for three ionization stages of argon as a function of peak laser intensity. The previous data has been well refined to clearly identify multi-electron effect near the threshold of ionization start. Three different theoretical curves are plotted as explained in the text

case where only NS double ionization is included. It is clear that when the ionization becomes deeper, the multi-electron effect becomes important. This suggests that one electron TDSE in (2.1.6) is not enough, and multi-electron TDSE should be solved for realistic calculation.

2.2 Laser as Electromagnetic Waves

Lasers are a bunch of mono-energetic photons in quantum physics and electromagnetic field in classical physics. Depending on a problem, it is better to choose one of two views convenient for analysis and understanding. Most of the analysis of the coupling of lasers and matters in the present text is carried out by assuming electrons are classical point charge, and the laser electric and magnetic fields are given by Maxwell equations.

Under the force by oscillating electromagnetic field, electrons make predominantly the local electric current compared to the ions, because the electron mass is more than 1000 times smaller than the ion mass, although they have almost the same charge. It is in general enough to consider the coupling between the laser fields and electron motion. Therefore, we are asked to solve the coupled system of Maxwell equations and the electron motions. It is, however, not so easy to solve the system in general, since the charge density and current density induced by many electrons in plasma alter the electric and magnetic fields.

2.2.1 Maxwell Equations

Let us start with revisiting Maxwell equations.

Most of the students involved in science and/or engineering have learnt the Maxwell equations. Maxwell's predecessor Michael Faraday was a genius in the frontier of electricity and magnetism as an experimentalist at his era. Given his poor family background, he was restricted to receiving higher education. But his talent was flowering in the UK Royal Institution where he was a technical official in the laboratory of Prof. H. Davy. During the 1830s, he was diligent in his responsibilities conducting experiments on electricity and magnetism and also accustomed to tabulating daily reports on his experimental results. One of his great achievements is Faraday's Law indicating physical quantities of electricity and magnetism are related to each other.

M. Faraday, however, was not able to express his experimental results by universal expressions due to lack of education in advanced mathematics. However, his accumulated results taken over to James C. Maxwell, and he resulted in one of the most important formulas of physics. Maxwell was educated at the University of Cambridge and demonstrated his scientific power as a student there. The experimental verification of his theories has to wait until the demonstration of electromagnetic waves by H. Hertz.

Maxwell read over Faraday's experiment note many times and wondered if there was a unified mathematical relation to explain Faraday's experimental results. At that time, there was also a time when fluid dynamics was comprehensive and studied vigorously. Fluid equation by Euler is expressed by the concept of density field, velocity field, and other "fields." Maxwell finally found that coupled partial differential equations like fluid equations can be derived by introducing concepts of electric field and magnetic field to electricity and magnetism, respectively. The concept of similarity (similar transfer) has led him to the great equations. Finally, in the book called "Theory of Electromagnetism," he published Maxwell equations that explain all the experimental results by Faraday in 1865.

Equation (2.2.1) is the induction relation invented by Faraday, and it indicates that the time variation of magnetic field produces electric field. This relation becomes important in plasma physics, for example, when we deal with particle acceleration in magnetic reconnection observed on the solar surface. The change of topology of magnetic field induces electric field which accelerates charge particles. The time variation of the electric field will induce the current in plasmas and induced magnetic field as seen in (1.3.2). Then, this magnetic field will return to (1.3.1), and the electromagnetic wave is generated to propagate in the plasmas. A typical collective phenomena in plasma are waves since charged particles in plasma interact each other over the long-distance coupling with many particles.

2.2 Laser as Electromagnetic Waves

From two Eqs. (1.3.1) and (1.3.2), the following relations are exactly derived:

$$\frac{\partial^2}{\partial t^2}\mathbf{E} - c^2\nabla^2\mathbf{E} + c^2\nabla(\nabla \cdot \mathbf{E}) = -\frac{1}{\varepsilon_0}\frac{\partial \mathbf{j}}{\partial t} \quad (2.2.1)$$

$$\frac{\partial}{\partial t}\left(\frac{\varepsilon_0}{2}\mathbf{E}^2 + \frac{1}{2\mu_0}\mathbf{B}^2\right) + \nabla(\mathbf{E} \times \mathbf{B}) = -\mathbf{j} \cdot \mathbf{E} \quad (2.2.2)$$

where the definition of the speed of light in vacuum is used.

$$c = (\varepsilon_0\mu_0)^{-1/2} \quad (2.2.3)$$

In deriving (2.2.1), the following mathematical relation is used:

$$\nabla \times (\nabla \times \mathbf{E}) = -\nabla^2\mathbf{E} + \nabla(\nabla \cdot \mathbf{E}) \quad (2.2.4)$$

Multiplying **B** to (1.3.1) and **E** to (1.3.2) and using the following mathematical relation:

$$\mathbf{B} \cdot \nabla \times \mathbf{E} - \mathbf{E} \cdot \nabla \times \mathbf{B} = \nabla(\mathbf{E} \times \mathbf{B}) \quad (2.2.5)$$

Equation (2.2.2) can be obtained. Equation (2.2.1) is general equation for the propagation of waves in any materials, and (2.2.2) is the field energy conservation relation.

It is convenient to rewrite Eq. (2.2.2) in the form:

$$\frac{\partial}{\partial t}W + \nabla \cdot \mathbf{S} = -\mathbf{j} \cdot \mathbf{E} \quad (2.2.6)$$

In (2.2.3), W is the energy density, and **S** is the energy flux of the electromagnetic wave. They are defined respectively in the forms:

$$W = \frac{1}{2}\left(\varepsilon_0|\mathbf{E}|^2 + \frac{1}{\mu_0}|\mathbf{B}|^2\right) \quad (2.2.7)$$

$$\mathbf{S} = \frac{1}{\mu_0}(\mathbf{E} \times \mathbf{B}) \quad (2.2.8)$$

Equation (2.2.4) represents the relation that the time change of electromagnetic wave energy density W balances the divergence of its energy flux, the **pointing vector S**, and the energy source term on RHS. When the plasma emits electromagnetic wave, the RHS of (2.2.3) is positive, but it is negative when the plasma absorbs the energy of electromagnetic field. RHS of Eq. (2.2.3) vanishes in vacuum, and such relation is called an **equation of conservation**. Readers will see many such forms in the present book.

Fig. 2.7 Geometrical relation for plane waves: (**a**) electromagnetic waves and (**b**) electrostatic waves. The electric field is stronger in Lorentz force than magnetic force in non-relativistic regime, and the wavenumber k is perpendicular to the electric field in (**a**), while it is parallel in (**b**). The electromagnetic waves have no density perturbation in plasmas, while the electrostatic waves are maintained by the electric field due to the charge separation of electrons from the ions

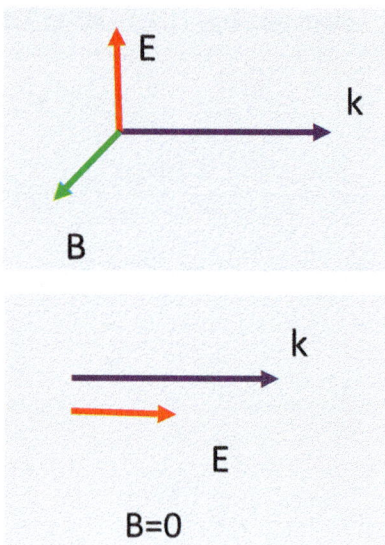

There are a variety of waves in plasma depending on how electron current is generated with electric and magnetic fields. In general, we can separate the type of such waves to two, namely, the **transverse waves** and **longitudinal waves**. They are defined by the following relations:

$$\text{Transverse waves: } \nabla \cdot \mathbf{E} = 0$$
$$\text{Longitudinal waves: } \nabla \times \mathbf{E} = 0$$

In the case of small amplitude electric and magnetic fields of the waves, we can expand the physical quantities in **Fourier-Laplace** components which allow to assume the space and time dependence of the quantities in the complex form pulse its complex conjugate:

$$\mathbf{E}(t, \mathbf{r}) \propto \sum_{\mathbf{k}, \omega} \mathbf{E}(\mathbf{k}, \omega) e^{-i\omega t + \mathbf{k} \cdot \mathbf{r}} + \text{c.c.} \qquad (2.2.9)$$

In (2.2.9), ω and **k** are the **angular frequency** and **wavenumber** vector, respectively.

In what follows, the complex expression is used for convenience of mathematics, and the real values can be obtained by taking the real components of the resultant physical quantities or assuming all components are the sum with its complex conjugate as shown in (2.2.9). In the form of (2.2.9), the transverse and longitudinal waves are characterized as shown in Fig. 2.7a and b. The vector **k** is the direction of the wave propagation, and the transverse waves have the electric field perpendicular

2.2 Laser as Electromagnetic Waves

to the propagation direction, but the longitudinal waves have the electric field in the direction of the propagation. In addition, (1.3.1) requires that the magnetic field is perpendicular to **k** and **E** for the transverse waves, but the longitudinal waves have no magnetic field. The typical transverse waves are electromagnetic wave, and they are also called **electromagnetic mode**; on the other hand, the longitudinal waves are sustained by the electric field due to charge separation given in (1.3.3), and they are also called **electrostatic mode**.

It is clear that from (2.2.1), the transverse waves are governed by the equation:

$$\frac{\partial^2}{\partial t^2}\mathbf{E} - c^2 \nabla^2 \mathbf{E} = -\frac{1}{\varepsilon_0}\frac{\partial \mathbf{j}_\perp}{\partial t} \quad (2.2.10)$$

where the current means the transverse component perpendicular to the electric field. On the other hand, the longitudinal waves are governed by the equation:

$$\frac{\partial^2}{\partial t^2}\mathbf{E} = -\frac{1}{\varepsilon_0}\frac{\partial \mathbf{j}_\parallel}{\partial t} \quad (2.2.11)$$

where the current represents the longitudinal component parallel to the electric field. It is clear that in order to study any waves in plasmas, the current response to the electric field evolution becomes the source or absorption of the waves.

2.2.2 Electromagnetic Waves in Vacuum

In the vacuum or rarefied gas like our atmosphere, the current of RHS in (2.2.10) can be neglected, and we obtain the following wave propagation equation:

$$\left(\frac{\partial}{\partial t} + c\nabla\right)\left(\frac{\partial}{\partial t} - c\nabla\right)\mathbf{E} = 0 \quad (2.2.12)$$

(2.2.12) shows two waves propagating along the opposite directions. The propagation velocity is the speed of light c. When the wave propagates in the x-direction with the electric field in y-direction, the solution of (2.2.12) is give in the form of a linear combination of two terms:

$$E_y = A e^{i(kx-\omega t)} + B e^{i(kx+\omega t)} \quad (2.2.13)$$

In (2.2.13), k is the absolute value of the wavenumber in the x-direction, and ω is the angular frequency. They have the following relations with **wavelength** λ and **frequency** f of the wave:

$$k = \frac{2\pi}{\lambda}, \omega = 2\pi f \quad (2.2.14)$$

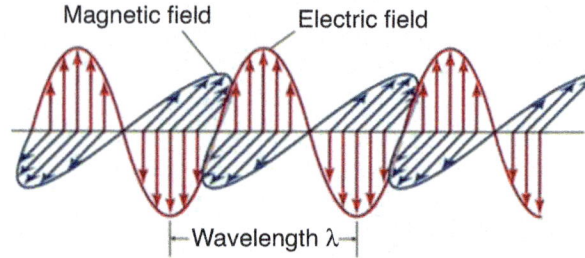

Fig. 2.8 Schematics of the electric and magnetic fields of plane electromagnetic fields propagating from the left to right

In addition, the following relation which is called in general a **dispersion relation** of waves is obtained:

$$\omega^2 = c^2 k^2 \qquad (2.2.15)$$

In Fig. 2.8, the schematics of a snapshot of the electromagnetic wave is plotted. It is noted that the magnetic field is perpendicular to the electric field, and **E**x**B** should be in the direction of **k**, the propagation direction. The wavenumber is a vector as seen in (2.2.9) and given to be in a vector form, **k**.

H. Hertz who heard the prediction by Maxwell has carried out the experimental proof of the electromagnetic wave in his laboratory with the use of high-voltage discharge. As suggested in (2.2.10), Hertz generated the electromagnetic wave by flowing discharge current in a high-voltage gap. Then, the electric field by the electromagnetic wave was detected, although it is not monochromatic, but it consists of a wide range of frequency.

Historically the electromagnetic waves have been called as a variety of names depending on the frequency. Radio waves have the longest wavelength, and gamma rays have the shortest. The visible region of light is very narrow in wavelength from about 350 nm to 750 nm (0.35–0.75 µm).

2.2.3 Lasers as Coherent Electromagnetic Waves

Let us introduce the vector potential of electromagnetic field of a laser beam:

$$\mathbf{A} = A_0(\omega)\boldsymbol{\varepsilon} \cos(\mathbf{k} \cdot \mathbf{r} - \omega t) \qquad (2.2.16)$$

In (2.2.16), $\omega = kc$ and $\boldsymbol{\varepsilon}$ is a unit vector showing the direction of the electric field. Then, the electric and magnetic fields of the wave are calculated with the use of the relation for electromagnetic waves:

$$\begin{aligned} \mathbf{E} &= -\frac{\partial}{\partial t}\mathbf{A} \\ \mathbf{B} &= \nabla \times \mathbf{A} \end{aligned} \qquad (2.2.17)$$

Equation (2.2.16) provides both fields as:

2.3 Electron Current Induced by Laser Fields

$$\mathbf{E} = E_0(\omega)\boldsymbol{\varepsilon} \sin(\mathbf{kr} - \omega t)$$
$$\mathbf{B} = E_0(\omega)\omega(\mathbf{k} \times \boldsymbol{\varepsilon}) \sin(\mathbf{kr} - \omega t) \quad (2.2.18)$$
$$E_0 = \omega A_0$$

Using (2.2.16), the energy density of (2.2.4) is obtained in the form:

$$W = \varepsilon_0 \omega^2 A_0^2(\omega) \sin^2(\mathbf{k} \cdot \mathbf{r} - \omega t) \quad (2.2.19)$$

Then, the time average of the energy density is obtained as:

$$\rho(\omega) = \frac{1}{2}\varepsilon_0 \omega^2 A_0^2(\omega)$$

The pointing vector of the energy flux density is

$$|\mathbf{S}| = I(\omega) = c\rho(\omega)$$

The energy flux density is equal to the laser intensity.

2.3 Electron Current Induced by Laser Fields

In order to evaluate RHS in (2.2.10), the current induced by the external fields should be calculated. In most of materials including plasma, the current is caused by the motion of electrons. In metal and plasma, there are a lot of free electrons carrying the current. The electrons are accelerated by electric and magnetic fields, and their motions are governed by the equation of motion with Lorentz force. It is given in general in the form:

$$\frac{d}{dt}\mathbf{p} = -e(\mathbf{E} + \mathbf{v} \times \mathbf{B}) \quad (2.3.1)$$

where $\mathbf{p}$ is the momentum of an electron, and v is its velocity. They have the relation in relativistic mechanics:

$$\mathbf{p} = m\gamma\mathbf{v} \quad (2.3.2)$$

where m is the electron mass and γ is Lorentz factor. Then, the current by single electron $\mathbf{j}_s$ at the position $\mathbf{r}$ and time t is defined as:

$$\mathbf{j}_s = -e\mathbf{v}\delta(\mathbf{r}, t) \quad (2.3.3)$$

where δ is the delta function.

In order to see the relativistic effect on RHS in (2.3.1) for an electron motion in laser field, assume that the laser field is approximated by an electromagnetic field in vacuum. Then, (1.3.1) requires the relation of the amplitude ratio of the form:

$$|\mathbf{E}| = c|\mathbf{B}| \tag{2.3.4}$$

This relation means that the magnetic force becomes important only in the case of relativistic motion. In the non-relativistic case, it is possible to treat the magnetic force as perturbation and Lorentz factor $\gamma = 1$.

Consider an electron motion in non-relativistic case for a linearly polarized laser in y-direction in (2.2.16), the oscillation motion only by electric field in y-direction is easily obtained from (2.3.1) as:

$$\frac{v_{os}}{c} = -a_0 \cos(\xi) \tag{2.3.5}$$

where ξ is the phase of the laser field defined as:

$$\xi = kx - \omega t \tag{2.3.6}$$

In (2.3.5), the nondimensional parameter is defined as:

$$a_0 \equiv \frac{eE_0}{m\omega c} = \frac{eA_0}{mc} \tag{2.3.7}$$

In (2.3.7), a_0 is a normalized amplitude of laser electric field and a very important parameter indicating the importance of relativistic effect in electron motion in strong laser field. It is called **strength parameter**. As shown below, non-relativistic condition is satisfied only for the case with $a_0 \ll 1$.

Evaluate the contribution of magnetic force as a perturbation force in non-relativistic case. Inserting the velocity of (2.3.5) into $\mathbf{v} \times \mathbf{B}$ term in (2.3.1), an approximate force to an electron is obtained in the form:

$$\frac{d}{\omega dt}\begin{pmatrix} v_x \\ v_y \\ v_z \end{pmatrix} = c \begin{pmatrix} \frac{1}{2}a_0^2 \sin(2\xi) \\ a_0 \cos(\xi) \\ 0 \end{pmatrix} \tag{2.3.8}$$

It is clear that for $a_0 \ll 1$, the dominant motion is a simple harmonic oscillation in the y-direction. On the other hand, the magnetic force is nonlinear force, and the oscillation frequency is 2ω. Since the x-component oscillates two times during the oscillation of y-direction, it makes **eight-figure motion** (figure "8").

It is, of course, clear that the solution of (2.3.8) cannot be used for relativistic regime. It is useful to relate the strength parameter to the laser intensity. It is calculated as a function of laser intensity I_L in W/cm^2 and its wavelength λ in cm:

2.3 Electron Current Induced by Laser Fields

$$a_0 = 8.5 \times 10^{-6} \lambda \sqrt{I_L} \qquad (2.3.9)$$

Since the conventional high-power laser has the wavelength around μm, the critical laser intensity giving $a_0 = 1$ is about 10^{18} W/cm^2. For the convenience to compare experimental conditions, (2.3.9) is written as a function of laser intensity in the unity of 10^{18}W/cm^2 as I_{18}, and laser wavelength in μm unit λμm.

$$a_0 = 0.85 \sqrt{I_{18} \lambda_{\mu m}^2} \qquad (2.3.10)$$

It will be derived in Chap. 8 that the exact solution in the relativistic motion for any value of a_0 is given as:

$$\frac{p_y}{mc} = a_0 \cos(\xi) \qquad (2.3.11)$$

$$\frac{p_x}{mc} = \frac{1}{2} p_y^2 \propto a_0^2 \qquad (2.3.12)$$

In this solution, the electron is assumed to be at rest before the laser comes, and due to the laser momentum, the electron is drifting to x-direction almost speed of light for the relativistic intensity case ($a_0 > 1$).

For highly relativistic intensity, the electron momentum is larger in the x-direction than in the y-direction by a factor of about a_0. It should be noted, however, that the energy increase of the electron is only due to the electric field. As shown in Chap. 5, an equation to the time evolution of electron energy can be derived from (2.3.1) to be:

$$\frac{d}{dt}(mc^2 \gamma) = -e\mathbf{v} \cdot \mathbf{E} \qquad (2.3.13)$$

Even in highly relativistic case, the electron energy increases only through the interaction with the laser electric field. The $\mathbf{v} \times \mathbf{B}$ force changes the direction of the momentum, namely, it is from y-direction to x-direction. It seems, however, difficult to image the physical reason why the momentum in the perpendicular direction becomes larger than that in the direction of electric field.

The finiteness of the speed of light enhances the dominant increase of the momentum in x-direction. As shown in (2.3.12), the electron obtains the momentum in the x-direction via $v \times B$ force, and its velocity approaches almost the speed of light. This means the electron can remain in the same phase of the laser field, and energy increases continuously from (2.3.13). However, the electron velocity cannot be the same as the speed of light, and de-phasing happens after a long time causing the change of sigh of electric field in (2.3.13). Such effect is proportional to a_0, and the x-momentum increases as increase of a_0.

Consider the case of a circularly polarized laser. The force due to electric field always works as centripetal force, and it keeps the rotating motion. Consequently, the magnetic field is always parallel to the electron velocity. The magnetic field force always vanishes and (2.3.1) leads a solution that the electron rotates with the velocity:

$$\frac{v_c}{c} = a_0 \quad (2.3.14)$$

The radius of the electron rotation can be obtained with the balance of the centrifugal force equal to the force by the electric field:

$$r_c = \frac{a_0}{k} \quad (2.3.15)$$

It is noted that the absolute values of v_c and r_c are the same as v_{os} and x_{os}, respectively. It should be noted that as you see below, the electron orbit changes to a spiral motion in the z-direction in the relativistic case. In addition, note that as far as a_0 is much smaller than unity, the oscillation amplitude is much shorter than the laser wavelength.

2.3.1 Antenna and Thomson Scattering

Before leaving this section, let us consider simple effects of an electron oscillation as emission and scattering of electromagnetic fields suggested in (2.3.10) intuitively.

It is obvious that an oscillating current generated at a certain point with oscillation frequency ω generates electromagnetic waves to propagate in space with the wavenumber $k = \omega/c$. This is the basic principle of antenna. **Yagi antenna** is well-known and widely used. A snapshot of the wave propagation by an oscillating current at the center is plotted in Fig. 2.9. It is well-known that the oscillating charge

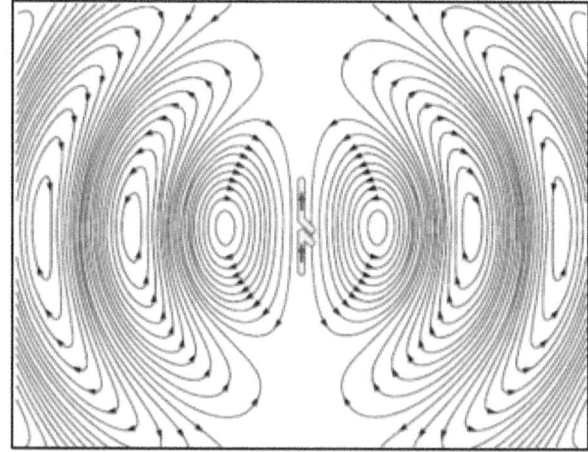

Fig. 2.9 A snapshot of the electric field generated by oscillating current located at the center is shown. An electron oscillating by external electromagnetic field emits such waves to scatter the electromagnetic waves

2.3 Electron Current Induced by Laser Fields

diploe emits the electromagnetic wave with its intensity dependence on angle proportional to:

$$I \propto \left(\frac{d\mathbf{j}}{dt}\right)^2 \cos^2\theta \tag{2.3.16}$$

where θ is the angle of the propagation direction from the horizontal axis. Time derivative of the current stems from RHS of (2.2.10) and is proportional to the square of the charge acceleration. This dependence represents **Larmor emission** of radiation.

The fact that the oscillating current generates electromagnetic field means that impinging electromagnetic field is scattered by free electrons in plasma. This scattering is **Thomson scattering**. The cross section of Thomson scattering σ_T doesn't depend on parameters of the electromagnetic field and constant:

$$\sigma_T = \frac{8}{3}\pi r_e^2 = 6.65 \times 10^{-25} \text{cm}^2 \tag{2.3.17}$$

where r_e is the **electron classical radius** defined as:

$$\frac{e^2}{4\pi\varepsilon_0 r_e} = mc^2 \quad \Rightarrow \quad r_e = \frac{e^2}{4\pi\varepsilon_0 mc^2} \tag{2.3.18}$$

The first relation above means that the electron-binding energy mc^2 can confine against Coulomb repulsion at the radius of r_e. It is noted that the electron density in solid metal is roughly of the order of 10^{22-24} cm^{-3}, and the scattering fraction of mm size sample is weak. With use of precise measurements of the scattered photons of cohenerent X-ray, say XFEL (X-ray- free electron lasers), from a dynamically compressed sample, it is possible to measure the column density [product of density and length], and the electron temperature from the spread of spectrum by Doppler shifts due to thermal electrons.

Since Tomson scattering gives a fraction of the momentum of the electromagnetic field ($\hbar \mathbf{k}$), a very bright light source repels the falling electron cloud to the source. Such situation may happen in the case of the birth of very massive stars like 100 times solar mass. Once the electrons obtain photon momentums, accompanying protons are also dragged by the ambipolar electric field generated by electrons. Assuming hydrogen plasmas are falling to the surface of a baby star with mass M and its total light energy flux L (luminosity), the maximum mass of the star is limited by the following force balance relation:

$$\sigma_T \frac{L}{c4\pi R^2} = G\frac{Mm_p}{R^2} \tag{2.3.19}$$

where R is the radius of the star, and G and m_p are the gravitational constant and the mass of a proton. In general, the luminosity L has strong dependence on the mass M as suggested in a stellar evolution theory; (2.3.19) provides the limit of the mass of a massive star M which can be a stationary star. This limiting mass is called **Eddington limit**.

2.3.2 Electron Current in Matters

To make the calculation more general, consider the finite resistivity even in conducting metals and plasmas. It is noted that the high-temperature plasma is of very good conductor, but still small resistivity remained due to electron binary collisions with heavy ions. This is a kind of viscosity fluid, and the frictional force should be included in (2.3.1).

Start with (2.3.1), neglect small force by magnetic field, and include the effect of electron collision by the background ions. The equation to derive the free electron current in non-relativistic condition is given to be:

$$m\frac{d}{dt}v = -eE - m\nu v \qquad (2.3.20)$$

where ν is the **collision frequency** of electron with the background ions. In order to obtain the induced current by laser field such as in (2.2.18), assume complex expression to (2.3.20) and assume that **E** is in the x direction inducing the velocity v. The following electron velocity is obtained:

$$v = -i\frac{e}{m\omega}\frac{1}{1+i\nu/\omega}E \qquad (2.3.21)$$

Assuming that all electrons have the same velocity, the following DC current is derived for $\omega = 0$:

$$j = -env = \frac{e^2 n_e}{m\nu}E = \frac{e^2 n_e}{m}\tau E$$

where n_e is the electron density, and the **collision time** $\tau(\equiv 1/\nu)$ is also introduced. It is familiar that metals are good conductor due to a lot of free electrons inside. The **electrical conductivity** σ is defined by:

$$j = \sigma E \qquad (2.3.22)$$

The DC conductivity σ_{DC} is obtained in the form:

2.3 Electron Current Induced by Laser Fields

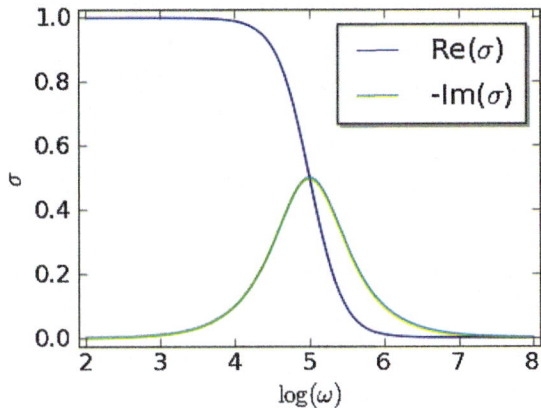

Fig. 2.10 The real and imaginary parts of the factor from DC to AC conductivity in Drude model defined in (2.3.18) is plotted. The imaginary part is at the maximum for the frequency $\omega\tau = 1$

$$\sigma_{DC} = \frac{e^2 n_e}{m}\tau \qquad (2.3.23)$$

It is easy to understand this relation, because the conductivity is higher in less collision in material, and higher density of free electrons. The well-known basic relation of Ohm law for an electric device is given in the form:

$$V = RI \qquad (2.3.24)$$

where V is an applied voltage, I is electric current, and R is resistivity in the unit of Ω. Comparing (2.3.22) to the relation of a device (2.3.24), the local **electrical resistivity** η should be:

$$\eta = 1/\sigma \quad [\Omega \cdot cm] \qquad (2.3.25)$$

In the case of alternating electric field like laser radiation in plasmas, the AC conductivity is given as **Drude model** in the form:

$$\sigma_{AC} = \sigma_{DC}\frac{1}{1+i\omega\tau} \qquad (2.3.26)$$

$$\frac{1}{1+i\omega\tau} = \frac{1}{1+\omega^2\tau^2} - i\frac{\omega\tau}{1+\omega^2\tau^2}$$

The real and imaginary parts of DC conductivity of (2.3.26) are shown in Fig. 2.10 as a function of frequency. It is noted that the real part of the conductivity drops near the frequency equal to the collision frequency. In the limit of no friction, the conductivity becomes pure imaginary in the form:

$$\sigma = -i\frac{e^2 n_e}{m\omega} \tag{2.3.27}$$

In this book, plasmas are frequently assumed collisionless; then the conductivity is pure imaginary, and no net energy loss takes place for waves in plasmas. Such media is called **reactive media** in comparison with the **dissipative media**. The RHS of (2.2.2) is time averaged to give the energy loss rate of the wave in the form:

$$\langle jE \rangle = \frac{1}{2}\sigma_r E_0^2 \tag{2.3.28}$$

where σ_r is the real part of the conductivity.

When (2.2.2) is regarded as the energy propagation of the laser in plasmas, taking the time average of the fields and use the following relations:

$$\langle W \rangle = \frac{1}{2}\varepsilon_0 E_0^2$$
$$\langle S \rangle = \langle W \rangle c = I_L \tag{2.3.29}$$

then one can obtain the following transport equation to laser intensity I_L:

$$\frac{\partial I_L}{\partial t} + c\frac{\partial I_L}{\partial x} = -\frac{\omega_{pe}^2}{\omega}\frac{\omega\tau}{1+\omega^2\tau^2} I_L \tag{2.3.30}$$

In (2.3.30), a very important physical value ω_{pe} is introduced. It is defined as:

$$\omega_{pe} = \left(\frac{e^2 n_e}{\varepsilon_0 m}\right)^{1/2} \tag{2.3.31}$$

This is the **electron plasma frequency** or simply called **plasma frequency** and one of the most important physical values in plasma physics. It is noted that in deriving (2.3.22), it is assumed that $\omega/k = c$, but it is not satisfied in plasmas in general as seen later.

It is clear that the coefficient of the RHS in (2.3.30) is **absorption rate** ν_{ab}:

$$\nu_{ab} = \frac{\omega_{pe}^2}{\omega}\frac{\omega\tau}{1+\omega^2\tau^2} = \frac{\omega_{pe}^2}{\omega^2}\frac{\nu}{1+\nu^2/\omega^2} \tag{2.3.32}$$

It has the following limiting dependence:

$$\omega\tau \gg 1 \quad \rightarrow \quad \nu_{ab} = \frac{\omega_{pe}^2}{\omega^2}\nu \tag{2.3.33}$$

$$\omega\tau \ll 1 \quad \rightarrow \quad \nu_{ab} = \frac{\omega_{pe}^2}{\nu} = \tau\omega_{pe}^2 \qquad (2.3.34)$$

It should be noted that for the given collision and plasma frequencies, the dependence of the absorption rate becomes opposite one to the collision frequency in two limiting regimes of the frequency of the external fields. It is strange intuitively the absorption rate of (2.3.34) is proportional to the collision frequency inversely. This is the case of DC conductivity, and electrons obtain large amount of energy before the energy is converted to the thermal energy, random velocities, before each collision.

2.4 Electron Coulomb Collision by Ions in Plasma

It is clear that in order to know how efficiently laser heats the plasmas, it is necessary to obtain the collision frequency of the plasmas.

The gas theory can be applicable not only to neutral gas but also plasmas both of which consist of thermal particles. Most properties of electron collision in plasmas can be derived according to the particle collisions in gas. As long as, therefore, the plasma is locally in thermodynamic equilibrium, the thermodynamics properties, statistical behavior, and hydrodynamics can be formulated from the analogy of gas. In what follows, the resemblance and difference in case of plasma compared to neutral molecular gas are explained by comparing the mean free paths due to collisions.

Although molecules in air cannot be seen by the naked eyes, it is made of a group of molecules with 80% nitrogen and 20% oxygen. The molecules interact with each other by molecular polarization force and collide like billiard balls. The radius at which the molecule force becomes strong is the effective distance of **Van der Waals force**, and for oxygen molecules, it is about 1.5 Å. Since the number density of molecules in air is about 3×10^{19} cm^{-3}, the mean free path of air is roughly evaluated as:

$$\ell_{air} = \frac{1}{\pi r^2 n} = 5 \times 10^{-5} \quad [\text{cm}] \qquad (2.4.1)$$

where $r = 1.5$ Å is assumed. The average molecular thermal velocity is roughly the speed of sound. Setting average molecular thermal velocity is $v = 330$ [m/s]; then it is calculated easily that the each molecule collides another molecules 6.6×10^8 times per second. This means that since the molecules very frequently exchange energy and momentum with other molecules, air is in thermodynamic equilibrium at any time much less than a second. Therefore, it is reasonable to use Maxwell distribution to the velocity distribution of the group of air molecules. In such molecular gas or liquids, collisional mean free path is too short, and, for example, the diffusion time due to the collisions is very long as see later. As the result, not

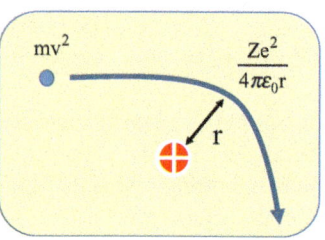

Fig. 2.11 A schematic trajectory of an electron scattered due to the Coulomb force of an ion at the center

molecular diffusion but convective and turbulent diffusion are more effective in general. This physics, however, is very complicated and out of the scope of this chapter. Readers may see the physics of turbulence in Vol. 2.

For the case like molecule gas, the mean free path is the same for any particles with different velocity if the collision cross section σ is constant. In this case, it is easy to imagine the collision process like hard ball collision, and transport coefficients are easily calculated. However, plasmas are complicated because it is made of ions and electrons with mass ratio of about 2000, and collision is due to Coulomb force, and its effective cross section is inversely proportional to the square of the thermal energy as seen below. The main idea of collisional process is the same as molecule gas, and the mean free path is derived after calculating the Coulomb collision cross section based on **Rutherford scattering** to be shown in Appendix-1.

It is instructive to evaluate the Coulomb cross section intuitively without invoking to the detail calculation with Rutherford scattering. Assume that an electron impacts an ion with the ionic charge Z and impact parameter r as shown in Fig. 2.11, then strong Coulomb collision may work if the following relation is satisfied:

$$\frac{Ze^2}{4\pi\varepsilon_0 r} = \frac{1}{2}mv^2 \tag{2.4.2}$$

Here v is the velocity of the impacting electron. The Coulomb collision cross section obtained from this radius is:

$$\pi r^2 = \pi \left(\frac{2Ze^2}{4\pi\varepsilon_0 mv^2}\right)^2 = 4\pi b_0^2 \tag{2.4.3}$$

Here, b_0 is the impact parameter providing the 90 degree scattering in the calculation by Rutherford [see (A1-9)], and the 90 degree scattering cross section is given as follows:

$$b_0 = \frac{Ze^2}{4\pi\varepsilon_0 mv^2}$$
$$\sigma_{90}^R = \pi(b_0)^2 \tag{2.4.4}$$

The impact parameter constant b_0 is only a function of Z and electron kinetic energy and

2.4 Electron Coulomb Collision by Ions in Plasma

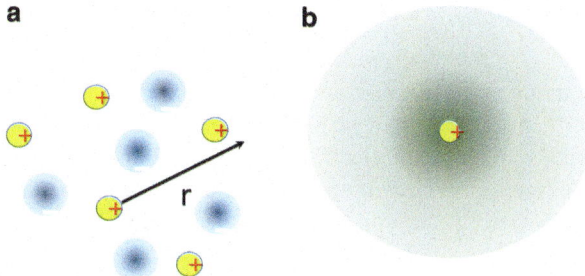

Fig. 2.12 The image of Debye shielding. (**a**): Schematic image of a snapshot of electrons and ions near a certain ion. The electrons are attracted by the Coulomb force of the ion, while ions are repulsed by this force. (**b**): The image of the Debye sphere where a huge number of electron and ions are distributed statistically. The green color means the distribution of negative charge. At the radius of Debye sphere, the 1/r Coulomb charge in vacuum is shielded, and the central ion has no effect outside of the Debye sphere

$$b_0 = 1.1 \times 10^{-8} \frac{Z}{\varepsilon_{eV}} \quad [\text{cm}] \quad (2.4.5)$$

where ε_{eV} is the kinetic energy of the impacting electron. It is noted that scattering cross section like molecules of about 1 Å corresponds to the collision in plasma with electron thermal energy of about 1 eV, and the cross section decreases in proportion to the inverse of square of the energy of impacting electron.

The 90 degree scatter mean free path is consequently proportional to the square of energy in the form.

$$\ell_{90}^R = \frac{1}{n_i \sigma_{90}^R} = 2.6 \times 10^{15} \frac{(\varepsilon_{eV})^2}{Z n_e} \quad [\text{cm}] \quad (2.4.6)$$

where n_e in [cm^{-3}]. Note that this mean free path is about 80 times longer than Coulomb mean free path as seen soon later. For example, in magnetic confinement fusion, the plasma with the density of 10^{14} cm^{-3} and the temperature of about 10 keV are confined in a torus device. The electron mean free path of (2.4.6) is roughly 6500 km or 100 km. It is clear that such high-temperature fusion plasma is collisionless plasma.

In order to evaluate the transport properties by Coulomb collision in plasma, small-angle scattering whose impact parameter is larger than (2.4.5) should be considered quantitatively. Rutherford scattering differential cross section is used to integrate all contribution of different angle scattering and the contribution by integrating over all velocities. However, a new problem appears, that is, the integration over the total impact parameter leads the divergence of the total scattering cross section. The Coulomb charge shielding schematically shown in Fig. 2.12a around a certain ion should be taken into account for small-angle scattering as described.

2.4.1 Debye Shielding

Electrolyte that helps to make water electrically conducting is salt (NaCl) which is decomposed Na^+ and Cl^- in water. This is a good example of shielding media. Inserting electrodes in such an electrolyte, Debye found that when applying a voltage, negative ions gather around the positive electrode, and positive ions gather around the negative electrode; consequently, the electric field is shielded in the electrolyte. This is **Debye shielding**.

Since plasma is a mixture of positive charge ions and negative charge electrons, the same phenomenon is expected. In the case of electrons and ions, the masses and thermal speeds are greatly different, so the shielding distance and time response of shielding are not the same. The shielding distance is called **Debye length** and is one of the most fundamental physical quantities in plasma physics. The picture of electron cloud around the central ion is shown in Fig. 2.12b. Debye length is obtained as follows mathematically.

Let's see how the electrostatic potential due to an ion charge located at the center spreads in the plasma. The equation to be solved is Poisson Eq. (2.2.3), and it is enough to assume spherical symmetry. Since electrons and ions are in Boltzmann distribution to the potential $\phi(r)$, the following equation is obtained for $r > 0$:

$$\varepsilon_0 \frac{1}{r^2} \frac{d}{dr}\left(r^2 \frac{d}{dr}\phi\right) = -e\{Zn_i(r) - n_e(r)\}$$
$$= -en_0\left[\exp\left(-Z\frac{e\phi}{T_i}\right) - \exp\left(\frac{e\phi}{T_e}\right)\right] \quad (2.4.7)$$

where n_0 is the average density of electrons. This is a nonlinear differential equation for the potential $\phi(r)$. This is called the **Debye-Huckel equation**. Although this cannot be solved analytically, it can be assumed in the high-temperature plasma that the energy due to the Coulomb force is sufficiently smaller than the kinetic energy. Then, the exponential function of (2.4.7) can be approximated by Taylor expansion:

$$\exp\left(-Z\frac{e\phi}{T_i}\right) - \exp\left(\frac{e\phi}{T_e}\right) \approx -\frac{e\phi}{T_e} - Z\frac{e\phi}{T_i}$$

Note that RHS of (2.4.7) cancels for $\phi(r) = 0$ and change the variable as:

$$\phi(r) = \frac{Ze}{r} y(r)$$

It is easy to find the following solution of the Debye shielded potential from (2.4.7):

2.4 Electron Coulomb Collision by Ions in Plasma

$$\phi(r) = \frac{Ze}{4\pi\varepsilon_0 r} e^{-r/\lambda_D} \tag{2.4.8}$$

Here, λ_D is **Debye length**

$$\frac{1}{\lambda_D} = \sqrt{\frac{Ze^2 n_0}{\varepsilon_0 T_i} + \frac{e^2 n_0}{\varepsilon_0 T_e}} \quad (= k_D) \tag{2.4.9}$$

It is useful to see the image of charge distribution given in (2.4.7) as shown in Fig. 2.7b. Such sphere with the radius of Debye length is called **Debye sphere**. The electrons gather around the central charge to shield the electric field generated by at the center. In (2.4.9), the **Debye wave number** k_D is also introduced for the convenient in expression. In the case of rapid changing phenomena like electron wave oscillation, the ions cannot follow the rapid change and cannot contribute to such shielding. It is appropriate to define the Debye length only by electrons:

$$\lambda_D = \sqrt{\frac{\varepsilon_0 T_e}{e^2 n_0}} \tag{2.4.10}$$

The dependence of Debye length on temperature and density is intuitively clear. At high temperature, electrons cannot be bend by the Coulomb force of the central ion charge because of larger kinetic energy; consequently, the distance need for shielding becomes long. In high- density case, many electrons contribute the shielding, and its distance becomes short as (5.9).

2.4.2 Yukawa Potential

It is noted that the radial dependence of potential (2.4.8) is well-known as **Yukawa potential** relating to his theory of meson, the particle which is the origin of nuclear force. Yukawa thought that the relativistic wave equation for meson particle replaced the energy and momentum in relativistic mechanics:

$$E^2 = c^2 p^2 + m^2 c^4$$

whose corresponding operator in quantum mechanics to obtain **Klein-Gordon** equation:

$$-\hbar^2 \frac{\partial^2}{\partial t^2} \psi = -\hbar^2 c^2 \nabla^2 \psi + m^2 c^4 \psi$$

Assuming steady state, Yukawa shows the following equation to meson wave function:

$$\hbar^2 \nabla^2 \psi = m^2 c^2 \psi \qquad (2.4.10')$$

Writing this equation spherically symmetrically would be the same as the equation resulting Debye shielding. Namely, correspondence of ψ and $mc/\hbar$ in (2.4.10') to ϕ and k_D in (2.4.7) in spherically symmetric condition provides the same mathematical solution to the wave function ψ. This is the reason why (2.4.8) is called Yukawa potential.

The effective distance, however, is much shorter than Debye length so that it is about the distance between nuclei.

$$\Delta x = \frac{\hbar}{mc} \qquad (2.4.11)$$

Yukawa evaluated the mass of meson so that (2.4.11) is equal to the nuclear distance. It is clear when one insets the radius of the distance of nuclei as 1 fm ($=10^{-13}$ cm), the meson mas becomes about 200 times electron mass.

It is also useful to note that since the electromagnetic force is mediated by photon; however, photon does not have any mass, and as a result, the force can extend to infinity by assuming the speed of light is infinite.

2.4.3 Coulomb Logarithm (Log)

The Coulomb potential by a charged particle cannot extent more than Debye length due to the exponential decay of the potential shown in (2.4.8). Consequently, the differential cross section obtained by Rutherford scattering in Appendix-1 cannot be applicable to the radius farer than Debye length.

It is usual to calculate the collision cross section mathematically with the use of Rutherford scattering. In what follows, derive it by using the fact that in Coulomb potential, the effective scattering occurs as the accumulation of small-angle scattering. After N times such as small-angle scattering (random walk in angle space), the following relation is obtained by the random walk model:

$$\sqrt{\langle \theta^2 \rangle} = \sqrt{N} |\Delta \theta| \qquad (2.4.12)$$

$$N = \frac{\langle \theta^2 \rangle}{|\Delta \theta|^2} \qquad (2.4.13)$$

Here $\Delta \theta$ is small angle by each scattering. In order to see substantial angle change, say around 90 degree $\langle \theta^2 \rangle \sim O(1)$, $N \approx |\Delta \theta|^{-2}$ time scattering is required. That is, for the scattering with impact parameter $\lambda_D > b >> b_0$, the differential cross section for (b, b + db) is given as:

2.4 Electron Coulomb Collision by Ions in Plasma

$$d\sigma_C = 2\pi b db |\Delta\theta|^2 \quad (2.4.14)$$

Here, the small-angle scattering angle is roughly evaluated as the ratio of impulse to momentum:

$$\Delta\theta \approx \frac{F \cdot \Delta t}{mv} = \frac{Ze^2}{4\pi\varepsilon_0 b^2} \frac{b}{mv^2} = \frac{b_0}{b} \quad (2.4.15)$$

Here F is Coulomb force at $r = b$ and b_0 in (2.4.4) is used. The relation of (2.4.15) is also obtained in the precise calculation with Rutherford scattering (Appendix A1).

Integrating (2.4.14) as follows, Coulomb collision cross section is obtained:

$$\sigma_C = 8\pi b_0^2 \int_{b_{min}}^{b_{max}} \frac{1}{b} db = 8\pi b_0^2 \ln\left(\frac{b_{max}}{b_{min}}\right) \quad (2.4.16)$$

Within the classical plasma assumption, the following evaluation may be used:

$$\begin{aligned} b_{min} &= b_0 \\ b_{max} &= \lambda_D \end{aligned} \quad (2.4.17)$$

The cut of integral at the minimum of b_0 is usually called **Landau cut**. Then, scattering cross section (2.4.16) is given:

$$\sigma_C = 8\ln(\Lambda)\sigma_{90}^R \quad (2.4.18)$$

where (2.4.4) is used and

$$\Lambda = \left\langle \frac{b_{max}}{b_{min}} \right\rangle = 4\pi n_e \lambda_D^3$$

In (2.4.18), $\langle\,\rangle$ means taking average to thermal electrons and the following relation was used:

$$\left\langle \frac{1}{2} mv^2 \right\rangle = \frac{3}{2} T_e$$

It is noted that the large-angle scattering cross section is roughly πb_0^2 and the **Coulomb logarithm**

$$\ln\Lambda \quad (2.4.19)$$

which is much larger than unity in normal plasmas. Such plasmas are called **ideal plasma**. In the ideal plasma, Debye sphere with radius λ_D contains a huge number of particles, and it is the reason why statistical model of (2.4.7) can be applied. This ideal plasma condition is:

$$\Lambda = 4\pi n \lambda_D^3 \sim \frac{4\pi}{3} n \lambda_D^3 \gg 1$$

The sphere made of the radius of Debye length is called **Debye sphere**. The typical values of $\ln \Lambda$ are around 10 and change very slowly to density and temperature. In the region where relatively low temperature and high-density region, the present ideal plasma assumption cannot be applied and (2.4.18) should be modified by taking account of the strongly coupling in nonideal plasma.

The b_{min} in the Coulomb logarithm has been evaluated classically while impacting electron has a finite size given by the uncertain principle, namely, **de Broglie length**. So, it is reasonable to think that the electron minimum impact should be larger than the de Broglie length:

$$b_{min}^{qm} = \frac{\hbar}{2mv}$$

It is better to define an effective b_{min} as:

$$b_{min} = \max \left\{ b_{min}^{cl}, b_{min}^{qm} \right\}$$

where $b_{min}^{cl} = b_0$ in (2.4.17). This condition is calculated to be for the thermal velocity $v = (T_e/m)^{1/2}$:

$$T_e \geq 20Z^2 \ [eV]$$

It is very clear that the quantum effect is dominant in most of plasma.

The electron collision frequency averaged over Maxwell distribution is obtained as:

$$\nu_{ei} = \langle n_i \sigma_C v \rangle = \frac{Z}{2\pi} \frac{\ln \Lambda}{n_e \lambda_D^3} \omega_{pe} \qquad (2.4.20)$$

The electron mean free path due to Coulomb scattering by ions in plasma is given from (2.4.18) by taking the average for Maxwell distribution:

$$\ell_{ei} = \frac{1}{n_i \langle \sigma_C \rangle} = \frac{2\pi}{Z \ln \Lambda} n_e \lambda_D^3 \ \lambda_D$$

where the mean free path of an electron with the velocity v is given with (2.4.6) in the form:

2.4 Electron Coulomb Collision by Ions in Plasma

$$\ell_{ei}(v) = \frac{1}{8\ln\Lambda}\ell_{90}^R \qquad (2.4.20')$$

As long as the plasmas are ideal plasmas and the number of electrons in Debye sphere is much larger than unity, the mean free path is much longer than the Debye length. This means the phenomena taking place over the distance of the order of Debye length are collisionless ones. From (2.4.20'), this effective mean free path is $8\ln(\Lambda)$ times shorter than that roughly estimated with only 90 degree scattering given in (2.4.6).

2.4.4 Collision Frequency and Electrical Resistivity

Using the mean free path, the collision frequency of free electrons scattering by heavy ions in the plasma is obtained. Since the collisions between electrons do not change the electron current, the collisions of electrons by ions in plasma determine the electrical conductivity. In the plasma, neutral atoms before ionization also contribute to the resistivity, because electron collision with neutral atoms works as drag to the electron flow. It is, however, neglected to deal with the case of the ideal plasmas.

First of all, the **collision time** τ_{ei} (the reciprocal of the collision frequency ν_{ei}) of the electrons due to the ions is defined by:

$$\tau_{ei} = \frac{1}{\nu_{ei}} = \frac{1}{n_i\langle\sigma_C v\rangle} = \frac{\ell_{ei}}{v_e} \qquad (2.4.21)$$

The collision time τ_{ei} is the time over which the electrons completely lose their initial momentum. This is also the frictional force to the electron motion, and the electrical conductivity σ defined in (2.3.15) is calculated with (2.4.21) directly:

$$\sigma_{DC} = \frac{e^2 n}{m\nu_{ei}} = \frac{8\pi}{Z\ln\Lambda}n_e\lambda_D^3\varepsilon_0\omega_{pe} \qquad (2.4.22)$$

It is easily found that the electric conductivity is high in the ideal plasmas, and the **electrical resistivity** of plasma η is defined to be:

$$\eta = \frac{1}{\sigma} = 3.2\times 10^{-11}\left(\frac{1\text{keV}}{T_e}\right)^{3/2} \quad [\Omega\cdot\text{cm}] \qquad (2.4.23)$$

The properties of the plasma resistivity in (2.4.23) are enumerated.

1. The resistivity does not depend on density.
2. In very high-temperature plasma such as fusion plasma, plasma responds like super conductor, and the plasma can be assumed to be collisionless. Collisionless

plasma should be treated kinetic theory, and the collisionless property makes plasmas very different from neutral gases and fluids.
3. Compare the resistivity of plasma to the other conducting metals in the same unit of (2.4.23). For example, cupper is 2×10^{-10} and stainless steel is 7×10^{-9}. In high-temperature plasmas, the temperature goes to more than 1 keV, and conductivity becomes much higher than such metals used in our life.
4. Tokamak plasma for fusion is heated up by the joule heating initially. It is, however, difficult to increase the temperature to initiate fusion reaction (10 keV), because of abrupt decrease of resistivity with increase of temperature. Additional heating such as wave heating (RF heating) or neutral beam injection heating is carried out.
5. Since the mean free path is electron energy dependent as shown in (2.4.20′), high energy component of electrons predominantly accelerated without collision in a strong electric field. Such electrons called **run-away electrons,** and the distribution function becomes to deviate from Maxwellian. The high-energy tail component is observed in high-temperature plasma production with electric field acceleration.

2.4.5 Relaxation Time to Thermal Equilibrium

Consider the case where the velocity distribution functions of electrons and ions are not in Maxwell distribution. According to the statistical mechanics, the particles exchange their momentum and energy through collisions and eventually become the Maxwell distributions with the same temperature after substantial time passes. It is well-known that the temperature equilibrium between ions and electrons takes a long time because of large mass ratio; consequently two Maxwellian with T_i and T_e are usually assumed.

In the case of molecular gas with same collision cross section σ and the same mass, the relaxation time to Maxwell distribution τ_M can be estimated

$$\tau_M = \frac{\ell}{\langle v \rangle} = \frac{1}{n\sigma \langle v \rangle} \quad (2.4.24)$$

Inserting the thermal velocity $<v>$ to (2.4.24), it is reasonable to regard that the particle velocity distribution relaxes to Maxwellian with the time scale of (2.4.24). In fact, the relaxation time is proportional to $1/v$, and it takes long time for particles with low velocity to tend to Maxwell distribution; however, it is only proportional to $1/v$ and not so strong dependence on the velocity compared to the case of plasma.

In charged particle collision, it is characteristic that the collisional relaxation time τ is proportional to the velocity as follows:

2.4 Electron Coulomb Collision by Ions in Plasma

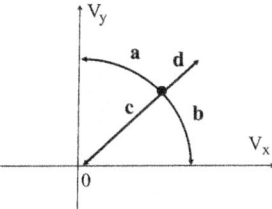

Fig. 2.13 In the velocity space of electrons, the scattering with ions changes the angle by keeping the absolute value of the velocity (energy), and almost no energy transfer is expected (the change is the direction a or b), while the collision with other electrons provides not only the angle scattering but also the energy change shown with the direction c or d

$$\tau \propto v^3 \qquad (2.4.24')$$

Since the relaxation time strongly depends on the velocity of plasma particles, there should be some unique characters in transport phenomena in plasma.

Consider the physical process where the non-Maxwellian velocity distribution finally becomes Maxwellian for electron and ion groups. It is important to take account of the fact that the electron-electron Coulomb collision allows exchange of momentum and energy in each collision and the same as between ion-ion collision. However, it is not the case of the collision of electrons by ions because of the big difference of the masses. The electrons can change momentum easily through the collision with ions, while energy does not change enough in this case. The ions do not exchange the momentum and energy substantially at the collision with electrons. Such fact indicates that the electron and ion groups become Maxwellian distribution keeping each total energy almost constant, namely, with different temperatures, and finally after a long time, both gases will be in the Maxwell distributions with the same temperature.

Now, consider the electron motion via Coulomb collision in the x-y 2-dimensional velocity space. As shown in Fig. 2.13, each electron moves like "a" or "b" after the collision with an ion, since the electron momentum predominantly changes after the collision with ion and the energy is almost kept constant. On the other hand, an electron changes momentum and energy after the collision with another electrons, namely, its position in Fig. 2.13 will move to any directions including special cases like a, b, c, and d.

Let us evaluate the relaxation times of electron and ion gases to their Maxwellian distributions only through the collisions with the same particles, electron gas τ_e and ion gas τ_i. The Rutherford scattering cross σ in (2.4.18) should be applied to the center of mass of binary collision, and the following relaxation times can be obtained:

$$\tau_e = \frac{1}{32\pi n_e} \left(\frac{4\pi\varepsilon_0 m_e}{e^2}\right)^2 \frac{\langle v^3 \rangle}{\ln\Lambda}$$
$$\tau_i = \frac{1}{32\pi n_i} \left(\frac{4\pi\varepsilon_0 m_i}{Z^2 e^2}\right)^2 \frac{\langle v^3 \rangle}{\ln\Lambda} \quad (2.4.25)$$

From (2.4.25), it is found that the relaxation times of electron gas and ion gas for the fully ionized hydrogen plasma with $Z = 1$ have the following relation:

$$\tau_e : \tau_i = 1 : \left(\frac{m_i}{m_e}\right)^{1/2} \quad (2.4.26)$$

It is noted that this root mass ratio is very large. If the velocity distribution of initial plasma is very far from Maxwellian, at first, electron gas becomes Maxwellian via Coulomb collision among electrons, and then the ion gas tends to Maxwellian via Coulomb collision among ions after the time more than 40 times.

Then, how long is the Coulomb collision relaxation time between electron gas and ion gas in fully ionized plasma? As mentioned above, the energy exchange between an electron and an ion in Coulomb collision is very inefficient. If the time scale of the plasma phenomenon is fast, it is reasonable to assume both charged gases have different temperatures, T_e and T_i. After the time scale of electron relaxation time τ_e and ion relaxation time τ_i, two gases relax their energy and momentum to finally become in thermodynamic equilibrium state with the same temperature. The exact derivation of this time scale is not so easy, but it is known to have the following simple relation including the electron-ion **energy relaxation time** τ_{ei}^e as briefly proofed soon below:

$$\tau_e : \tau_i : \tau_{ei}^e = 1 : \left(\frac{m_i}{m_e}\right)^{1/2} : \frac{m_i}{m_e} \quad (2.4.27)$$

As seen later in the case of laser heating of plasmas, most of the heating process or production process of plasmas is due to the input energy from outside like laser to deposit the energy to electrons. In the case where the main purpose is to produce high-temperature ions for the purpose like driving hydrodynamic phenomena or nuclear fusion, it takes time to transfer the electron energy to the ion energy.

The fact that energy relaxation between electron gas and ion gas takes long time as shown in (2.4.27) can be understood by a simple consideration. Assume that the time has passed and electron gas and ion gas are already in Maxwellian velocity distribution. This means the velocity distribution is isotropic, spherically symmetric in velocity space. Consider the following one-dimensional collision model to evaluate the energy transfer fraction via one collision event:

2.5 Lasers in Plasmas

$$m_e v_e + m_i v_i = m_e v'_e + m_i v'_i$$
$$m_e v_e^2 + m_i v_i^2 = m_e (v'_e)^2 + m_i (v'_i)^2 \qquad (2.4.28)$$

(2.4.28) are the momentum and energy conservation relations for electron-ion head-on collision. The LHS is before the collision, and RHS is after the collision. Eq. (2.4.28) can be solved approximately with using the relation $m_e \ll m_i$ and $v_e \gg v_i$. In such case, using the assumption that $v'_e \approx -v_e$, the ion momentum change after the collision can be obtained for both cases where the ion is moving in the same direction or opposite direction as the colliding electron:

$$m_i v'_i \approx m_i v_i \left(1 \pm 2 \frac{m_e}{m_i} \frac{v_e}{v_i}\right)$$

where (+ sign) means the ion moving in the same direction as the electron, while (− sign) is the ion moving with the opposite direction to the electron. Since the velocity distribution function is isotropic and it is expected that both collisions occur with the same probability, the following approximated relation can be obtained:

$$\left\langle \Delta\left(\frac{1}{2} m_i v_i^2\right)_L + \Delta\left(\frac{1}{2} m_i v_i^2\right)_R \right\rangle / \left\langle \frac{1}{2} m_i v_i^2 \right\rangle \approx \left\langle \frac{m_e}{m_i} \frac{v_e}{v_i} \right\rangle^2 \approx \frac{m_e}{m_i} \qquad (2.4.29)$$

After one collision between an electron and ion, energy is transferred by the fraction of electron and ion mass ratio which is very small. Namely, in order for both particles exchange energy substantially to finally become thermodynamic equilibrium, more than 1000 times Coulomb collisions are required. This is the reason why in plasma collision, large difference on the relaxation time appears in the plasma consisting of heavy ions and light electrons. These mass difference properties can be seen even in collision-less phenomena.

2.5 Lasers in Plasmas

Using the relations (2.3.27) and (2.3.22), the RHS of (2.2.10) is written in the form:

$$-\frac{1}{\varepsilon_0} \frac{\partial \mathbf{j}_{ind}}{\partial t} = \omega_{pe}^2 \mathbf{E} \qquad (2.5.1)$$

where the current is separated to the induced one by the electric field and the other external current, $\mathbf{j} = \mathbf{j}_{ind} + \mathbf{j}_{ext}$. The equation of electromagnetic wave propagation (2.2.13) reduces to

Fig. 2.14 The dispersion relation of electromagnetic waves in plasmas. Due to the induced current in plasmas, the low-frequency waves are forbidden to propagate in the plasmas

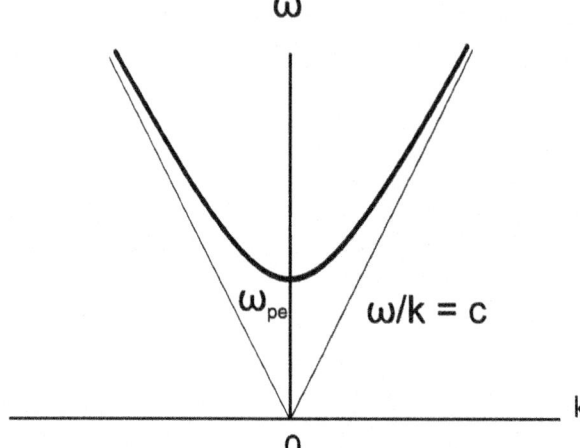

$$\frac{\partial^2}{\partial t^2}\mathbf{E} - c^2\nabla^2\mathbf{E} + \omega_{pe}^2\mathbf{E} = -\frac{1}{\varepsilon_0}\frac{\partial \mathbf{j}_{ext}}{\partial t} \quad (2.5.2)$$

In the case without the external current in plasma, the plane electromagnetic field in the form (2.5.2) propagating in a uniform plasma should satisfy the following relation:

$$\omega^2 = c^2 k^2 + \omega_{pe}^2 \quad (2.5.3)$$

This is **dispersion relation** of the electromagnetic waves in plasmas, and given k or ω, the other one is not independent and should satisfy (2.5.3) in order for the wave to propagate as a plane wave.

The dispersion relation (2.5.3) is plotted in (k, ω) space in Fig. 2.14. It is clear that in the high-frequency limit, it asymptotically becomes the relation of electromagnetic waves in vacuum. It is new to find that the wave cannot propagate with a frequency below the plasma frequency. The frequency

$$\omega = \omega_{pe} \quad (2.5.4)$$

is called **cut-off frequency**. It should be noted that the cut-off frequency is only the function of the plasma electron density, and the corresponding density is called **critical density** (n_{cr}). This cut-off property explains why X-rays penetrate our body, why mirror reflects the image, why metal box shield the electric noise, and so on. It is informative to know the phenomena happening at the cut-off point. Inserting the current (2.3.22) into (1.3.2), it is clear that:

2.5 Lasers in Plasmas

Fig. 2.15 The schematic picture showing the propagation of radio waves emitted from grand station to the space. The plasmas in the ionosphere reflect the waves emitted obliquely. Using this property, long-distance communication was done before satellite communication becomes popular

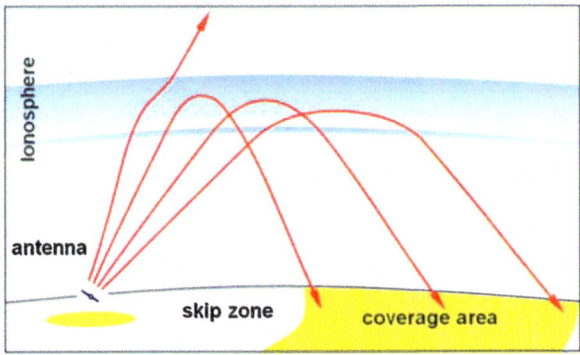

$$\frac{1}{\mu_0} \nabla \times \mathbf{B} = \mathbf{j} - \varepsilon_0 \frac{\partial \mathbf{E}}{\partial t} = 0 \quad \text{at } \omega = \omega_{pe}$$

The electromagnetic field can propagate thanks to the induction of the magnetic field by the time variation of the electric field. However, the induced current by the electric field becomes equal to the displacement current at the critical point, and the magnetic field is not induced any more.

In the situation where the plasma density gradually increases in space, consider how the electromagnetic field propagates. This has been studied relating to the radio wave communication in space in the early twentieth century. E. V. Appleton who carried out comprehensive study was appreciated as a winner of Nobel Prize in 1947. It is found that the electromagnetic wave approaching the cut-off density point is reflected, and the reflecting point gets lower density for oblique incident to the density gradient as seen in Fig. 2.15. In Fig. 2.15, trajectories of radio waves emitted from an antenna are shown in red arrows, when radio waves are emitted from the surface of the earth to the space. In space, the plasma layers are observed as the **ionosphere**. For the case of oblique incidence to the ionosphere with angle θ, the wave momentum in the perpendicular direction to the density gradient is conserved:

$$k \sin \theta : \text{const.}$$

Then, an effective wavenumber k_x in the direction of the density gradient becomes:

$$c^2 k_x^2 = \omega^2 - \omega_{pe}^2 - c^2 k^2 \sin^2 \theta \tag{2.5.5}$$

This simple relation explains the trajectories of the radio waves in Fig. 2.15. The waves are reflected at the point where $k_x = 0$ is satisfied. It is noted that the trajectory of the wave is described by the same equation of a point mass in a potential in Newton equation. In the past, one used such reflection by the ionosphere to send radio waves far away.

For the convenience of further study, introduce the **dielectric constant** of the plasmas. It is a function of k and ω in general; it is called **dielectric function**. In

Maxwell equation in matters, it is usual to introduce the polarization **P** by electrons, and (1.3.2) is shown in the form:

$$\frac{1}{\mu_0} \nabla \times \mathbf{B} = \frac{\partial \mathbf{E}}{\partial t} + \frac{\partial \mathbf{P}}{\partial t} = \frac{\partial \mathbf{D}}{\partial t}$$

where the displacement field is shown as:

$$\mathbf{D} = \varepsilon_0 \varepsilon \mathbf{E} \tag{2.5.6}$$

where ε is (relative) **plasma dielectric constant** and given to be:

$$\varepsilon(k, \omega) = 1 - \frac{\omega_{pe}^2}{\omega^2} \tag{2.5.7}$$

This means that the induced electron current is always opposite to the direction of displacement current in (1.3.2), and both cancel at the cut-off density. The dielectric constant has a direct relation with conductivity in the form:

$$\varepsilon(k, \omega) = 1 + i \frac{1}{\varepsilon_0 \omega} \sigma(k, \omega) \tag{2.5.8}$$

It should be noted that the imaginary part of ε is equal to the real part of σ. Therefore, the imaginary part of ε gives the net power of RHS in (2.2.2), and it is clear that

$$\text{Im}(\varepsilon) \begin{cases} > 0 & \text{(dissipation)} \\ < 0 & \text{(generation)} \end{cases} \tag{2.5.9}$$

In discussing the properties of any waves in plasmas, deriving the imaginary part of the dielectric constant and identifying (2.5.9), it is clear that the plasmas or any media are absorbing energy of the wave or emitting waves in the plasmas. It is informative to show an example of both in plasmas. Plasma is dissipative in **Landau damping** shown in Vol. 3, and wave energy decreases, and electrons gain the energy. However, if the electron velocity distribution has a bump on tail, certain waves gain energy from the particles. The latter is called **inverse Landau damping** to be explained in Vol. 3. They will be explained in later chapter. The XFEL is the amplifier of X-ray by converting the relativistic electron beam kinetic energy to the coherent X-ray energy.

2.5.1 Classical Absorption of Laser Energy

Before calculating the absorption rate of laser energy to electrons in plasma, discuss about the effect of Coulomb collision between two electrons. It is clear that the total energy and momentum of two electrons are conserved after the collision. Even if the momentum change of one electron like the case of the collision by ions, but the changed amount goes to the other electron. As the two electron systems, the momentum is conserved, and the energy of two electron system is conserved. This represents that the electron-electron Coulomb collision does not contribute to the absorption of laser energy to the electrons in plasma.

It is reasonable to regard (2.3.28) the absorption rate of lasers propagating in plasmas. By the use of the total collision frequency of electrons by ions, the velocity-dependent collision frequency ν_{ei} is written again in the form:

$$\nu_{ei}(v) = n_i \sigma v = \frac{Ze^4}{8\pi m^2} \frac{n_e}{v^3} \ln \Lambda \qquad (2.5.10)$$

The Coulomb log is reasonable to be evaluated by assuming electrons are in Maxwell distribution with temperature T. Then, the average collision frequency is obtained as:

$$\langle \nu_{ei}(v) \rangle = \frac{Z}{8\pi} \omega_{pe} \frac{1}{n_e \lambda_{De}^3} \ln \Lambda \qquad (2.5.11)$$

Inserting (2.5.11) as ν in (2.3.32), it is possible to obtain the absorption rate of laser in plasmas.

(2.3.33) has the following relation:

$$\frac{\langle \nu_{ei}(v) \rangle}{\omega_{pe}} \sim \frac{1}{n_e \lambda_{De}^3} \ln \Lambda \qquad (2.5.12)$$

As long as the plasmas are ideal plasmas and the plasma size is not large enough, the collisional absorption is weak.

It is useful to know the condition of the ideal plasma from microscopic view. The average distance between electrons r_a is defined to be:

$$\frac{4\pi}{3} n_e r_a^3 = 1 \qquad (2.5.13)$$

Then it is possible to define the Coulomb **coupling parameter** Γ as:

$$\Gamma_{ee} = \frac{e^2/(4\pi\varepsilon_0 r_a)}{T_e} = \frac{1}{(4\pi\sqrt{3})^{2/3}} \left(\frac{1}{n_e \lambda_{De}^3}\right)^{2/3} \quad (2.5.14)$$

(2.5.14) represents average Coulomb coupling energy divided by electron thermal energy. The ideal plasma also means the Coulomb interaction of electrons is much weaker than the freely running kinetic energy. In case of electron-ion coupling, just Z factor appears in (2.5.14).

Calculate the absorption fraction in an inhomogeneous plasmas, say laser-produced ablating plasma whose density continuously changes from the solid density to the vacuum. Since the laser propagates in plasmas, it is assumed that the laser intensity is stationary and possible to find the solution of (2.5.3). In order to obtain the absorption length locally, we assume that the change of density is very slow compared to the wavelength of laser, and the dispersion relation is satisfied locally. Then, the solution of the dispersion relation is obtained by assuming

$$\varepsilon = \varepsilon_r + i\varepsilon_i, \quad \omega = \omega, \quad k = k_r + ik_i \quad (2.5.15)$$

and Taylor expanding (2.5.3) to small imaginary components. The imaginary part of k provides the absorption rate in space:

$$k_i = \frac{1}{2} \frac{\nu}{c} \frac{\omega_{pe}^2}{\omega^2} \frac{1}{\sqrt{\varepsilon_r}} \quad (2.5.16)$$

The fraction of the absorption of laser from the incidence to plasmas and back to the vacuum is obtained by the following path integration:

$$\alpha_{abs} = 1 - \exp\left(-2 \int k_i dl\right) \quad (2.5.17)$$

It is given in Ref. [7] that the fraction of absorption is calculated for the density profile of exponential and linear ones with the scale length L, respectively, as:

$$\alpha_{abs}^{exp} = 1 - \exp\left(-\frac{8}{3} A \cos^3\theta\right) \quad (2.5.18)$$

$$\alpha_{abs}^{lin} = 1 - \exp\left(-\frac{32}{15} A \cos^5\theta\right) \quad (2.5.19)$$

where

$$A = \frac{\nu_{ei}^* L}{c}$$

2.5 Lasers in Plasmas

where ν_{ei}^* is the value at the cut-off density. It is useful how the fraction of absorption depends on the physical parameter of laser heating. When intense laser is absorbed at the surface of a solid target, abrupt increase of the pressure expands the heated material toward the vacuum, and this ablation plasma expands roughly with a speed of sound C_s. The sound velocity is a function of electron temperature Te, and after a while, stationary expansion wave is formed by the electron heat conduction to the over-dense region. The heat flux is sustained by laser heating so that the following relation is satisfied:

$$\alpha_{abs} I_L = f n_e T_e v_e \quad (2.5.20)$$

where f is a factor and assumed to be constant, about $\sqrt{m_e/m_i}$. The scale length of plasma increases in time, and it can be about the sound velocity times the laser pulse duration τ_L.

$$L = C_s \tau_L \quad (2.5.21)$$

It is enough for rough estimate to used Taylor expanded form of () giving an elation

$$\alpha_{abs} \frac{\tilde{\nu}_{ei}^* L}{c} \quad (2.5.22)$$

Combining (2.5.20), (2.5.21), and (2.5.22) and assuming that the typical density in these relations is given by the critical density, it is easy to find the following relation:

$$\alpha_{abs} \propto \frac{f^{0.4} \left(Z^{3/2} \tau_L\right)^{0.6}}{I_L^{0.4} \lambda_L^2} \quad (2.5.23)$$

where I_L and λ_L are the laser intensity and laser wavelength, respectively. This simple relation suggests that

1. Absorption is efficient for a long pulse lasers.
2. Increase of the laser intensity reduces the absorption fraction.
3. Shorter wavelength laser is better for higher absorption.

It is noted that with shorter wavelength laser, it deposits energy at higher critical density, and more electrons are heated with less temperature. It is obvious that a long pulse laser produces a long- scale plasma, and the laser can deposit its energy over a long traveling distance. In Fig. 2.16, many of experimental data are plotted as a function of laser intensity for four different wavelengths [8]. As suggested in (2.5.23), absorption fraction reduces as intensity increases, and the absorption is higher for shorter wavelength lasers. Material dependence and pulse duration dependence are also shown in Fig. 2.16 [8]. Except the detail values of the absorption rate, the dependence of (2.5.23) is well proofed in these figures.

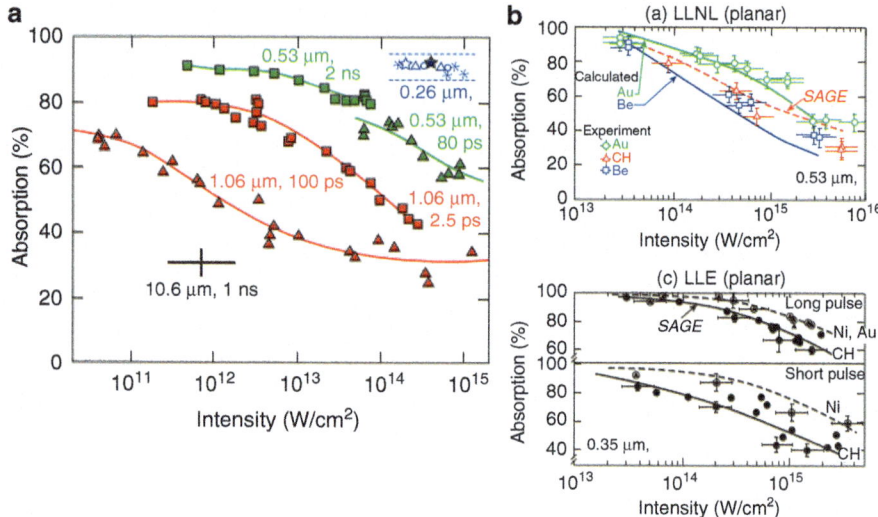

Fig. 2.16 (**a**): Many of experimental data of absorption rate are plotted as functions of laser intensity for the fundamental wavelength (1.06 μm), the second, third, and fourth harmonic wavelengths of glass lasers. Pulse duration is also shown in the figure. This figure drives the research direction that the shorter wavelengths have advantage to heat targets efficiently by classical absorption, namely, higher harmonic conversion is useful to produce idealistic hydrodynamic plasmas, such as ablations, shock waves, implosions, and so on. (**b**): Target material dependence (top) and pulse duration dependence (bottom) are also shown for plane targets. The data with SAGE are from computational results [8]

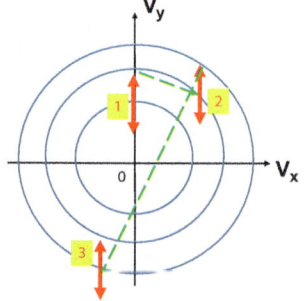

Fig. 2.17 Schematics of an electron orbit in velocity space under laser oscillation in y-direction. The electron position jumps by a pulse force due to Coulomb scattering by the fixed ions. This is a random walk in velocity space of electron by nonadiabatic force

2.6 Laser Absorption in Plasma

2.6.1 Physical Image of Classical Absorption

Before going to mathematics, consider how electrons oscillating under laser field can get energy from laser via Coulomb collision with fixed ions. For example, an electron running to y-direction is located at (1) in Fig. 2.17 with the oscillation shown with red arrow by the electric field in y-direction, linear polarized in

2.6 Laser Absorption in Plasma

y-direction. When the electron has pulse force due to the Coulomb scattering at the time with maximum velocity, it jumps to position (2) while preserving the kinetic energy, meaning the path in the same circle as seen in Fig. 2.17. Then, the pulse force should satisfy the nonadiabatic condition (Appendix 2). Then, strong force changes the momentum drastically like the jump to (3). If the jump happens when the electron has the maximum energy at (2), the average energy of the electron increases.

The laser absorption via Coulomb collision process can be described qualitatively as follows. Electron motion is the repeat of increase and decrease of kinetic energy by laser oscillation field. The Coulomb collision due to the fixed ions works as nonadiabatic force to electrons while keeping the kinetic energy. Change of the motion direction of electron allows the increase or decrease of energy after the short time nonadiabatic force. This is nothing without the famous random walk statistical problem. If we assume that the energy change by each collision is given by the change of absolute value of the velocity Δv, a kind of diffusion equation is obtained in velocity space. Then, the random number Δv should satisfy the relation

$$-v_{os} \leq \Delta v \leq v_{os}$$

where v_{os} is the amplitude of oscillation define din (1.3.10).

2.6.2 Simple Diffusion Model

It is useful to derive the absorption rate based on an intuitive model for collisional absorption. For simplicity, only the velocity space of electrons is considered. Define a velocity distribution function of electron in velocity space $f(v)$ then assume that the velocity distribution change in time because of random chance to obtain an additional velocity toward forward and backward directions in colliding with the background ions. At the time step n and the velocity point k, the following finite difference relation is obtained:

$$f_{n+1,k} - f_{n,k} = -f_{n,k} + \frac{1}{2}(f_{n,k+1} + f_{n,k-1}) \quad (2.6.1)$$

This is the same method to obtain diffusion phenomena in space based on random walk model. Then, the time step of n and the step of velocity jump k in (2.6.1) are assumed simply to be

$$\Delta v = v_{os}, \quad \Delta t = v_{ei}^{-1} \quad (2.6.2)$$

where v_{os} is the oscillation velocity by laser electric field defined in (2.3.5). It is reasonable to assume that an electron loses or gain the velocity of oscillation velocity in colliding with an ion, where it is assumed that the collision event happens in very short time compared to the laser oscillation time. Remain the lowest order terms of (2.6.1) after Taylor expansion:

$$f_{n+1,k} = f_{n,k} + \Delta t \frac{\partial f}{\partial t}$$

$$f_{n,k\pm 1} = f_{n,k} \pm \Delta v \frac{\partial f}{\partial v} + \frac{(\Delta v)^2}{2} \frac{\partial^2 f}{\partial v^2}$$

Assume that the finite difference form of (2.6.1) can be approximated with a partial differential equation. The following equation of diffusion type is obtained:

$$\frac{\partial}{\partial t} f = \frac{\partial}{\partial v}\left(D \frac{\partial}{\partial v} f \right) \qquad (2.6.3)$$

$$D = \frac{\Delta v^2}{2\Delta t} \Rightarrow v_{os}^2 \nu_{ei} = \frac{\omega_{pe}^2}{\omega^2} \frac{\nu_{ei}}{n} \frac{\varepsilon_0}{4} \langle E^2 \rangle \qquad (2.6.4)$$

It is possible to obtain the absorption fraction from (2.6.3). The increase of the kinetic energy of electrons in a unit volume is the deposited energy by laser:

$$\frac{\partial}{\partial t} \int \frac{1}{2} m v^2 f dv = -\int m v D \frac{\partial}{\partial v} f dv \approx \frac{\omega_{pe}^2}{\omega^2} \frac{\nu_{ei}}{2} W \qquad (2.6.5)$$

In deriving RHS of (2.6.5), the velocity in the collision frequency is set the thermal velocity, and W is the energy density of the laser electric field. This simple and intuitive random walk model also leads to the correct absorption rate of laser energy in plasma except the factor 1/2.

It is better to discuss here an appropriateness of using the assumption (2.6.2). It is assumed that at each collision of an electron against an ion, the electron always obtains the quivering velocity. This is possible only when the collision time is much shorter than the oscillation period, and the collision is nonadiabatic case (see Appendix-2). The collision time with the impact parameter b with impact velocity v should satisfy the following condition:

$$\frac{b}{v} \ll \omega^{-1} \qquad (2.6.6)$$

It is doubtful for small-angle scattering to satisfy this condition. Inserting Debye length and thermal velocity, (2.6.6) becomes

$$\frac{\omega}{\omega_{pe}} \ll 1$$

2.6 Laser Absorption in Plasma

This condition is in general not satisfied in normal plasmas, and the calculation of Coulomb log should be improved for the case of laser absorption. It will be discussed later.

2.6.3 Kinetic Derivation by Dawson and Oberman

Precise calculation of the inverse Bremsstrahlung absorption was done by Dawson and Oberman [9]. It is very important to know their theory for laser absorption. They start with Vlasov equation:

$$\frac{\partial f}{\partial t} + \mathbf{v} \cdot \frac{\partial f}{\partial \mathbf{r}} + \frac{\mathbf{F}}{m} \cdot \frac{\partial f}{\partial \mathbf{v}} = 0$$

Electron distribution function $f(t,\mathbf{r},\mathbf{v})$ is perturbed by two electric fields. One is laser electric field $\mathbf{E} = \mathbf{E}_0 \sin \omega_0 t$, and the other is the electrostatic fields by randomly located ions. The basic equation is given in the form:

$$\frac{\partial f}{\partial t} + \mathbf{v} \cdot \frac{\partial f}{\partial \mathbf{r}} - \frac{e}{m}(\mathbf{E}_0 \sin \omega_0 t - \nabla \Phi) \frac{\partial f}{\partial \mathbf{v}} = 0 \qquad (2.6.7)$$

where the self-consistent electrostatic potential Φ is determined by Poisson equation:

$$\varepsilon_0 \nabla^2 \Phi = e \left[\int f d\mathbf{v} - Z \sum_i \delta(\mathbf{r} - \mathbf{r}_i) \right] \qquad (2.6.8)$$

In (2.6.8) $\mathbf{r}_i$ is the position of the i-th ion. The space and velocity coordinates are transformed to those in the moving frame with the oscillation velocity by laser field. In this frame, the electrons are stationary fluid in oscillating ions. Then, the response of the electrons by the random ion fields is obtained as a small deviation from this stationary state. The transformation is:

$$\boldsymbol{\rho} = \mathbf{r} + \boldsymbol{\varepsilon} \sin \omega_0 t, \quad \mathbf{u} = \mathbf{v} + \omega_0 \boldsymbol{\varepsilon} \cos \omega_0 t$$

where

$$\boldsymbol{\varepsilon} = -e\mathbf{E}_0/m\omega_0^2$$

Then, the distribution function as a function of $(t, \boldsymbol{\rho}, \mathbf{u})$ is newly defined as $g(t, \boldsymbol{\rho}, \mathbf{u})$, and it is expanded with small perturbation in the form:

$$g(t, \boldsymbol{\rho}, \mathbf{u}) = g_0 + g_1$$

It is reasonable to assume that g_0 is given with Maxwell distribution for the velocity $\mathbf{u}$.

Vlasov equation to the linear perturbation g_1 is:

$$\frac{\partial g_1}{\partial t} + \mathbf{u} \cdot \frac{\partial g_1}{\partial \boldsymbol{\rho}} + \frac{e}{m} \frac{\partial \phi}{\partial \boldsymbol{\rho}} \cdot \frac{\partial g_0}{\partial \mathbf{u}} = 0 \qquad (2.6.8a)$$

and Poisson equation becomes:

$$\frac{\partial^2 \phi}{\partial \boldsymbol{\rho}^2} = 4\pi e \left\{ n_0 \int g_1 d^3 u - Z \sum_i \delta(\boldsymbol{\rho} - \boldsymbol{\varepsilon} \sin \omega_0 t - \mathbf{r}_i) \right\} \qquad (2.6.8b)$$

Carrying out Fourier-Laplace transformation of (2.6.8a) and (2.6.8b), they become:

$$g_{\mathbf{k},\omega} = \frac{e/m}{\omega - \mathbf{k} \cdot \mathbf{u}} \mathbf{k} \cdot \frac{\partial g_0}{\partial \mathbf{u}} \phi_{\mathbf{k},\omega} \qquad (2.6.9)$$

$$\phi_{\mathbf{k},\omega} = -\frac{n_0 e}{\varepsilon_0 k^2} \int g_{\mathbf{k},\omega} d\mathbf{u} + S_{\mathbf{k},\omega} \qquad (2.6.10)$$

where $S_{\mathbf{k},\omega}$ stems from the electric field by randomly located ions. It is calculated to be:

$$S_{\mathbf{k},\omega} = -\frac{Ze}{8\pi^3 \varepsilon_0 k^2} \sum_{n=-\infty}^{\infty} (-1)^n \frac{J_n(\mathbf{k} \cdot \boldsymbol{\varepsilon})}{\omega - n\omega_0} \sum_j \exp(-i\mathbf{k} \cdot \mathbf{r}_j) \qquad (2.6.11)$$

In obtaining (2.6.11), Fourier transform is done to obtain at first:

$$\int d\boldsymbol{\rho} \delta(\boldsymbol{\rho} - \boldsymbol{\varepsilon} \sin \omega_0 t - \mathbf{r}_i) e^{i\mathbf{k} \cdot \boldsymbol{\rho}} = (8\pi)^{-1} e^{i\mathbf{k} \cdot (\boldsymbol{\varepsilon} \sin \omega_0 t - \mathbf{r}_i)}$$

Then, the following mathematical relation is used:

$$e^{iz \sin \omega t} \equiv \sum_{n=-\infty}^{\infty} (-1)^n J_n(z) e^{in\omega t}$$

It is noted that (2.6.11) has many poles at $\omega = n\omega_0$ which means that there are **higher harmonic** oscillation components in the electric field.

Substituting (2.6.9) to (2.6.10) leads

2.6 Laser Absorption in Plasma

$$\phi_{k,\omega} = \frac{S_{k,\omega}}{D_{k,\omega}} \qquad (2.6.12)$$

$$D_{k,\omega} = 1 + \frac{\omega_{pe}^2}{k^2} \int \frac{1}{\omega - \mathbf{k} \cdot \mathbf{u}} \mathbf{k} \cdot \frac{\partial g_0}{\partial \mathbf{u}} d\mathbf{u}$$

where the denominator represents the dynamic screening effect by electron clouds. It is noted that in the limit $\omega \to 0$, one can obtain the Debye screening given in (2.4.8). Dawson and Oberman derived the absorption coefficient by calculating the joule heating from the electron current derived by the equation of motion of electrons:

$$\frac{\partial}{\partial t} \mathbf{j} = \frac{\omega_{pe}^2}{4\pi} \mathbf{E}$$

Then, time average of Joule heating $\langle \mathbf{j} \cdot \mathbf{E} \rangle$ gives the absorption rate of laser by plasmas. Finally the energy absorption rate ν_E relating to the collision frequency ν_{ei} is obtained in the form:

$$\nu_E = \frac{\omega_{pe}^2}{\omega_0^2} \nu_{ei} \qquad (2.6.13)$$

$$\frac{\nu_{ei}}{\omega_{pe}} = \frac{Z}{3} \left(\frac{1}{2\pi}\right)^{3/2} \frac{1}{n\lambda_{De}^3} \ln \Lambda \qquad (2.6.14)$$

$$\Lambda = \sqrt{2} (k_{max} \omega_{pe} / k_{De} \omega_0)$$

In deriving (2.6.13), the following assumption has been used:

1. Higher harmonic components in (2.6.11) are neglected.
2. Bessel function is approximated $J_1^2(\mathbf{k} \cdot \boldsymbol{\varepsilon}) \approx (\mathbf{k} \cdot \boldsymbol{\varepsilon})^2$ by assuming that the quivering distance is much shorter than the laser wavelength.
3. Laser intensity is assumed week enough to satisfy $v_{os}/v_e \ll 1$.

It is informative to compare the collision frequency in (2.6.14) with one obtained previously in (2.4.20):

$$\frac{\nu_{ei}(2.4.20)}{\nu_{ei}(2.6.14)} = \frac{3}{2} \left(\frac{\pi}{2}\right)^{1/2} = 1.89 \qquad (2.6.15)$$

The relation (2.6.15) is except Coulomb log definition. It is noted that the evaluation of $\ln \Lambda$ is also another discussion point.

2.6.4 Quasi-Linear Model of Absorption

Quasi-linear theory is widely used to include the first-order nonlinear effect to the time evolution of the electron distribution function. The detail explanation is given in, e.g., in [10]. The time average of the nonlinear product provides the diffusion term with the energy density of fluctuating electric fields. Let us evaluate the absorption rate with the use of the **quasi-linear diffusion theory** to the perturbation due to laser oscillations of electrons.

Assuming that the collision frequency is given and starting with the Boltzmann equation with **Krook collision** operator, try to derive the case including strong laser intensity. Then, the basic equation in uniform density is given:

$$\frac{\partial f}{\partial t} - \frac{e}{m}(\mathbf{E}_0 \sin \omega t) \cdot \frac{\partial f}{\partial \mathbf{v}} = -\nu(f - f_M) \tag{2.6.16}$$

where f_M is Maxwell distribution and a function of time. In this case, we assume that the laser field is small perturbation, and the velocity distribution function is given as the sum of Maxwellian and the perturbation f_1:

$$f(t, \mathbf{v}) = f_0(t, \mathbf{v}) + f_1(t, \mathbf{v}) \tag{2.6.17}$$

In addition, we assume that the zero order function is Maxwellian

$$f_0(t, \mathbf{v}) = f_M(t, \mathbf{v}) \tag{2.6.18}$$

Then, it is easy to obtain the following relation for the linear terms:

$$f_1 = i\frac{e}{m\omega}\frac{1}{1 + i\nu/\omega}(\mathbf{E}_0 \sin \omega t) \cdot \frac{\partial f_0}{\partial \mathbf{v}} \tag{2.6.19}$$

Inserting (2.6.19) into the second term in (2.6.16) and remaining only the component slowly varying in time, we obtain the following **quasi-linear diffusion** equation to the time evolution of the background electron Maxwell distribution:

$$\frac{\partial f_0}{\partial t} = \frac{1}{2}\mathbf{v}_{os} \cdot \frac{\partial}{\partial \mathbf{v}}\left\{\frac{\nu}{1 + (\nu/\omega)^2}\mathbf{v}_{os} \cdot \frac{\partial}{\partial \mathbf{v}}f_0\right\} \tag{2.6.20}$$

Note that time average $\langle \sin^2(\omega t)\rangle = 1/2$ and $\mathbf{v}_{os} = e\mathbf{E}_0/m\omega$. It is easy to consider the case of linear polarization. It is clear in (2.6.16) that the distribution function changes in time only the direction of laser polarization. If we can assume that $\nu/\omega << 1$ and ν can be replaced by some averaged value $\langle \nu \rangle$ we obtain:

$$\nu_{\text{abs}} = \frac{\omega_{\text{pe}}^2}{\omega^2} \langle \nu \rangle \tag{2.6.21}$$

It should be noted that the simple calculation results the same mathematical form of the laser absorption rate as (2.6.13). Then, the final problem becomes how accurately obtain the collision frequency given in (2.6.14). This can be also done with relatively easy way as done in Sect. 2.4. It is, of course, obvious to keep in mind that the precise theory given by Dawson and Oberman is the best way to calculate the classical absorption of intense lasers.

2.7 Bremsstrahlung and Collisional Absorption

It is useful to note the relation between radiation emission from plasma due to the Coulomb collision and the collisional absorption seen so far in plasma. The thermal radiation from the optically thin plasma is called **Bremsstrahlung**. It is well-known that the X-rays are generated when electron beams are irradiated on materials. The physical process of Bremsstrahlung is that when a free electron orbit is modified by an ion like Fig. 2.11, time change of electron current, namely, acceleration of orbit, is induced. From (2.3.16), Larmor emission of radiation with broad-spectrum is emitted from the electron, and the electron loses the corresponding energy after the radiation emission.

It is shown in Fig. 2.18a with Feynman diagram. The electron with momentum p interacts with an ion via virtual photon and scattered with less momentum p' to emit a photon (γ). Its reverse process is shown in Fig. 2.18b, where an electron interacts with a photon near the ion with a distance so that the virtual photon acts. In this case, the photon energy is absorbed by electron to be scattered with the photon energy. In these processes, the ion is assumed to have no reaction by such process. The process

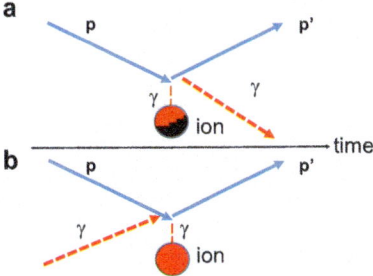

Fig. 2.18 (**a**) A Feynman-like diagram indicating Bremsstrahlung process. The electron with momentum p interacts with an ion via a virtual photon (γ) and scattered with less momentum p' to emit a photon (γ). The reverse process is shown in (**b**). An electron interacts with a photon near the ion with a distance so that the virtual photon acts. The process (**b**) is nothing without the collisional absorption, and it is also called inverse Bremsstrahlung (IB) absorption

(b) is nothing without the collisional absorption, and it is also called **inverse Bremsstrahlung (IB) absorption**.

If the system is optically thick plasma such as inside the sun, the photons generated by Bremsstrahlung process are soon absorbed by IB process. In such optically thick system, it is well-known the radiation is in Planckian energy distribution. It is said that in such LTE system, detail balance of each elementally process is established. Then, the photon distribution is Planckian distribution for a given temperature. Planckian distribution as a function of photon frequency is in the form:

$$f_P(\nu) = \frac{8\pi h}{c^3} \nu^3 \frac{1}{e^{h\nu/T}-1} d\nu \qquad (2.7.1)$$

In the laser absorption case, the photon distribution at laser frequency is much higher than (2.7.1) for plasma temperature, and the IB absorption is much larger than the Bremsstrahlung emission process.

Note that (2.7.1) has a form of (1.2.2) and g is proportional ν^3 and f is given in the term including exponential function in (2.7.1). If the density of state does not decrease so fast as ν, the distribution diverges near ε (=$h\nu$) = 0. This is Bose-Einstein condensation. However, the density of state of photon quanta is small near $\varepsilon = 0$, Planckian distribution has a peak near $h\nu = 2.8$ T.

References

1. T. Makabe, Z. Petrovic, *Plasma Electronics* (Taylor & Francis, 2006)
2. C.J. Joachain, N.J. Kylstra, R.M. Potvliege, *Atoms in Intense Lasers* (Cambridge University Press, 2012). M. Protopapas, C. H. Keitel and PO. L. Knight, Rep. Prog. Phys. **60**, 389 (1997). C. J. Joachain, EPL, **108**, 44001 (2014)
3. L.D. Landau, E.M. Lifshitz, *Quantum Mechanics* (Elsevier Science, 2003)
4. K. Mishima et al., Phys. Rev. A **66**, 033401 (2002)
5. S. Augst et al., J. Opt. Soc. Am. B **8**, 858 (1991)
6. S. Augst et al., Phys. Rev. A **52**, R917 (1995)
7. W.L. Kruer, *The Physics of Laser Plasma Interactions* (Addison-Wesley, New York, 1988)
8. R.S. Craxton et al., Phys Plasmas **22**, 110501 (2015)
9. J. Dawson, C. Oberman, Phys. Fluids. **5**, 517 (1962). Phys. Fluids. **6**, 394 (1963)
10. T.J.M. Boyd, J.J. Sanderson, *The Physics of Plasmas* (Cambridge University Press, 2003). Chap. 10

Chapter 3
Ultra-Short Pulse and Collisionless Absorption

3.1 Ultra-Short Pulse in Non-relativistic Intensity

After the appearance of ultra-short pulse technology in the 1980s as described in Chap. 1, the interaction of laser pulse of sub-picosecond duration with many solid matters has been studied intensively. The physics of the ablation by ultrafast intense laser, the removal of matter from solid surface or bulk, is of great fundamental and practical interest. Many papers have been published regarding experimental results and theoretical studies. The surface ablation of condensed matter under the irradiation of sub-picosecond laser pulses has a number of peculiar properties which distinguish this process from the ablation induced by nanosecond and longer laser pulses. In the absorption analysis, the penetration of laser field into solid density due to skin effect should be considered, and consequently the properties of conductivity in metal, insulator, and semiconductor become key physics to study laser absorption and subsequent ablation phenomena.

Ultra-short laser interaction with solids opens interesting opportunities for the study of optical and thermodynamic properties of matter with electron temperature higher than the solid matter, namely, the state called **warm dense matter (WDM)**. The diagram showing the important phenomena in the considered intensity range is given in Fig. 3.1 where pathways of the material from excitation, melting, and phase transition to ablation are shown around the relevant timescale and intensity ranges [1]. The intuitive image of Fig. 3.1 covers from about 10^{10} W cm^{-2} to above 10^{14} W cm^{-2}. At these intensities, a variety of phase transitions are observed near solid surface. Excitation of the solid by electron heating takes place due to the laser heating in pulse duration shown by red in the time scale in Fig. 3.1. Depending on excitation strength, melting occurs roughly on a picosecond timescale. In semiconductors and insulators irradiated with high laser intensities, the loss of crystalline order is possible within less than 1 picosecond. Laser-solid interaction in such

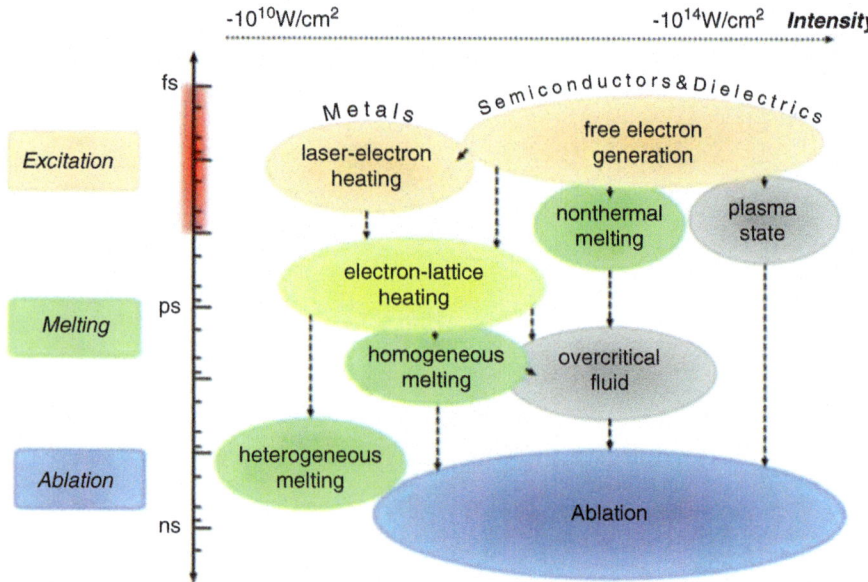

Fig. 3.1 Typical timescales and intensity ranges of several phenomena and processes occurring during and after irradiation of a solid with an ultra-short laser pulse of about 100 fs duration. Excitation takes place in the range of femtoseconds (duration of the laser pulse). The timescale of melting may vary for different processes and lies roughly in the picosecond regime. Material removal, i.e., ablation, lasts up to the nanosecond regime [1]

intensity regime is studied intensively relating to the laser cutting, material processing, and other applications [1].

The intensity of ultra-short pulse has reached around 10^{22} W/cm². As already mentioned, the relativistic effect and nonlinear physics become essential to study the physics of laser- plasma interaction with such ultra-intense lasers as seen in Chaps. 5, 6, 7, and 8, mainly focusing on the interaction of ultra-intense and ultra-short pulse with matter. Here, in contrast, the physics of interaction of ultra-short pulse and solids with sharp boundary is studied for the case of non-relativistic intensity.

When ultra-short laser pulse of laser wavelength $\lambda_L = 0.4\mu m$ and pulse duration 120 fs is irradiated onto a variety of solid materials with normal incident, the laser absorption fraction is observed over the wide range of laser intensity as shown in Fig. 3.2 [2]. As suggested above, the absorption fraction strongly depends on the difference of materials, metal, or insulator. The absorption is different for different types of the same metal in the range of laser intensity up to 10^{15} W/cm² ($\sim I_L \lambda_L^2 = 10^{14}$ W/cm²µm²). It is seen that insulator quartz has high absorption (~90%), but all materials show high reflectivity at high intensity, and they can be said to be in **universal plasma mirror**. The maximum temperature of the laser-irradiated material was also studied with **LASNEX** simulation code in LLNL with precision laser propagation program with solid conductivity model. The resultant values are plotted in Fig. 3.3 [2]. It is clear that the materials are at very high

3.1 Ultra-Short Pulse in Non-relativistic Intensity

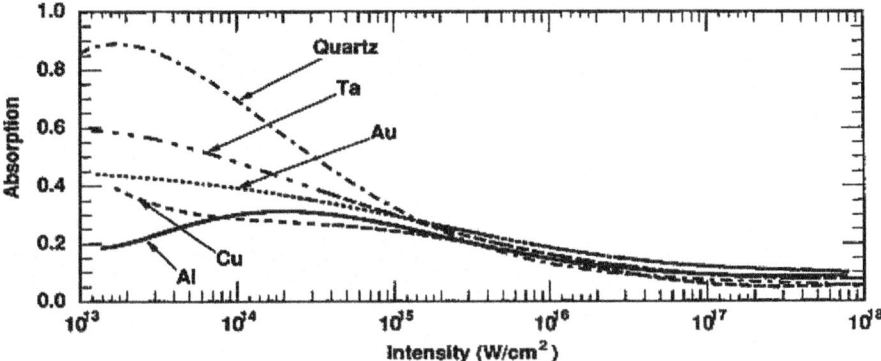

Fig. 3.2 Absorption fraction vs. peak laser intensity for aluminum, copper, gold, tantalum, and quartz targets. The laser intensity is the peak value. When ultra-short laser pulse of laser wavelength 0.4 mm and pulse duration 120 fs is irradiated onto a variety of solid materials with normal incident, the laser absorption fraction is observed over the wide range of laser intensity as shown in the figure. All materials show high reflectivity at higher intensity, and they can be said to be in universal plasma mirror. [Fig. 1 in Ref. 2]

Fig. 3.3 The predicted spatial maximum electron temperature occurring at the temporal and spatial peak of the laser pulse plotted as a function of the peak laser intensity. [Fig. 4 in Ref. 2]

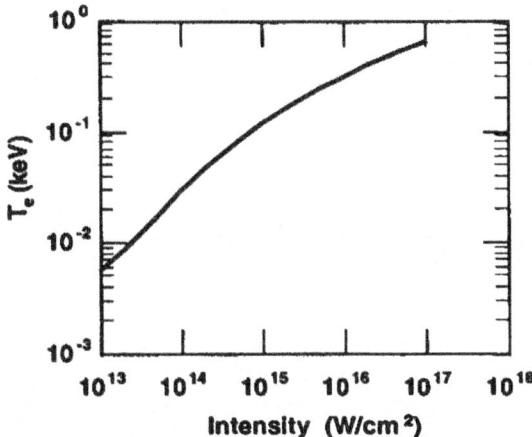

temperature near solid density, and lasers are reflected by such high-temperature and high-density state of matters.

This property is widely used as **plasma mirror** for focusing ultra-high intensity and ultra-short laser pulse, where no conventional material available for final focusing optics. Recent experimental data for designing the plasma mirror is shown in Fig. 3.4 for 200 fs laser (red) and 500 fs lasers (black) [3]. This also shows that substantial reflection of intense lasers is obtained over the laser intensity of 10^{15} W/cm^2.

Light absorption in ultra-short-scale length plasma has been studied by precisely solving the laser propagation Eq. (2.5.2) numerically in the complex space [4]. As long as the density profile is given, (2.5.2) is a linear differential equation to the laser

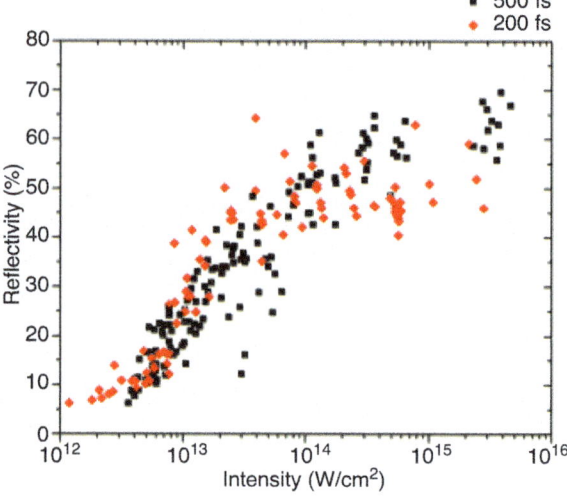

Fig. 3.4 Experimental data of laser reflectivity for designing the plasma mirror is shown for 200 fs laser (red) and 500 fs lasers (black) [3]

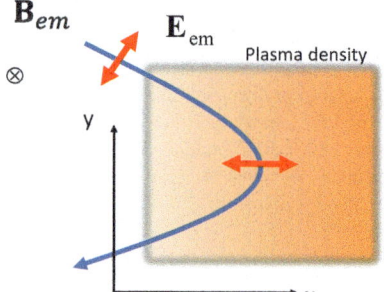

Fig. 3.5 Schematic picture of laser propagation for oblique incidence in an inhomogeneous plasmas. The red shows oscillating electric field for the case of p-polarization. It induces charge separation near the turning point due to the electric field in the direction parallel to the density gradient

electric field **E** or magnetic field **B**, and it is simply solved because of no external current. In order to solve (2.5.2), however, the induced current should be calculated from (2.3.12). Propagation Eq. (2.5.2) becomes equation to complex valuables. In solving (2.5.2), the two cases should be solved separately. The density profile is given of inhomogeneous in x-direction as shown in Fig. 3.5. In the laser propagation in **s-polarized** case, the equation to E in z-direction is solved, while in **p-polarized** case, the equation to B in z-direction should be solved for convenience of mathematics.

In the s-polarization, the electric field is always in z-direction, and it is convenient for numerically integrating the following **Helmholtz equation** for laser irradiated with the incident angle θ into the plasma with density gradient.

$$\frac{d^2}{dx^2}E + k_0^2[\varepsilon(x) - \sin^2\theta]E = 0 \qquad (3.1.1)$$

3.1 Ultra-Short Pulse in Non-relativistic Intensity

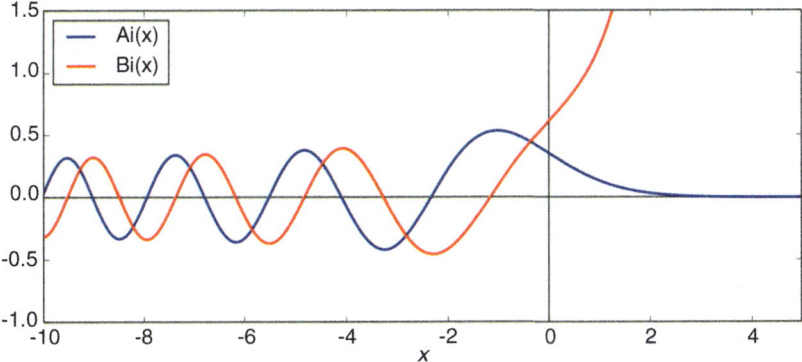

Fig. 3.6 Two independent solutions of Airy function. The laser waves penetrate over the turning point (x = 0) due to the tunneling effect typical for any wave phenomena

In the case of p-polarization since the magnetic field is always in z-direction, the Helmholtz equation to B is given as:

$$\frac{d^2}{dx^2}B + k_0^2[\varepsilon(x) - \sin^2\theta]B = -\frac{1}{\varepsilon}\frac{d\varepsilon}{dx}\frac{dB}{dx} \quad (3.1.2)$$

In (3.1.1) and (3.1.2), the location of the density satisfying the following condition is the **turning point**:

$$\varepsilon(x) - \sin^2\theta = 0 \quad (3.1.3)$$

In the case of a linear density profile, assume the turning point is at x = 0, (3.1.1) has the following form after stretching x-coordinate properly to a dimensionless coordinate ζ:

$$\frac{d^2}{d\zeta^2}E + \zeta E = 0 \quad (3.1.4)$$

(3.1.4) is well-known equation whose solution is **airy functions** with two independent solutions as shown in Fig. 3.6. The waves coming from $\zeta < 0$ region are reflected, and the wave penetrates into $\zeta > 0$ region evanescently due to the tunneling effect. Therefore, only the blue curve in Fig. 3.6 is physically meaningful solution. As the result of solving the wave equation, the penetration of the electric field in the over-dense region is obtained. For the normal incident, the oscillating laser field can penetrate the density higher than the cutoff density. It is noted that this is the same as the **tunneling effect** in Schrodinger equation for quantum particle wave function. Geometrical optics corresponds to the classical mass point motion governed by Newton equation, while the wave optics treated here is mathematically same as the wave function of quantum particles.

In the case of p-polarization, RHS of (3.1.2) has a singularity at the critical density ($\varepsilon = 0$), and the solution of B should satisfy the condition dB/dx = 0 at the critical point for avoiding nonphysical solution. It is known that the finite value of B is possible at the critical point, and as a result, the electric field in the x-direction remains at this point, namely, resonant point, as shown below.

Taking into account a finite collisional effect, the singularity of RHS in (3.1.2) can be avoided, and (3.1.2) becomes easier to be solve. Given the dielectric constant (2.5.8) with the complex conductivity (2.3.26) in the form:

$$\varepsilon = 1 - \frac{\omega_{pe}^2}{\omega^2}\frac{1}{1+\nu^2/\omega^2} + i\frac{\omega_{pe}^2}{\omega^2}\frac{1}{1+\nu^2/\omega^2}\frac{\nu}{\omega}$$

then, the complex variable of E in (3.1.1) is decomposed to a function with two real variables

$$E(x) = E(x)\exp[-i\phi(x)] \qquad (3.1.5)$$

Inserting this into (3.1.1), the following coupled equations are derived:

$$\frac{d^2}{dx^2}E - \left(\frac{d\phi}{dx}\right)^2 E + k_0^2[\cos^2\theta - \xi]E = 0$$
$$\frac{d^2\phi}{dx^2}E + 2\frac{d\phi}{dx}\frac{d}{dx}E - k_0^2\frac{\nu}{\omega}\xi E = 0 \qquad (3.1.6)$$

where

$$\xi = \frac{\omega_{pe}^2}{\omega^2}\frac{1}{1+\nu^2/\omega^2}$$

For p-polarization case, the same kind but more complicated coupled differential equations are obtained [4]. Both cases can be solved numerically for given density profile and the normalized resistivity ν/ω.

From the calculated E and B profile in s- and p-polarization cases, the magnetic field and electric fields are calculated, respectively, from Maxwell equations:

$$\mathbf{B} = -\frac{i}{\omega}\nabla \times \mathbf{E}$$
$$\mathbf{E} = \frac{c}{\omega\varepsilon(x)}\nabla \times \mathbf{B} \qquad (3.1.7)$$

In the case of p-polarization, the electric field in the x-direction at the resonance point is given with the solution of (3.1.2) as:

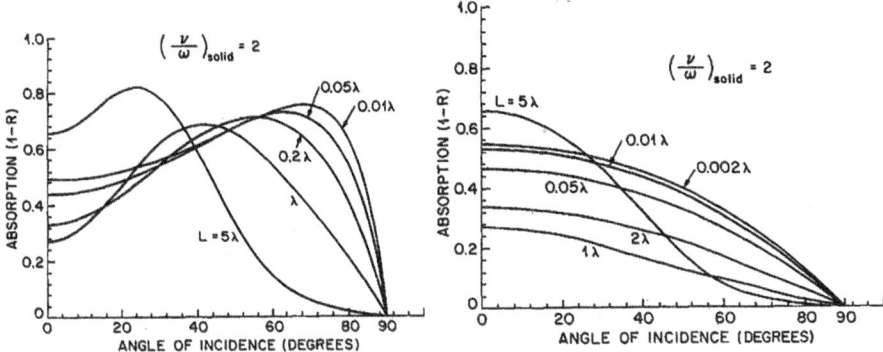

Fig. 3.7 Angular dependence of P-light (**a**) and S-light (**b**) absorption for $(\nu/\omega)_{\text{solid}} = 2$ and scale lengths $L/\lambda = 5, 1, 0.2, 0.05, 0.01$ with $\lambda = 308$ nm. As $L \to 0$ the dependence converges to that predicted by the wave equation for p-polarization. [Figs. 3 and 4 in Ref. 5]

$$E_x(x) = i\frac{\sin\theta}{\varepsilon(x)}B(x) \qquad (3.1.8)$$

It is clear that the case with small ν/ω, the electric field in x-direction is resonantly enhanced near the critical density. Given density scale length, it is seen that there is an optimum angle for the highest E_x value.

For a given value of the dissipation ν/ω, the Eqs. (3.1.6) are solved for s-polarization case, and the corresponding equations from (3.1.2) are also solved for p-polarization case numerically. Since assuming sub-picosecond pulse, the density scale length L is varied from very sharp case to 5λ, where λ is the laser wavelength in vacuum. The resultant angular dependence of absorption is shown in Fig. 3.7(a) and (b) for p- and s-polarizations, respectively. In Fig. 3.7, a large number of the dissipation is assumed $\nu/\omega = 2$. Since the density is very high and the temperature is low for such short pulse, Λ in (2.6.14) is of order of unity, and the ideal plasma formula (2.6.14) cannot be applied. The calculation results show that the absorption has the maximum angle for each scale length in p-polarization, while it is monotonic in s-polarization case. In both cases, substantial absorption is obtained.

In contrast, the absorption fraction decreases dramatically in the case of sharp density profile and small value of ν/ω for p-polarization as shown in Fig. 3.8. It is concluded that for the case with relatively long-scale length $L > 0.2\ \lambda$, the absorption profiles are resemble except for the shift of the angle at the peak absorption, and about 60% absorption is obtained at the peak, while the absorption fraction reduces rapidly as the decrease of the density scale length, and almost no absorption is obtained for the sharp boundary. It is clear in such cases that the modeling of the resistivity of the materials is very important issue to predict the laser absorption fraction.

Intensity dependence of the absorption is not determined within the present model, because (3.1.1) and (3.1.2) are linear equations to E and B, respectively,

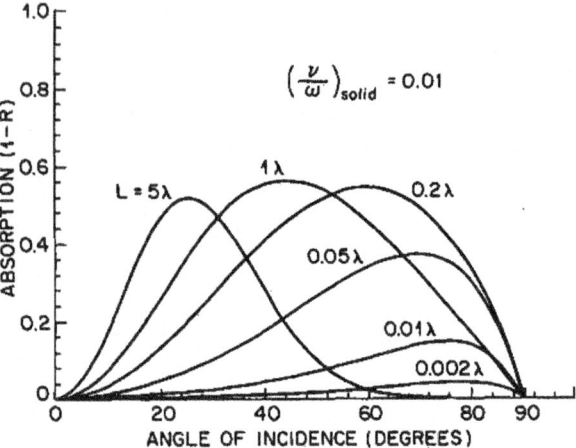

Fig. 3.8 Angular dependence of P-light absorption for $(\nu/\omega)_{solid} = 0.01$ and scale lengths $(L/\lambda) = 5, 1, 0.2, 0.05, 0.01,$ and 0.002 with $(n_e)solid = 2 \times 10^{23}$ cm^{-3}, $\lambda = 308$ nm. [Fig. 2 in Ref. 4]

for the given density profile and the constant collision frequency ν/ω. It is obvious that these physical values are the functions of the temperature of matter and the expanding density profile. They should be determined self-consistently as mentioned previously. At least hydrodynamic simulation with appropriate collision frequency modeling should be coupled with laser propagation Eqs. (3.1.1) or (3.1.2) for consistent approach. In the next session, let us see such work done in the early stage of short pulse interaction with solid materials.

3.2 Self-Consistent Analysis of Short Pulse Absorption

Modeling of the collision frequencies and related microphysics of the matter from solid in room temperature to plasma at high temperature has been done to couple with 1-D hydrodynamic simulation for sub-picosecond pulse lasers in a wide range of laser intensity 10^{11}–10^{17} W/cm^2 [5]. The accuracy of the modeling is proofed by comparing with the experimental data of aluminum solid shown in Fig. 3.2. It is pointed out that the following three points are well modeled in the simulation code:

(1) Precise treatment of laser propagation in steep gradient matter
(2) Precise modeling of collision frequency of electrons in solids to plasmas continuously
(3) Realistic electron ion temperature relaxation with separated equation of state (EOS) for electron fluid and ion fluid

The requirement (1) is done by solving (3.1.1) and (3.1.2) with complex dielectric constants as described in the previous section. For this purpose, the each Lagrangian fluid mesh is divided to sub-grids to keep the precision in numerical integration of Maxwell equations. To realize (2), the electron collision frequency for plasmas given in (2.6.14) is only used for the case where the electron temperature is higher than its

3.2 Self-Consistent Analysis of Short Pulse Absorption

Fermi temperature, and the collision frequency is calculated through electron-phonon interaction in the lower temperature. In the case of electron-phonon interaction, the electron collision frequency depends on the ion temperature determining the phonon amplitude, and the electron-ion collisional temperature relaxation time should be well modeled in non-plasma state. Since the phonon amplitude increases in proportion to the ion temperature, the collision frequency increases at first as the increase of ion temperature and then decreases when the collision frequency as plasmas becomes dominant as the form (2.6.14). In addition, the electron and ion temperatures are very different, and **equation of state (EOS)** should be modeled independently.

The collision of electrons by phonon in dense matter has been studied relating to the electron collision in the partially degenerated matter in white dwarfs and at the surface of neutron stars. When the electron temperature is below the Fermi temperature, the free electrons are predominantly scattered by the phonons or lattice vibrations, and the collision frequency in such case has been derived in the approximated form [5]:

$$\nu_{ep} = 2k_s \frac{e^2 T_i}{\hbar^2 v_F} \qquad (3.2.1)$$

where k_s is a numerical constant and v_F is the Fermi velocity. It is noted that the solid aluminum **Fermi temperature** ($= m v_F^2$) is 11.7 eV, and (3.2.1) is applicable for the lower temperature than about 10 eV. In contrast, (2.6.14) is good for higher temperature than the Fermi temperature in the dense aluminum. Both models, however, becomes nonphysical if the mean free path obtained with these collision frequencies becomes shorter than the average ion distance r_0 defined by:

$$\frac{4\pi}{3} r_0^3 n_i = 1$$

Note that this radius r_0 is usually called **ion sphere radius**. Taking account of such physics, the collision frequency from the solid material to ablating plasmas can be modeled as shown in Fig. 3.9 [5]. In Fig. 3.9, the thick solid curve is used for simulation, while the thin solid curve obtained from the interpolation formula of (2.6.14) and (3.2.1):

$$\nu^{-1} = \nu_{ei}^{-1} + \nu_{ep}^{-1} \qquad (3.2.2)$$

In Fig. 3.9, the collision frequency in the ambiguous region sandwiching the Fermi temperature is given by a simple relation that the mean free path is the ion sphere radius, namely, $\nu = v_e/r_0$. In addition, a collision frequency derived from the surface reflectivity at room temperature is used as the limiting value of collision frequency in the lowest temperature region as shown in Fig. 3.9.

Fig. 3.9 Collision frequency of solid aluminum as a function of the temperature $T_e = T_i$ (thick solid line). The thin solid line is the result of the interpolation of (3.2.2); the dashed line the upper limit of the collision frequency given by the requirement that the mean free path should be longer than the ion mean distance. [Fig. 1 in Ref. 5]

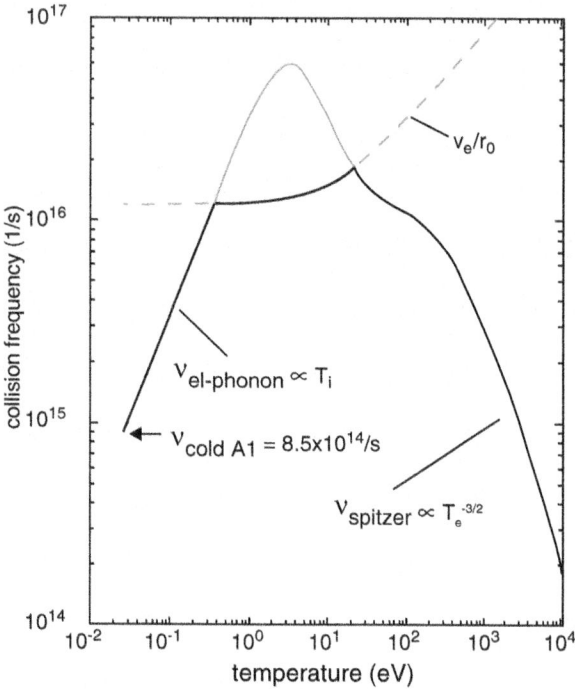

At the beginning of laser absorption, the temperature increases only at the solid surface and the heated energy is carried into the inner solid region via electron thermal conduction. This electron thermal conductivity K_e should be modeled in the electron energy equation in the form:

$$\frac{d}{dt}\varepsilon_e = \nabla(K_e \nabla T_e) \quad (3.2.3)$$

where ε_e is the internal electron energy per electron particle. Since the electrons transfer their energy by random walks via scattering by the background ions, the thermal conductivity can be also derived with the collision frequency and is known to have the form (Chap. 2, Ref. [1]):

$$K_e = K_0 \frac{T_e}{m\nu} \quad (3.2.4)$$

In (3.2.4), K_0 is a nondimensional constant and is given in the form $K_0 = 320/(3\pi)\delta\varepsilon$, where δ and ε are correction factors weekly dependent on the charge state, since the electron scattering also contributes to the thermal conductivity (Chap. 2, Ref. [1]). For the fully ionized aluminum, $K_0 = 10$.

3.2 Self-Consistent Analysis of Short Pulse Absorption

It is useful to note that the thermal conduction is the diffusion process of the temperature, and by the use of the same sort of random walks as (2.6.1) in the real space, the diffusion of temperature is roughly described as:

$$\frac{\partial T_e}{\partial t} = \frac{\partial}{\partial x}\left(D\frac{\partial}{\partial x}T_e\right), \quad D = \frac{(l_e)^2}{\tau} = \frac{v_e^2}{\nu} \quad (3.2.5)$$

Equation (3.2.5) is an approximate form of (3.2.3), since the internal energy of a particle in the ideal plasma is $\varepsilon_e = 3/2T_e$.

The requirement (3) is not at the moment well modeled, since it is very hard to obtain the electron-ion temperature relaxation time in the low-temperature and near-solid density state. As already explained in (2.4.27), the energy relaxation time in the Coulomb collision is about mass ratio longer than the electron collision time by ions. According to the detail calculation it is given to be with a factor 1/2:

$$\tau_{ei}^e = \frac{m_i}{2m_e\nu}$$

Compared to ν plotted in Fig. 3.9, this value is about 10^{-13} s, namely, the temperature relaxation time is about 10 ps. Setting this value as τ_R, it is used in the solid aluminum in simulation. The contribution to the electron and ion temperature internal energies by this relaxation effect is given as:

$$\frac{d}{dt}\varepsilon_e = -\frac{T_e - T_i}{\tau_R}, \quad \frac{d}{dt}\varepsilon_i = \frac{T_e - T_i}{\tau_R} \quad (3.2.6)$$

It is noted that the specific heat is defined by:

$$C_V^\alpha = \frac{\partial \varepsilon_\alpha}{\partial T_\alpha} \quad (\alpha = i, e)$$

The specific heat is almost order of unity for ions in low-temperature region and 3/2 at high temperature limit, while it is very small in the low-temperature degenerate state for electron and proportional to T_e. This is also the reason why the electron temperature increases abruptly in the partial degenerate state, while more energy is required to heat the ions. Since the temperature relaxation is sensitive to the electron-phonon collision to determine T_i via (3.2.6) and not so important for T_e higher than the Fermi temperature, τ_R was assumed constant, and the measured value of $\tau_R = 10$ or 20 ps at the solid state was used [5].

Finally, the EOS is modeled with SESAME table [6], which is mainly based on Thomas-Fermi EOS with the other correction, same as the concept of **quotidian EOS (QEOS)** [7].

It is demonstrated that such hydrodynamic model well explains the experimental results of short pulse over the wide range of laser intensities [5]. The aluminum data in Fig. 3.2 [2] is compared to the simulation results as shown in Fig. 3.10 [5]. The

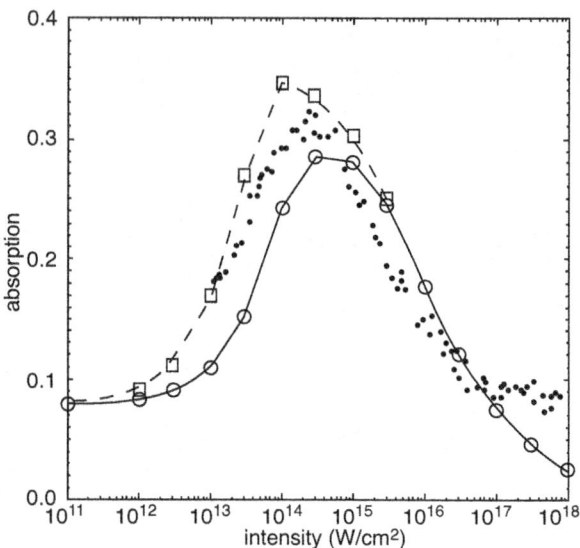

Fig. 3.10 Absorption as a function of the laser intensity. Pulse duration 150 fs, wavelength 400 nm. One-dimensional fluid simulations are done with different electron-ion relaxation times (solid line: 20 ps, dashed line: 10 ps). The experimental points are taken from Ref. 5. [Fig. 3 in Ref. 5]

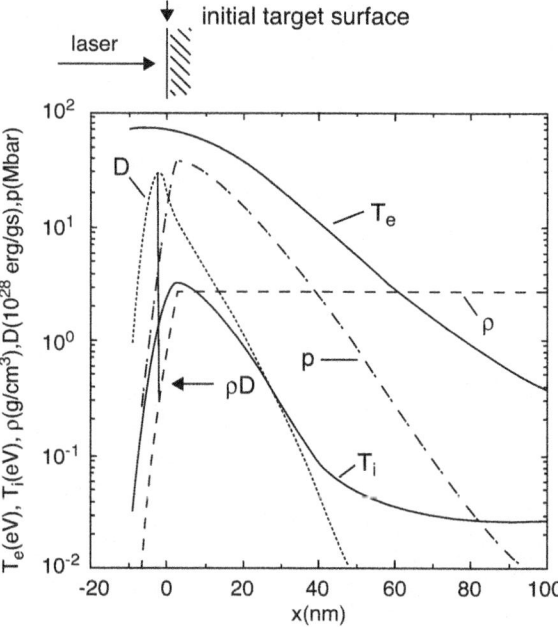

Fig. 3.11 Spatial dependence of the electron temperature (T_e), the ion temperature (T_i), the mass density (ρ), the pressure (p), and the laser energy deposition (D). Laser intensity is 10^{15} W/cm^2. [Fig. 5 in Ref. 5]

laser pulse is 150 fs and wavelength 0.4 μm, and laser is normal incident. The solid and dotted lines are data with $\tau_R = 20$ ps and 10 ps, respectively. Except for the intensity more than 10^{17} W/cm^2, the simulation well reproduces the experimental data. The physical values' profiles are plotted in Fig. 3.11 at the time of intensity

3.2 Self-Consistent Analysis of Short Pulse Absorption

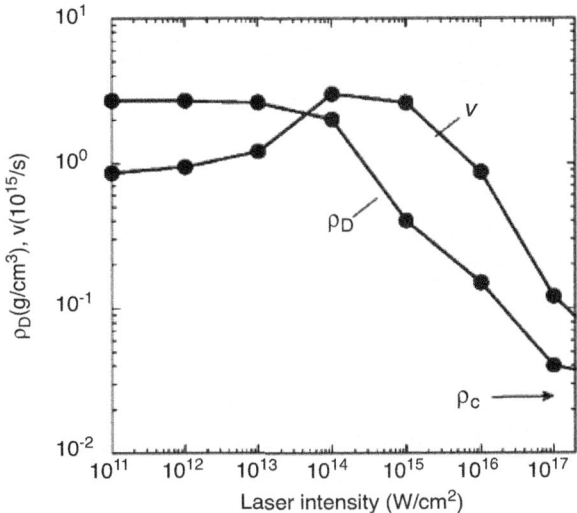

Fig. 3.12 Intensity dependence of the density ρ_D of the maximum laser deposition (see Fig. 3.11) and the collision frequency ν at this point. Laser conditions are the same as in Fig. 3.10. [Fig. 8 in Ref. 5]

peak for the case of laser intensity 10^{15} W/cm^2, where the peak absorption is observed.

In Fig. 3.11, the electron temperature (T_e), density (ρ), pressure (p), ion temperature (T_i), and the laser energy deposition rate (D) are plotted. The density ρ_D indicates the density where the peak of laser energy deposition is obtained. It is observed that at lower laser intensity, surface ablation is not dominant, and the laser is mainly absorbed in the over-dense solid region, while with increase of the intensity to about 10^{14} W/cm^2, ablation becomes dominant, and the volume of laser absorbing plasma increases. Such fact is seen in Fig. 3.12, where the density at the peak absorption and the corresponding collision frequency are plotted as functions of laser intensity. At the intensity of 10^{17} W/cm^2, the electron temperature increases to about 500 eV same as shown in Fig. 3.3, and the classical absorption becomes very less efficient as in Fig. 3.9.

It is also noted that the difference in 10^{17-18} W/cm^2 range seen in Fig. 3.10 is caused by several reasons: one is that the radiation pressure becomes higher than the pressure of laser absorbing region, the second is the laser oscillation velocity defined in (2.3.5) becomes larger than the thermal velocity, and so on. Of course, collective physics to be studied later in this chapter will couple to the irradiated laser and enhance the absorption rate, although most of such energy goes to the generation of hot electrons. Such physics is the main topics in the later part in this chapter.

This physical modeling is also compared to other experimental data. The laser polarization dependence of absorption rate is experimentally observed [8]. It clearly shows in Fig. 3.13 that enhanced absorption of p-polarization compared to s-polarization at oblique incidence. It is surprising to know that the present simulation well reproduces the experimental data for the intensity range of 10^{11-15} W/cm^2. In the experiment and simulation, the laser is pulse duration 400 fs and wavelength 308 nm with the incident angle 45°. The solid and dotted lines are simulation results

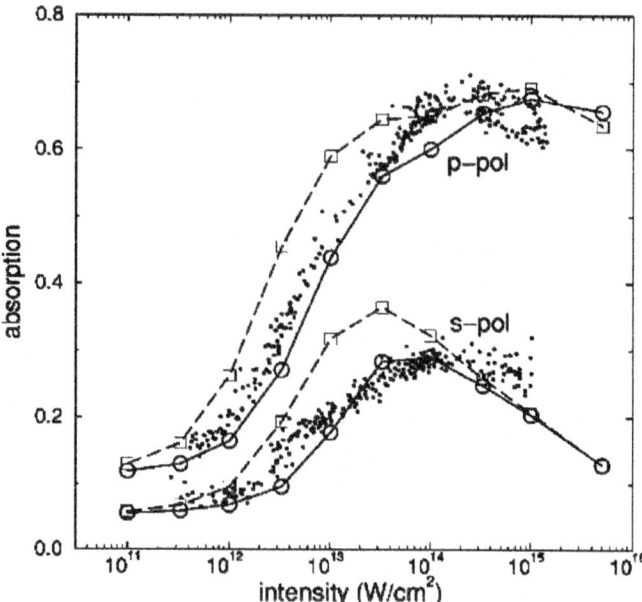

Fig. 3.13 Absorption at 45° as a function of intensity. Pulse duration 400 fs, wavelength 308 nm. The solid and dashed lines are calculated with the temperature relaxation time of 20 ps and 10 ps, respectively. The experimental data points (the small dots) are taken from Ref. 10. [Fig. 9 in Ref. 5]

with $\tau_R = 20$ and 10 ps, respectively. It is seen that even at sharp density gradient for less than 10^{13} W/cm^2, p-polarization is better absorption than s-polarization case. Angle dependence of absorption at intensity 10^{14} W/cm^2, pulse duration 250 fs at wavelength 248 nm is also studied experimentally [9]. Both experimental data and the corresponding computational result are compared about angular dependence of absorption rate for s- and p-polarization irradiation. It is demonstrated that this simulation can reproduce well the angle dependence of absorption rate for both polarizations and obtains the maximum absorption of about 60% at around 60° of irradiation.

It is important to discuss about the rough ratio of the collision frequency around the dominant absorption region to the laser frequency ν/ω which was a key nondimensional value in absorption calculation in Figs. 3.7 and 3.8. In Fig. 3.12, the density of the maximum absorption point ρ_D and the corresponding collision frequency ν are plotted along the same simulations resulting Fig. 3.11 [5].

It is interesting to know that the ν for less than 10^{15} W/cm^2 is about $1 \sim 3 \times 0^{15}$ s^{-1}, namely, $\nu/\omega \sim 0.2$–0.6 for wavelength 400 nm. This is far from the ideal plasma case as pointed out in Session 2.4. The density of the maximum absorption point is about 10–100 times the critical density, and it is also over the critical density even for the intensity where the ablation plasma formation is important in laser absorption. These facts indicate that the electron collision modeling by ion scatter in over critical density is essential to study the laser-matter interaction of sub-picosecond pulse with its peak intensity of wide range.

In the other words, laser tunneling effect into over-dense region is very essential in such short pulse absorption, and since the tunneling is determined by the precise properties of the dielectric constant, namely, complex form of electric conductivity, just a simple model in Fig. 3.9 in the range of about $T_e < 100$ eV doesn't give us any satisfaction in understanding the micro physics. We have to invoke to quantum theory to obtain the collision frequency consistently as seen below.

3.3 Quantum Theory of Electric Conductivity in Dense Plasmas

The intuitive upper limit of the collision frequency introduced in Fig. 3.9 with $\nu < v_e/r_0$ can be consistently derived with the use of quantum mechanical analysis for Fermi-Dirac distribution of electrons [11]. The procedure to obtain the fluctuating electrostatic fields by ions is the same as that used in obtaining (2.6.13) (Chap. 2, Ref. [9]). However, the **Fermi-Dirac distribution** function should be used, since the low temperature and Fermi degeneracy affect the electron collision process. After numerous mathematics, the following collision frequency of electrons by ions is derived for arbitrary amplitude of laser field; Eq. (33) in Ref. [10]:

$$\nu_{ei} = \frac{Zm}{\pi^2 \hbar} \frac{\omega^2}{\varepsilon_0 E_0^2} \int_0^\infty dq S_{ii}(\mathbf{q}, T_i) \\ \times \sum_{n=1}^\infty n\omega \mathrm{Im}\varepsilon_{ee}(\mathbf{q}, -n\omega, T_e) \int_{-1}^1 dz J_n^2\left(\frac{eE_0 q}{m\hbar\omega^2}z\right) \qquad (3.3.1)$$

In (3.3.1), the imaginary part of the dielectric constant of electrons $Im(\varepsilon_{ee})$ is given in Eq. (26) of Ref. [10]:

$$Im\varepsilon_{ee}(\mathbf{q}, -n\omega, T_e) = \frac{\pi e^2 \hbar^2}{\varepsilon_0 q^2} \int \frac{d^3 p}{(2\pi\hbar)^3} \delta(\hbar n\omega + E(\mathbf{p}) - E(\mathbf{p}+\mathbf{q})) \\ \times \{ f(\mathbf{p}) - f(\mathbf{p}+\mathbf{q})\} \qquad (3.3.2)$$

where the δ-function represents the n-photon absorption (**multiphoton absorption; MPI**), and the distribution function f is Fermi-Dirac distribution for a given electron temperature T_e. Note that above, MPI is the formulation of dielectric media, although MPI in Sect. 2.1 was based on a single atom. The difference is that (3.3.2) is MPI by free electrons.

At first, the delta function in (3.3.2) allows only the multiphoton absorption satisfying the energy conservation relation:

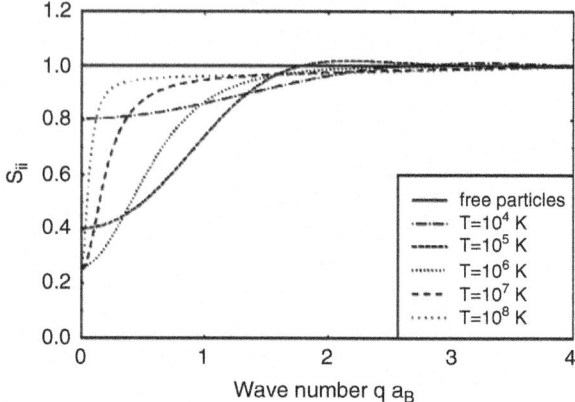

Fig. 3.14 The temperature dependence of the static structure factor in (3.3.1). Ion-ion structure factor for aluminum at solid-state density, $\rho = 2.7$ g/cm^3 (corresponding to an ion density of $n_i = 6 \times 10^{22}$ cm^{-3}) from HNC (hyper-net-chain) calculations for several temperatures. [Fig. 2 in Ref. 10]

$$E(\mathbf{p} + \mathbf{q}) = E(\mathbf{p}) + n\hbar\omega$$

The electron energy after the collision has to increase the discrete value of the n times the photon energy.

At second, it is clear that (3.3.2) is proportional to the difference of $f(\mathbf{p})$ before and after the energy increase and becomes small compared to Boltzmann distribution near the Fermi energy. If the final state $\mathbf{p} + \mathbf{q}$ is almost occupied in the Fermi distribution, the collisional absorption is forbidden because of quantum statistical reason. This is the real reason of the saturation of the collision frequency in the case where T_e approaches Fermi temperature. Then, the rough cut of $\nu < v_e/r_0$ condition in Fig. 3.9 is not necessary.

The $S_{ii}(\mathbf{q}, T_i)$ in (3.3.1) is called the **static structure factor** of ions with temperature T_i. It is the Fourier transformation of the **pair distribution function** $g_{ii}(\mathbf{r})$, which gives the probability of the surrounding ions at the radius $\mathbf{r}$ from the central ion. It is noted that in deriving (2.6.14), $S_{ii} = 1$ has been assumed, since the ideal plasma is assumed. $S_{ii}(\mathbf{q}, T_i)$ depends on the ion temperature as shown in Fig. 3.14. The ion sphere radius of the solid aluminum is $r_0 = 0.054$ nm $= 1.05$ a_B, where a_B is the Bohr radius.

In (2.6.11), the contribution from n = 1 is a linear absorption due to one photon absorption, and in Sect. 2.6, only n = 1 contribution is left, and higher n components are neglected. In addition, Taylor expansion was used to Bessel function $J_1^2(x) \approx x^2$ in (2.6.11), and as a result, the collision frequency (2.6.13) is obtained in the form independent of the laser field strength E_0. Since Bessel function shown in Fig. 3.15 is an oscillating function and $J_1^2(x) < x^2$, the above assumption always gives overestimated value of the n = 1 absorption contribution in the strong laser field case. It should be noted that at the same time, the multiphoton absorption increases at strong fields and the collision frequency become dependent on the intensity of lasers. This is called **nonlinear classical absorption** or **nonlinear inverse Bremsstrahlung absorption**.

3.3 Quantum Theory of Electric Conductivity in Dense Plasmas

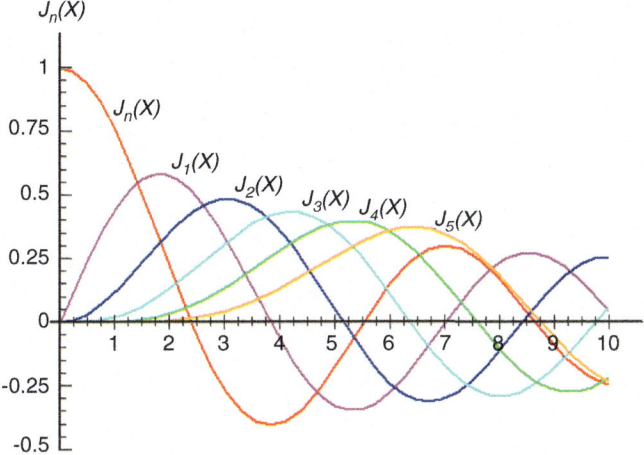

Fig. 3.15 Profiles of the n-th order Bessel function for n = 1 ~ 5

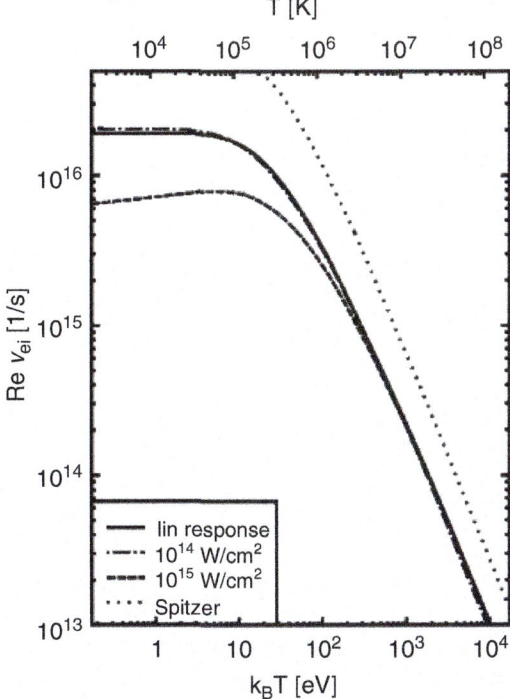

Fig. 3.16 Real part of the electron-ion collision frequency vs. electron temperature for aluminum ($n_i = 6 \times 10^{22}$ cm^{-3}, $T_i = T_e$) for laser wavelength 800 nm. The limiting case for high T (Spitzer's formula [1]) is also shown. [Fig. 6 (a) in Ref. 10]

Numerically solving (3.3.1) to the solid density aluminum plasma shown in Fig. 3.9, the resultant collision frequency is plotted in Fig. 3.16 [10]. In Fig. 3.16, the laser wavelength is 800 nm, and two cases are shown for the laser intensity of

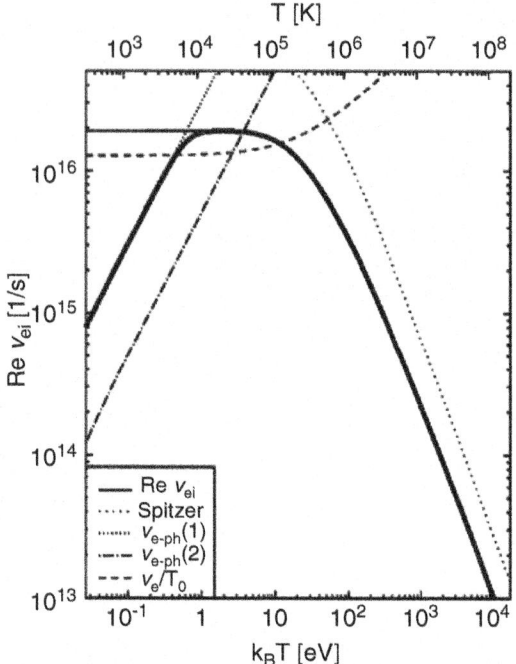

Fig. 3.17 Dynamic collision frequency vs. electron temperature for aluminum, $T_i = T_e$ after physical improvement of the solid line of Fig. 3.9. The thick solid line denotes the final appropriate collision frequency by taking into account of the quantum effects in cold solid matter [Fig. 7 in Ref. 10]

10^{14} W/cm^2 and 10^{15} W/cm^2 with dash-dotted line and dashed line, respectively. Although it was mentioned above that at higher intensity, the multiphoton ionization will commit the laser absorption, the collision frequency is smaller in higher intensity than the lower one in Fig. 3.16. This is due to another type of nonlinear effect in the collisional absorption, when the electron mean velocity is smaller than the electron quivering velocity by laser field and will be explained in the next session.

In Fig. 3.16, the Spitzer's collision frequency is also plotted with thin dotted line, and it is clear that it is overestimated at low temperature. In addition, even in the ideal plasma, the **Spitzer's collision frequency** is about factor three higher than the present nonlinear one. This is resulted after taking account of the higher velocity moment more precisely as shown in [10].

The authors of [10] proposed to model the collision frequency as the solid thick line in Fig. 3.17. In the low-temperature limit, the collision is dominated by the electron-phonon scattering, and (3.2.1) is also used in the interpolation formula like (3.2.2) with the present numerical result. The dash-dotted line is the electron-phonon collision frequency from Ref. 14 in [10].

3.4 Nonlinear Inverse Bremsstrahlung (IB) Absorption ($v_e < v_{os}$)

In the case where a strong laser field is irradiated on matters with cold temperature, the electron quivering velocity v_{os} defined in (2.3.5) is larger than the electron thermal velocity v_e. In such case, it is intuitively understand that the collision cross section σ_L should be evaluated with the transit velocity at the time of nonadiabatic collision with a scattering ion. In obtaining Maxwell distribution average of the collision frequency, it is reasonable to replace the velocity dependence in the denominator in (2.5.10) as:

$$\langle v^3 \rangle \rightarrow \left(v_e^2 + v_{os}^2 \right)^{3/2} \qquad (3.4.1)$$

Then the absorption rate becomes nonlinear function of the laser intensity. Based on the derivation of the classical absorption by **Dawson-Oberman** (Chap. 2, Ref. [9]), Decker et al. have numerically integrated (2.6.11) by including n up to 10 in order to taking into account the oscillating property of J_1 and other J_n (n = 2–10) correctly [11]. They have clearly shown that the absorption rate decreases for the region of $v_{os}/v_e > 1$. This nonlinear effect has been previously derived by Silin analytically as follows [12]:

$$\frac{\nu_{ei}^{IB}}{\omega_{pe}} = \frac{Z}{3} \left(\frac{1}{2\pi} \right)^{3/2} \frac{1}{n_e \lambda_{De}^3} \ln \left(\frac{k_{max}}{k_{min}} \right), \qquad v_{os} \ll v_e \qquad (3.4.2)$$

$$\frac{\nu_{ei}^{IB}}{\omega_{pe}} = \frac{Z}{\pi^2} \frac{1}{n_e \lambda_{De}^3} \left(\frac{v_e}{v_{os}} \right)^3 \left[\ln \left(\frac{v_{os}}{2v_e} \right) + 1 \right] \ln \left(\frac{k_{max}}{k_{min}} \right), \qquad v_{os} \gg v_e \qquad (3.4.3)$$

where

$$k_{min} = \omega/v_e, \quad k_{max} = b_0(v = v_e) \qquad (3.4.4)$$

In the Coulomb log, b_0 is **Landau cut** defined in (2.4.4). It is surprising that (3.4.2) is the same as (2.6.14) except for the definition of Coulomb log, although Silin published in 1964 [12] independently from the work of Dawson-Oberman in 1962 (Chap. 2, Ref. [8]) and pointed out the nonlinear effect of (3.4.3) before Ref. [11].

An explicit derivation of the nonlinear collision frequency based on the quantum mechanical approach is carried out for fully ionized hydrogen with standard parameters $n_e = 10^{22}$ cm^{-3}, T = 30 eV and 100 eV and $\omega/\omega_{pe} = 5$ [13]. The energy loss rate for IB absorption is finally found to have the following form:

$$\langle \mathbf{j} \cdot \mathbf{E} \rangle = \frac{1}{\sqrt{2}\pi^{3/2}} \frac{1}{n_e \lambda_{De}^3} Z^2 n_i m \sum_{n=1}^{\infty} (n\omega)^2 \int_0^{\infty} \frac{dk}{k^3} \frac{1}{|\varepsilon_{RPA}(k, n\omega)|^2}$$
$$\times \exp\left[-(n\omega)^2/(2v_e^2 k^2)\right] \exp\left[-(\hbar k)^2/(8m^2 v_e^2)\right] \quad (3.4.5)$$
$$\times \frac{\sinh(n\hbar\omega/2T)}{(n\hbar\omega/2T)} \int_0^1 dz J_n^2\left(\frac{eE_0 k}{m\omega^2} z\right)$$

The collision frequency by the IB process ν_{ei}^{IB} is derived using the relation (2.6.13) as:

$$\nu_{ei}^{IB} = \frac{\omega^2}{\omega_{pe}^2} \nu_E, \quad \nu_E = \frac{\langle \mathbf{j} \cdot \mathbf{E} \rangle}{\langle \varepsilon_0 \mathbf{E}^2 \rangle} \quad (3.4.6)$$

In the classical mechanics, $\hbar \to 0$, this expression tends to the classical ones (3.4.2) and (3.4.3) in the both limits derived by Silin [12]. Decker et al. [11] have obtained such expression in the framework of the nonlinear Dawson-Oberman kinetic theory. However, the classical formulation has the well-known problem of a divergence at large k that is solved by Landau cut procedure. In contrast, the quantum formula of (3.4.5) has no divergence in the integral to k.

Quantum effects, indicated by $\hbar$, appear in the following three places in (3.4.5). The first place is one of the exponential functions in (3.4.5) describing the quantum diffraction effect at large momenta k. The exponential function ensures the convergence of the integral by k. It is noted that the first exponential vanishes at small k, and the second one does at large k, and the divergence is avoided at both limits. This is approximately expressed with Coulomb log in the classical calculation. The second place is the term with the *sinh* function that is connected with the Bose statistics of multiple-photon emission and absorption. Finally, the quantum effects are also included in the dielectric function derived with random phase approximation (RPA).

Equation (3.4.5) is solved numerically for hydrogen with the standard parameters given above and compared with the classical ones. In Fig. 3.18, the solid line is the result from (3.4.5) for T = 30 eV [13]. It is clear that the nonlinear effects become dominant for $v_{os}/v_e > 1$, and the collision frequency reduces roughly in proportion to $(v_{os}/v_e)^3$ as predicted by Silin [12]. It is constant in the linear collision phase for $v_{os}/v_e < 1$, and the difference of the value comes from how to evaluate the Coulomb log in the classical formulations. It is seen that the quantum formulation provides 3–5 times larger values compared to the two classical results.

In order to see explicitly the breakdown of Coulomb log obtained by the cuts at both k limits, the collision frequency dependence to the **coupling parameter** Γ defined in (2.5.14) is studied for a fixed $v_{os}/v_e = 0.2$ and 10 by varying the temperature T. In Fig. 3.19, the results for the case of $v_{os}/v_e = 10$ are plotted and compared to two classical models. The result from (3.4.5) continuously increases

3.4 Nonlinear Inverse Bremsstrahlung (IB) Absorption ($v_e < v_{os}$)

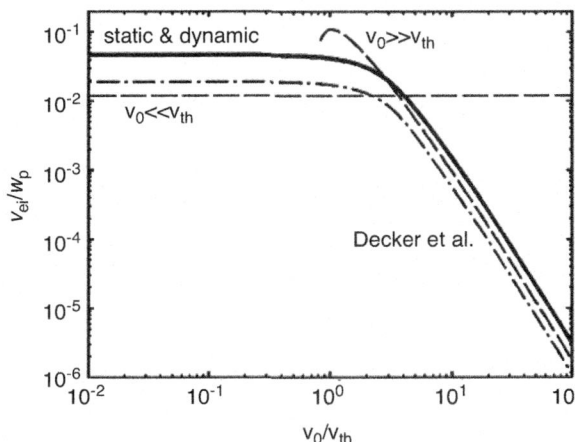

Fig. 3.18 Electron-ion collision frequency as a function of the quiver velocity (v_{os}/v_e) for a fully ionized hydrogen with parameters $n_e = 10^{22}$ cm^{-3}, $T = 30$ eV, and 100 eV and $\omega/\omega_{pe} = 5$. For comparison, results of Decker et al. (dash-dotted line) and of the asymptotic formulas of Silin (dashed line) are given. [Fig. 1 in Ref. 13]

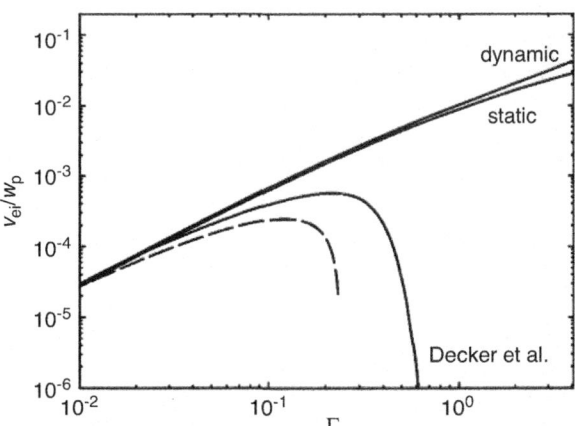

Fig. 3.19 Electron-ion collision frequency as a function of the coupling parameter Γ for a hydrogen plasma in a laser field ($Z = 1$, $v_{os}/v_e = 10$, $n_e = 10^{22}$ cm^{-3}, $\omega/\omega_{pe} = 5$.). Comparison is given with the theory of Decker et al. and with the asymptotic formula of Silin (dashed line). [Fig. 5 in Ref. 13]

with Γ, while the classical ones get to negative since the k-cutting model is not applicable in large Γ regime.

The contribution of each multiphoton component in (3.4.5) is shown in Fig. 3.20 for the case of $v_{os}/v_e = 10$ at $\Gamma = 0.1$ [13]. The solid line is obtained from (3.4.5). It is seen that there is a clear cut around n ~ 500, and the values decrease about 5 times at n = 10 compared to n = 1 contribution. In the classical limit, the multiphoton absorption is obtained as the dotted line in Fig. 3.20. The multiphoton absorption is underestimated in the classical model. The reason for the faster decreasing contribution in the classical approach is the constant Landau cut for maximum k, while the maximum of the integrand in (3.4.5) is shifted to higher k due to the exponential factor with n. This is because the multiphoton absorption occurs when the electron approaches nearer to the central ion at higher n-process. This is purely quantum effect, and it is very important relating to the higher harmonic emission in laser-

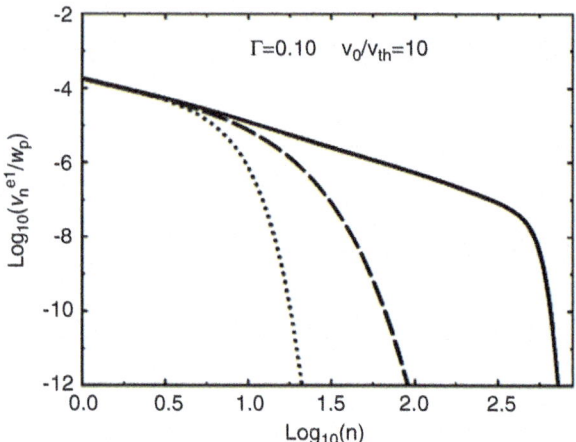

Fig. 3.20 Contribution of multiphoton absorption in a hydrogen plasma ($n_e = 10^{22}$ cm^{-3}; $\omega/\omega_{pe} = 5$; $\Gamma = 0.1$) for $v_{os}/v_e = 10$. Present theory (solid line), *sinh* term neglected (dashed line); classical dielectric theory (dotted line). [Fig. 7 in Ref. 13]

plasma interaction. In Fig. 3.20, the dashed line is the result obtained without *sinh* (x)/x term in (3.4.5).

It should be noted that even such quantum analysis of the collision physics the electron-phonon interaction in solid regime is not included, and the numerical result of (3.4.5) should be used like the interpolation form given in (3.2.2).

The nonlinear effect of IB in the strong field $v_{os}/v_e > 1$ could be modeled intuitively by replacing the thermal velocity v_e in (2.6.14) as:

$$v_e \rightarrow v_e \left(1 + v_{os}^2/v_e^2\right)^{1/2} \tag{3.4.7}$$

Then, ν_{ei} in (2.6.14) will be modified as:

$$\nu_{ei} = \nu_{ei}^{DO} \frac{1}{\left(1 + v_{os}^2/v_e^2\right)^{3/2}} \tag{3.4.8}$$

where ν_{ei}^{DO} is the collision frequency of (2.6.14) given by Dawson-Oberman. This can explain the strong field modification of (2.6.14) by Silin [12] and Decker et al. [11].

It is, however, pointed out that the absorption is better than (3.4.8) in the strong field region [14]. After numerical calculations, the resultant heating rate is found rather constant as shown in Fig. 3.21 in strong field region [14]. It is proposed that the fitting formula to such computational result is expressed as:

$$\nu_{ei} = \nu_{ei}^{DO} \frac{1}{\left(1 + 0.3 v_{os}^2/v_e^2\right)} \tag{3.4.9}$$

In the strong field limit, absorption rate is rather proportional to $1/I_L$ instead of $1/I_L^{3/2}$ so that the temperature increasing rate becomes independent of the laser intensity as

3.4 Nonlinear Inverse Bremsstrahlung (IB) Absorption ($v_e < v_{os}$)

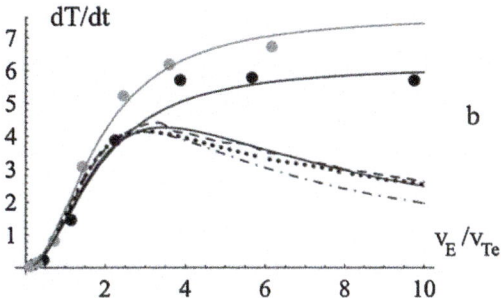

Fig. 3.21 Dependence of the absorption rate per electron dT_e/dt normalized to $\nu_{ei}T_e$ on the normalized oscillation velocity for a plasma with $Z = 10$ and for the Maxwellian electron distribution function with $T_e = 200$ eV (black dots) and 500 eV (gray dots). The continuous lines following large dots are calculated from (3.4.9). Numerical results are compared with the theoretical results for $T_e = 200$ eV: the Kroll-Watson approximation (dash dotted line), the Dawson-Oberman approximation (dashed line), and the classical approach (dotted line). The laser wavelength is 0.25 μm. [Fig. 11 in Ref. 14]

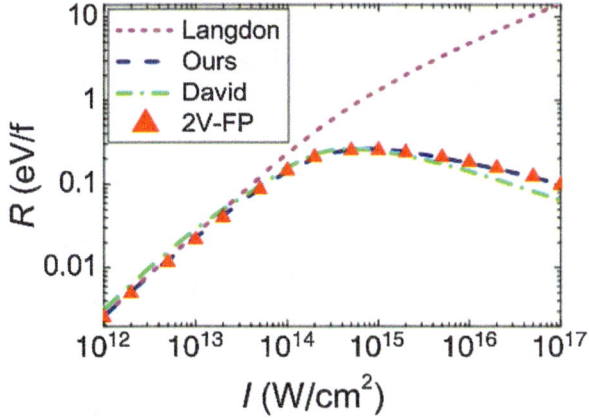

Fig. 3.22 The absorption rates R averaged over the first four laser cycles as functions of laser intensity obtained from two-dimensional velocity space Fokker-Planck (2VFP) code (triangle), conventional liner operator (short dashed line), the present IB operator (dashed line), and David's fitted formula from molecular dynamic method (dash dotted line) [16]. The plasmas and laser parameters are electron density $n_e = 10^{20}$ cm^{-3}, initial temperature $T_e = 10$ eV, ionization state $Z_i = 1$, and laser wavelength $\lambda = 1.06$ μm. [Fig. 2 in Ref. 15]

seen in Fig. 3.21. The change from the previous one shown intuitively in (3.4.8) is due to the anisotropy of the velocity distribution to the laser polarization direction and non-Maxwell effects. It is noted that the electron gains energy in IB process as the increase of velocity in the quivering velocity direction, and it is natural to assume that the electron has anisotropic distribution function.

This constant heating rate is confirmed with Vlasov-Fokker-Planck simulation as shown in Fig. 3.22 [15]. The authors concluded that their absorption rate dependence on the laser intensity is well described by a fitting formula given in [16] where Coulomb log is modified to include the field strength v_{os}/v_e dependence in v_{ei}^{DO} in (2.6.14).

3.5 Electron Plasma Waves and Collisionless Absorption

In the discussion so far, the absorption of laser energy stems from the collisions of electrons oscillating by the laser electric field, and the study of such classical laser absorption demanded the correct evaluation of the collision frequency of electrons by the ions in plasma or solid materials. When the plasma temperature is high enough, such collisions merely happen, and the classical absorption rate tends to inefficient. In such **collisionless plasmas**, the plasma waves are easily excited, and the coupling of laser wave and the plasma waves will become important in the energy conversion from laser to the plasma waves. In the limit of cold plasmas, the plasma waves are the so-called electron plasma oscillation as shown soon, while with finite temperature, this oscillation induces electron pressure perturbation and propagates in the form of waves. Before discussion of the wave-wave coupling, the plasma oscillations and plasma waves are derived. Since such oscillations and waves are induced and propagated in plasma due to the long-range Coulomb force and they are formed with the collective interactions among many electrons, such phenomena like waves are called **collective phenomena** in plasmas.

In the case where small amplitude longitudinal waves are excited in plasmas, the basic equation is the linearized equation of (2.2.11). The current by electron oscillations is obtained for the electrostatic field E in (2.3.3) by assuming one-direction x:

$$j_x = -en_0 v, \quad \frac{1}{\varepsilon_0}\frac{\partial j_x}{\partial t} = \omega_{pe}^2 E_x$$

Then, (2.2.11) reduces the form in the oscillation direction, x:

$$\frac{\partial E}{\partial t^2} + \omega_{pe}^2 E = 0 \qquad (3.5.1)$$

where ω_{pe} is the electron plasma frequency defined in (2.3.31), and it is constant with a constant density, $n_e = n_0$. (3.5.1) is the equation of a harmonic oscillator, while each oscillation of electron layer is interacting each other via electric field in the continuous plasma media.

The dispersion relation of (3.5.1) is simply obtained and shows the waves oscillating with the frequency ω in the form:

3.5 Electron Plasma Waves and Collisionless Absorption

$$\omega^2 = \omega_{pe}^2 \quad (3.5.2)$$

Relation (3.5.2) is the dispersion relation of **electron plasma oscillation** and means the plasmas have an eigenfrequency ω_{pe} and is a kind of resonator with the plasma frequency. Therefore, in the case where (2.2.11) has an external current in the x-direction, the plasma oscillations are excited in the plasmas.

For the convenience to study the collisionless absorption process, the effect of finite temperature in the dispersion relation (3.5.2) is also derived mathematically. The basic equations of fluid description of electron gas will be derived and explained in Vol. 2, and the equations of continuity and motions are used here without no derivation. They are:

$$\frac{\partial n_e}{\partial t} + \nabla(n_e \mathbf{u}_e) = 0 \quad (3.5.3)$$

$$mn_e \frac{d}{dt}\mathbf{u}_e = -en_e \mathbf{E} - \nabla P_e \quad (3.5.4)$$

where n_e and $\mathbf{u}_e$ are the electron density and flow velocity. In (3.5.4), the pressure force appears, and it is defined as $P_e = n_e T_e$ for fully ionized ideal plasmas.

Consider that the small amplitude perturbations of density, flow velocity, and electric field are excited in the uniform plasmas initially at rest. The ions can be assumed to be at rest due to their heavy mass. Then, they are assumed to be in the forms:

$$n_0 + n_{e1}, \quad \mathbf{u}_{e1}, \quad \mathbf{E}_1 \quad (3.5.5)$$

Inserting (3.5.5) into (3.5.3) and (3.5.4) and remaining only the perturbations, the following equations to the perturbed quantities are obtained:

$$\frac{\partial}{\partial t} n_{e1} + n_{e0} \nabla \mathbf{u}_{e1} = 0 \quad (3.5.6)$$

$$m_e \frac{\partial}{\partial t} \mathbf{u}_{e1} = -\frac{1}{n_{e0}} \nabla P_{e1} - e\mathbf{E}_1 \quad (3.5.7)$$

Such equations to small amplitude perturbations are called **linearized equations** in general. In order to make (3.5.6) and (3.5.7) closed relation, Poisson equation is also required:

$$\nabla \mathbf{E}_1 = -\frac{e}{\varepsilon_0} n_{e1} \quad (3.5.8)$$

In addition, we assume the electron oscillating motion is adiabatic, and the pressure perturbation is proportional to the electron density:

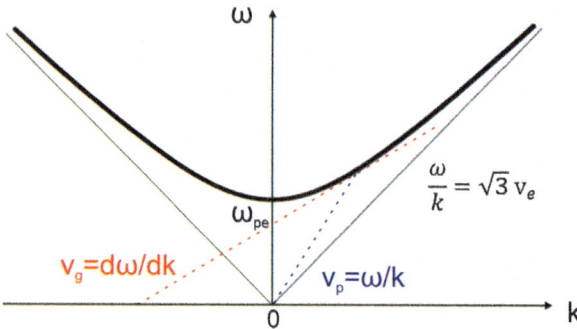

Fig. 3.23 Dispersion relation of the electron plasma waves. The definition of phase velocity and group velocity is also shown graphically

$$P_{e1} = \left(\frac{\partial P_e}{\partial n_e}\right) n_{e1} = \gamma T_e n_{e1} \tag{3.5.9}$$

In Eq. (3.5.9), γ is the adiabatic constant. In the case of one-dimensional wave, namely, the degree of freedom $N = 1$; then, it is known that:

$$\gamma = \frac{N+2}{N} = 3$$

The three basic Eqs. (3.5.6), (3.5.7), and (3.5.8) are homogeneous equations. After the Fourier-Laplace transformation, (3.5.6), (3.5.7), and (3.5.8) provide the dispersion relation to the **electron plasma wave**.

$$\omega^2 = \omega_{pe}^2 + 3v_e^2 k^2 \tag{3.5.10}$$

where v_e is the electron thermal velocity. The dispersion relation Eq. (3.5.10) is plotted in Fig. 3.23. Compared to the dispersion relation of electromagnetic waves in plasma (2.2.17), the plasma wave has the maximum phase velocity of $\sqrt{3}v_e$ instead of the speed of light. In the limit of the cold electron plasmas, the wave cannot propagate, and (3.5.10) tends to the dispersion relation obtained in (3.5.2).

Consider the phase and group velocities of the electron plasma waves. Operating Fourier transform for perturbation quantities decomposes any perturbations in space with the sum of Fourier components. For a given wavenumber k, a physical quantity of the wave "a(**r**,t)" can be written as the sum of all Fourier components:

$$a = \sum_{\mathbf{k}} A(\mathbf{k}) e^{i\mathbf{k}\mathbf{r} - i\omega t} \tag{3.5.11}$$

Assume that the Fourier component A has the peak at $\mathbf{k} = \mathbf{k}_0$. Taking the first-order Taylor expansion of Eq. (3.5.11), it can be written as:

$$a = e^{i\mathbf{k}_0\mathbf{r}-i\omega_0 t}\sum_{\Delta\mathbf{k}}A(\mathbf{k}_0+\Delta\mathbf{k})e^{i\Delta\mathbf{k}\mathbf{r}-i\Delta\omega t} \qquad (3.5.12)$$

where ω_0 is the solution of Eq. (3.5.12) for $k = k_0$. It is shown in (3.5.12) that the first term represents the oscillation of the phase of the wave, while the second term in the summation means the amplitude in the form:

$$\Delta\mathbf{k}\mathbf{r} - \Delta\omega t = \Delta\mathbf{k}\left(\mathbf{r} - \frac{\partial\omega}{\partial\mathbf{k}}t\right)$$

It is clear that the waves have the two different kinds of wave velocities, namely, the phase velocity and group velocity shown below:

$$\mathbf{v}_p = \frac{\omega}{\mathbf{k}}, \quad \mathbf{v}_g = \frac{\partial\omega}{\partial\mathbf{k}} \qquad (3.5.13)$$

Any information is carried in the form of the amplitude, namely, with the group velocity. The wave packet propagates with the group velocity, and the wave propagation is sustained by the phase velocity. Both velocities are plotted in Fig. 3.23.

The characteristics of the electron plasma wave are easily found from Fig. 3.23 that

1. Electron plasma frequency depends only on the electron density, and the wave with the frequency below the electron plasma frequency cannot propagate in plasmas.
2. The wavenumber dependence of the electron plasma wave stems from the electron temperature, and the resultant group velocity is due to the electron thermal motion.
3. The group velocity is about the electron thermal velocity in the limit of high frequency. The electron plasma wave, in principally, cannot be treated by fluid approximation when the wavelength is approaching to the Debye length. In order to describe the plasma wave with shorter wavelength, the plasma kinetic theory to be explained in Volume 2 is required. It will be found that such short wavelength plasma wave is damped through wave-particle interaction, **Landau damping**, and cannot exist as stationary waves.

3.6 Resonance Absorption

Consider the case of the p-polarization in Fig. 3.5 and use (3.1.2) as the basic equation to obtain the electric field profile **E** in the x- and y-directions. As seen below, the electromagnetic field couples resonantly with the plasma oscillation via the x-component of **E** near the cutoff density. The electric field in the x-component penetrates to the critical density by the tunneling effect after the turning point (3.1.3). And the electric field oscillates the electron, and charge separation appears since the density is inhomogeneous in x-direction. Due to the resonance at the cutoff point, the induced electrostatic field is enhanced, and large amplitude current is induced to increase the $\langle \mathbf{j} \cdot \mathbf{E} \rangle$ term in (2.2.2). In the case of s-polarization, there is no x-component of the laser electric field in the plasmas, and no resonance is induced.

In order to study the energy conversion rate from laser to the plasma waves, a **driver model** is usually assumed. In this model, the x-component of laser field is calculated by solving (3.1.2) approximately, and its value at the critical point is used to solve the induction of the plasma oscillation around the critical point. Given the strength of laser field at the critical point, the electrostatic field E in the x-direction should satisfy Maxwell relation (1.3.2):

$$\frac{\partial}{\partial t}E = j + \frac{1}{\mu_0}[\nabla \times \mathbf{B}]_x, \quad j = -en_e V \tag{3.6.1}$$

In (3.6.1), we remain only the electron current by the electrostatic wave. The Poisson Eq. (1.3.3) is:

$$\varepsilon_0 \frac{\partial}{\partial x}E = e(n_0 - n_e) \tag{3.6.2}$$

The equation of the electron fluid velocity V in electrostatic field is:

$$m\frac{d}{dt}V = -eE \tag{3.6.3}$$

Taking the sum of (3.6.1) and the velocity time (3.6.2) to use the relation:

$$\frac{d}{dt}E = \left(\frac{\partial}{\partial t} + V(t,x)\frac{\partial}{\partial x}\right)E = -\frac{m}{e}\frac{d^2}{dt^2}V \tag{3.6.4}$$

The total derivative to E is obtained. Taking the total derivative of (3.6.3), the following simple forced oscillation equation is obtained:

$$\frac{d^2}{dt^2}V + \omega_{pe0}^2 V = -\omega^2 V_d \sin(\omega t)$$
$$V_d = \frac{e}{m\omega}E_d, \quad E_d = -B_z(x_{cr})\sin\theta \tag{3.6.5}$$

In the driver model, the external force due to B_z is obtained by solving (3.1.2) approximately. The propagation Eq. (3.1.2) was solved analytically, and the solution is given in a book by V. Ginzburg, in which he discuss about the propagation of electromagnetic waves in the space, especially radio waves in the ionosphere [17]. His book has been widely used in laser plasma in its early time of research,

3.6 Resonance Absorption

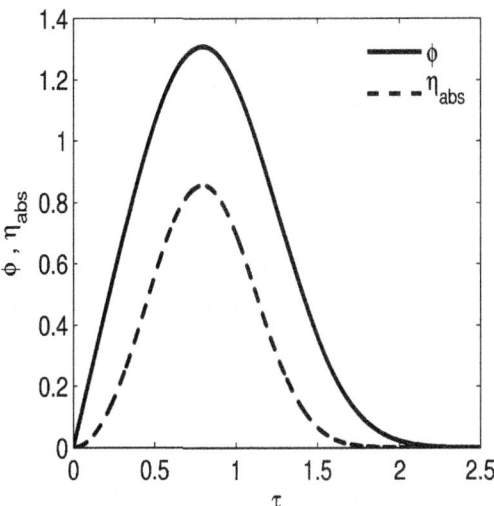

Fig. 3.24 The Ginzburg curve defined in (3.6.8) (solid line) and the absorption fraction given in (3.6.14) (dashed line)

and Kruer showed how to obtain approximately the functional dependence of the driver filed in his textbook (Chap. 2, Ref. [2]).

Kruer assumed the magnitude at the turning point of (3.1.2) can be approximated with the solution of RHS = 0 in (3.1.2), and in addition, he assumed $k_0 L \gg 1$. Then, with use of the airy function in Fig. 3.6, the following relation is obtained:

$$cB_z(\varepsilon = \sin^2\theta) = 0.9E_0(k_0 L)^{1/6} \quad (3.6.6)$$

Then, this value at the turning point is extending by taking account of the attenuation to the critical point by the tunneling effect. The following attenuation rate $e^{-\beta}$ is obtained by integrating k of (3.1.2):

$$e^{-\beta}: \quad \beta = \frac{2}{3} k_0 L \sin^3\theta \quad (3.6.7)$$

By the use of (3.6.6) and (3.6.7), we obtain the following function:

$$E_d = \frac{E_0}{\sqrt{2\pi k_0 L}} \Phi(\tau)$$
$$\Phi(\tau) = 2.3\tau \, \exp\left(-2/3\tau^3\right) \quad (3.6.8)$$
$$\tau \equiv (k_0 L)^{1/3} \sin\theta$$

The function $\Phi(\tau)$ is called **Ginzburg curve**.

The Ginzburg curve is plotted with solid line in Fig. 3.24. It is useful to note the property of this curve. As expected, the curve tends to zero at both limits of $\tau = 0$ and $\tau \gg 1$. It has the maximum at $\tau = 0.8$. This means if we expect a significant driving force in (3.6.5), the density scale L should satisfy the condition $L \approx 1/k_0 = \lambda/2\pi$ at a finite incident angle. In case of glass laser, it is about 0.1μm and may be possible to obtain such density scale length in sub-picosecond laser irradiation. However, it may be difficult to expect such density steepening within the driven hydrodynamic expansion of plasmas in the case of relatively long ns laser pulse. It will be explained that in such case, the laser pressure helps to keep steepened density profiled near the critical point, and substantial driver field is obtained.

It is noted that the driver field (3.6.5) is directly proportional to the magnetic field at the critical point, and the laser field induces the plasma oscillation by the electrostatic field. Through this process, the laser energy is transferred to the plasma wave oscillation energy. Such physical process is called **resonance absorption**. In order to estimate laser absorption rate by the resonant absorption, we have the following three different ways with different physical models. Such consideration is widely used in many cases.

3.6.1 Collisional Absorption Model

Assume that the displacement of electron fluid oscillations is small enough compared to the density scale length and laser wavelength. As in Sect. 3.1, assume that the electron plasma has a small collision frequency ν by the collision with the background ions. Then, (3.6.4) is modified like:

$$\frac{\partial^2}{\partial t^2}V + \nu\frac{\partial}{\partial t}V + \omega_{pe0}^2(x)V = V_d e^{-i\omega t} \qquad (3.6.9)$$

Although E and V are 90° difference for $\nu = 0$, imaginary part of (3.6.9) appears for finite ν case in the form:

$$\text{Im}(V) = -\frac{e}{m\omega}\frac{(\nu/\omega)}{(x/L)^2 + (\nu/\omega)^2}E_d \qquad (3.6.10)$$

where we assumed a linear density profile with the critical density at $x = 0$, although the conclusion doesn't depend on the profile of the density. Since the electron current in the x-direction is $j = -en(x)V$ and the time-averaged $\langle E \cdot j \rangle$ of (2.2.6) can be calculated as:

$$\langle \text{Re}(E \cdot j) \rangle = \frac{\omega_{pe0}^2(x)}{\omega}\frac{(\nu/\omega)}{(x/L)^2 + (\nu/\omega)^2}\frac{\varepsilon_0 E_d^2}{2} \qquad (3.6.11)$$

Integrate (3.6.11) in space x with assumption $\nu/\omega << 1$ so that the following integration can be applicable:

3.6 Resonance Absorption

$$\lim_{\delta \to 0} \int_{-\infty}^{\infty} \frac{\delta}{y^2 + \delta^2} dy = \pi \qquad (3.6.12)$$

With the use of (3.6.12), the resonant absorption rate is obtained:

$$\int_{-L}^{\infty} \langle \operatorname{Re}(E \cdot j) \rangle dx = \pi \omega L \frac{\varepsilon_0 E_d^2}{2} \qquad (3.6.13)$$

Inserting E_d of (3.6.8) to (3.6.13), the following absorption rate is obtained:

$$\eta_{\text{abs}} = \frac{1}{2} \Phi(\tau)^2 \qquad (3.6.14)$$

This absorption rate is plotted with dashed line in Fig. 3.24. It has about 80% at the maximum.

3.6.2 Singular Point Integral Model

It is important to know that the resonance absorption rate derived above can be obtained even without collisional process in (3.6.9). Although the laser absorption is very sensitive to the real value of ν in case of the classical absorption in Sects. 3.1 and 3.2, the absorption rate obtained in (3.6.14) is independent of the collision frequency ν in the resonance absorption. The reason is that the resonance absorption is essentially collisionless process, and the friction was introduced for convenience to avoid mathematical difficulty in integrating (3.1.2) in space.

It is important to know that the absorption rate can be calculated mathematically in really collisionless process. In the present case, there is a resonance at $x = 0$, and without ν, the integration is carried out for integral function with a singular point at $x = 0$:

$$\int \langle E \cdot j \rangle dx \propto \int \frac{f(x)}{x} dx \qquad (3.6.15)$$

The resonance absorption is a collisionless process physically, and it is found that the integration of (3.6.15) has a finite value as follows. The integration of a function with singular points is called **Cauchy integral** in mathematics and is given to be:

$$\int_{-\infty}^{\infty} \frac{f(x)}{x - x_0} dx = P \int_{-\infty}^{\infty} \frac{f(x)}{x - x_0} dx + iS\pi f(x_0) \qquad (3.6.16)$$

The first term is the **principal integral** to be explained more in Vol. 3 relating to Landau damping. The second term is pure imaginary, and S is +1 or −1 depending on the process is absorption or emission, namely, energy loss or gain in the present case. Using (3.6.16) in the x integration of $\langle E \cdot j \rangle$, the second term of RHS of (3.6.16) gives us the same result as (3.6.14).

Any dissipation is required to the derivation, and the finite value is obtained because of the phase change of the oscillation velocity across the resonance points as seen below.

It should be noted that the driver model uses the value of electric field at the critical point approximately obtained by solving the laser propagation Eq. (3.1.2). It should be, however, solved by coupling with the equation of electrostatic wave self-consistently. Namely, (2.2.10) and (2.2.11) should be solved consistently with proper evaluation of the current density by both waves. Then, an effective electric field at the resonance point is smaller than (3.6.8) because the laser pump to the electrostatic wave is depleted. This is called **pump depletion** and discussed later.

3.7 Resonance in Pendulum

It may be a question why laser energy is absorbed in plasmas without any dissipation process. In order to understand this physics, consider an oscillation of a pendulum in an external oscillating force. For example, (3.6.5) is a simple mathematical equation to an oscillation of a pendulum with an external force. For convenience, it is rewritten in the form:

$$\frac{d^2y}{dt^2} + \omega_0^2 y = F\cos(\omega t)$$
$$\omega_0 = \sqrt{g/l} \qquad (3.7.1)$$

It is possible to solve (3.7.1) analytically as well-known, and the solution to the initial condition $y = dy/dt = 0$ $(t = 0)$ is given as:

$$y(t) = \frac{F}{2\omega_0} t \sin(\omega_0 t) \qquad \text{for } \omega = \omega_0$$
$$y(t) = \frac{2F}{(\omega_0^2 - \omega^2)} \sin\left(\frac{(\omega_0 - \omega)t}{2}\right) \sin\left(\frac{(\omega_0 + \omega)t}{2}\right) \qquad \text{for } \omega \neq \omega_0 \qquad (3.7.2)$$

In the case of the pure resonance, the amplitude y continues to grow linearly in time and no saturation amplitude. In the other case, a beat wave structure appears.

3.7 Resonance in Pendulum

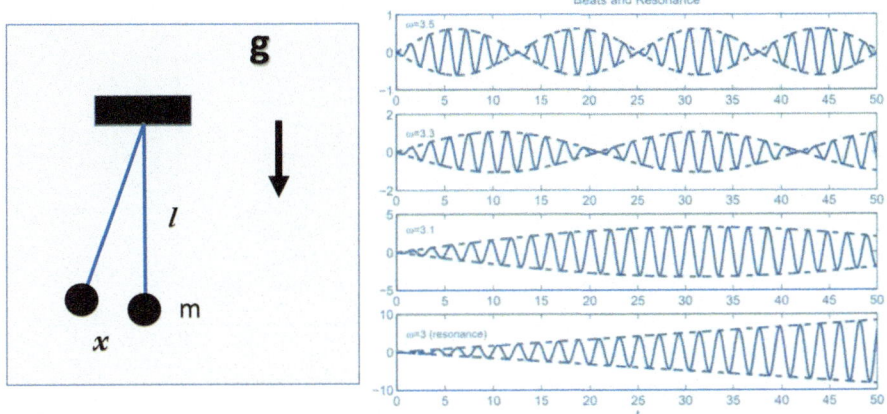

Fig. 3.25 The pendulum and the time evolution of the oscillation amplitude of the pendulum, when a forced oscillation is imposed to the pendulum. The right figure is the time evolution of the force oscillation with frequency matched case (bottom) and three mismatched cases. The mismatching of frequency is 1/30, 1/10, and 5/30 from the second bottom to the top. It indicates that even 10% mismatching, the oscillation amplitude increases to higher one

Examples of the time evolution given in (3.7.2) are plotted in Fig. 3.25 for the case of $F = 1$, $\omega_0 = 2\pi$. It is noted that for ultra-short pulse laser, the amplitude continues grows to very high amplitude even if the density is not the critical one but near the critical.

In the case with the collision frequency, (3.6.9) has the following stationary oscillating solution:

$$y(t) = \frac{F}{Z} \sin(\omega t + \varphi) \qquad (3.7.3)$$

where

$$Z = \left[(1 - \xi^2)^2 + (2\zeta\xi)^2\right]^{1/2}$$
$$\tan\varphi = \frac{2\zeta}{1 - \xi^2} \qquad (3.7.4)$$
$$\xi = \frac{\omega_0}{\omega}, \quad \zeta = \frac{\nu}{\omega_0}$$

In Fig. 3.26, the enhancement factor $1/Z$ in (3.7.4) is plotted for several cases with different values of ζ as a function of the frequency difference $1/\xi$. In the limit of $\omega = \omega_0$, there is no stationary solution, and the amplitude diverges. It is clear from Fig. 3.26 that the case of no dissipation or very small dissipation, the amplitude of the oscillation is very much enhanced compared to the amplitude of the force field,

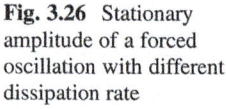

Fig. 3.26 Stationary amplitude of a forced oscillation with different dissipation rate

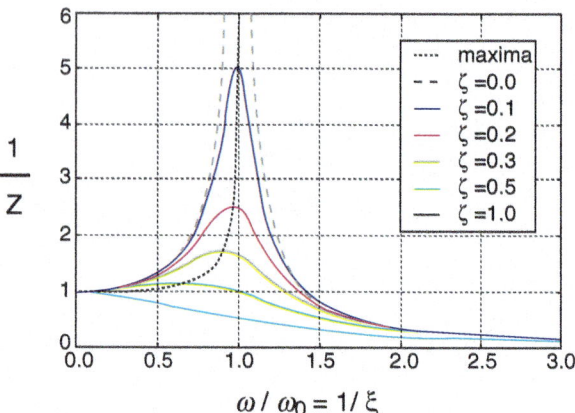

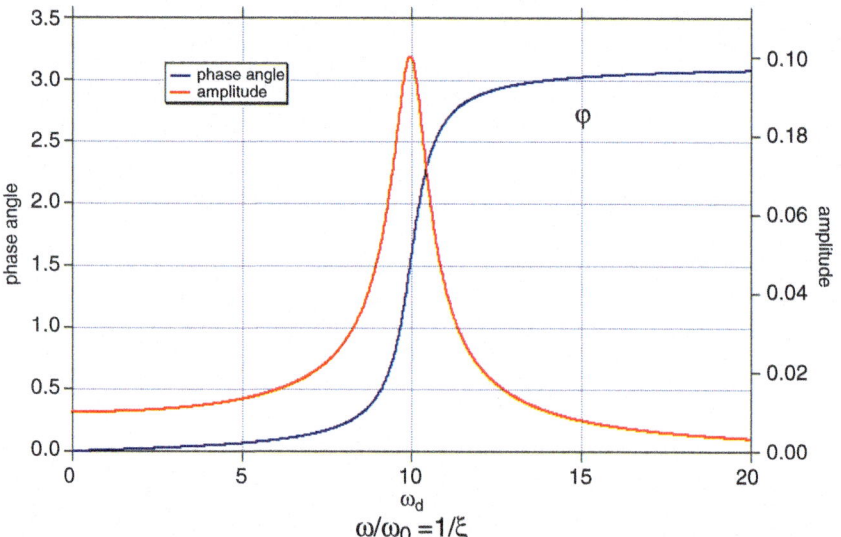

Fig. 3.27 One of the stationary solutions in Fig. 3.26 with the phase change from the force oscillator. It is important that the phase changes by π crossing the resonant frequency independent from the value of dissipation

and it is expected that high amplitude electrostatic fields are generated around the critical density.

Not only the amplitude but also the change of the phase shift φ is very important to understand the physics of resonance absorption. In Fig. 3.27, the phase change as the function of the frequency is plotted with blue line, and the amplitude is also plotted with red line for the case of $\zeta = 0.15$. It is clear that the phase changes by 180° over the region where the amplitude is enhanced. Given external force F, the oscillation of y makes local current, and the phase of current change across the

3.8 Linear Mode Conversion in Resonance Absorption

critical point by 180° means there should be real part of $\langle E \cdot j \rangle$. In the limit of no dissipation, the phase changes like a step function, and the velocity given by y in (3.7.3) becomes delta function. Then, integrating in space, the total $\langle E \cdot j \rangle$ is given in the form (3.7.3).

3.8 Linear Mode Conversion in Resonance Absorption

As shown in Fig. 3.23, the electron plasma waves can propagate in plasmas with finite temperature ($v_e > 0$). This fact means the plasma waves dominantly generated near the critical density propagate to the lower density region, and the electron oscillation energy given by the resonance is carried toward the low-density plasmas. Then, it is possible to assume stationary oscillating solutions in solving (2.2.10) for laser field and (3.5.10) with external current by laser field on RHS. This is the fourth-order differential equations, and the solutions are obtained numerically. The basic equation for a stationary propagation of the plasma waves is obtained from (3.5.10) by using (3.6.3) and (3.6.4) and assuming the wave amplitude is small enough to allow the linear wave model:

$$\frac{\partial}{\partial t^2} \mathbf{E} + \omega_{pe}^2 \mathbf{E} - 3k_{De}^2 \nabla^2 \mathbf{E} = -\frac{1}{\varepsilon_0} \frac{\partial \mathbf{j}_{ext}}{\partial t} \quad (3.8.1)$$

Assuming a stationary propagating plasma wave in x-direction, (3.8.1) is reduced to:

$$3k_{De}^2 \frac{\partial^2 E_x}{\partial x^2} - [\varepsilon(x) + i\nu_{pw}\omega] E_x = c\sin\theta B_z \quad (3.8.2)$$

Here, the friction to the plasma wave stemming from wave-particle interaction such as **Landau damping** is included. In (3.8.2), Debye wavenumber $k_{De}(=v_e/\omega_{pe})$ at the critical point is introduced. If it is able to solve (3.8.2) self-consistently with (3.1.2), the energy conservation of two-wave system is guaranteed.

Note that the typical wavelengths of laser wave and plasma waves are very different by the ratio of c/v_e, the former is much longer than the latter. The coupled Eqs. (3.1.2) and (3.8.2) are solved numerically for an appropriate boundary condition. One example is shown in Fig. 3.28 [18]. Laser impinges from right in the plasma with a linear density profile and reflected to the right. The laser field penetrates by the tunneling effect beyond the turning point as shown in Fig. 3.28 (a). Then, the plasma wave induced near the critical density propagates to the right as seen in Fig. 3.28(b). The plasma wave is evanescent to the over-dense region and propagates to the right.

Figure 3.28 shows the real and imaginary parts of B of the electromagnetic field, and the real part of electric field of the plasma wave is plotted. The normalized coordinate $\xi = 0$ is the critical density point. It is clear that the magnetic field has

Fig. 3.28 Spatial dependence of (**a**) the electromagnetic mode B (Z) and (**b**) the real part of the electrostatic field E(ζ) in a linear density profile for $\tau^2 = 0.5$ [τ defined in (3.6.8)], 3 T/mc^2 = 0.001, and $k_0L = 10$. It is noted that Z is normalized z axis by the laser wavelength, while ζ is normalized with a shorter length of 1/30 times the laser wavelength. The electromagnetic wave impinging from the right is converted to the plasma wave propagating to the right due to the linear mode conversion. The EM wave is reflected near the critical density (Z = 0). The beat wave structure is seen in (b) through the interference with the x-component of the EM wave electric field. [Figs. 4 and 5 in Ref. 18]

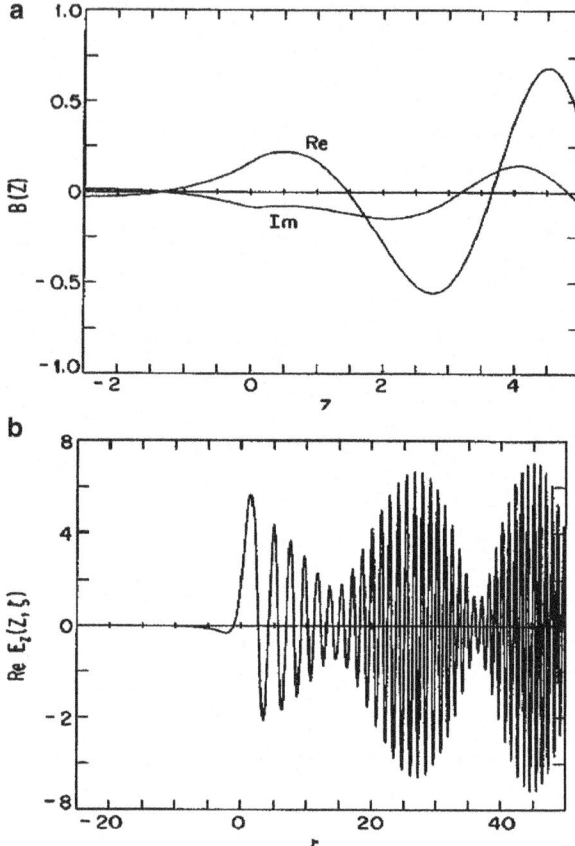

local maximum at ξ = 0. This is the case with q = τ^2 = 0.5 in (3.6.8). The plasma temperature about 330 eV and $k_0L = 10$ is assumed to study the case of ablating plasmas produced by the ns laser pulse. The numbers in the space axis are the same for the both figures, but it is noted that the normalized coordinate Z and ζ are ten times different [18]. Such wave conversion or generation near resonance point is called **linear mode conversion**. This idea can be used for heating any kinds of plasmas efficiently by adjusting the eigenfrequency of plasmas with the frequency of external waves, laser for dense plasma or radio microwaves for low-density plasmas.

It is informative to note that the coupled wave equations solved above have four simple waves, since there are governed by the fourth-order partial differential equations originally. They allow not only the solution mentioned above but also the case when the plasma waves are injected in the lower-density region to the critical density. Since no plasma wave can propagate from the over-dense region, the injected plasma waves are reflected at the critical point to return to the lower-density region. In such case, the coupled equations will allow the generation of electromagnetic wave near the critical density. In this case, the electromagnetic wave

3.8 Linear Mode Conversion in Resonance Absorption

propagates only from the critical point to vacuum, and it gains energy from the plasma wave. This is also resonant coupling of the plasma waves and electromagnetic waves.

3.8.1 Absorption Rate and Pump Depletion

The resonance absorption is precisely studied computationally [19], and the resultant absorption rates are shown in the solid lines in Fig. 3.29 for the case of the background electron temperatures being 2.5 keV and 50 keV, where the horizontal axis is $q = \tau^2$ of (3.6.8). They carried out **PIC simulation** for three different incident angles, and the results are plotted in the same figure, where the density scale is $k_0 L = 12.5$. It is clearly seen that the numerical result is insensitive to the plasma temperature, and PIC simulation well agrees with the numerical result of the liner mode conversion discussed above.

It should be noted, however, that the maximum of the absorption rate in the consistent calculations is about 50%, while the absorption rate shown in (3.6.14) and plotted in Fig. 3.24 is higher than that the solid line in Fig. 3.29. This is because of the issue of self-consistency in the case of the capacitor model. The evaluation of the B field from (3.1.2) without the resonance absorption effect is the reason of the discrepancy. It may be possible to include this effect approximately to the capacity model of (3.6.4). We assume the driver energy is depleted by the absorption and only a half of it is included to the strength of E_d in (3.6.8), then the absorption rate η_a is modified as:

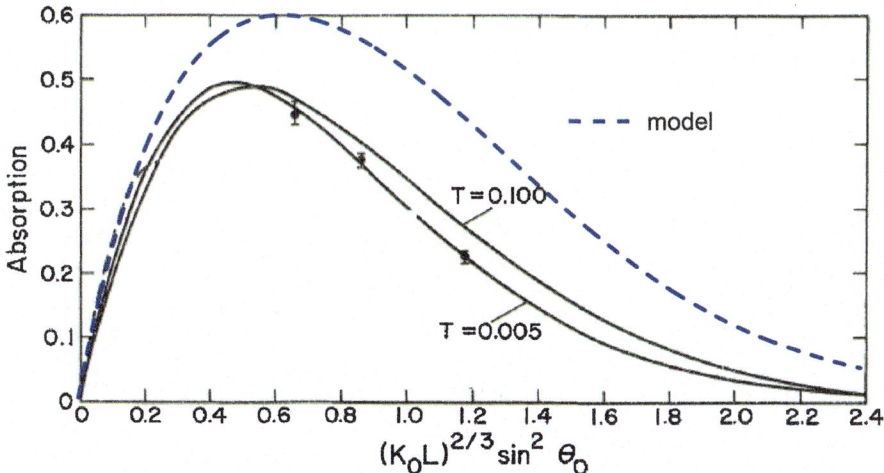

Fig. 3.29 The absorption curve obtained by numerical calculation (solid lines) and three points obtained with PIC simulations. The dashed line is the result given in (3.8.3) where the pump depletion is approximately taken into account in the driver model

$$\eta_a = \left(1 - \frac{1}{2}\eta_a\right)\frac{1}{2}\Phi^2$$

This equation simply gives us a modified absorption rate instead of (3.6.14) in the form:

$$\eta_a = \left(1 + \frac{1}{4}\Phi^2\right)^{-1}\frac{1}{2}\Phi^2 \tag{3.8.3}$$

This absorption rate is plotted as a function of τ in Fig. 3.29 with dotted line. It is noted that this simple modification gives almost the same result. Suck kind of feedback to an external parameter in a simplified model is called **pump depletion** in general.

3.9 Large Amplitude Electron Plasma Waves

Eliminating the driver term in Eq. (3.6.4), the equation tends to an exact equation for nonlinear plasma wave at $T = 0$:

$$\frac{d^2}{dt^2}V + \omega_{pe0}^2 V = 0 \tag{3.9.1}$$

Introducing the Lagrange coordinate x_0 defined by:

$$x_0 = x - \int_0^t V(x_0, \tau) d\tau \tag{3.9.2}$$

It is easy to obtain the nonlinear plasma wave structure. The plasma wave propagating to the positive direction in a uniform density has a solution of (3.9.1) in the form:

$$V = V_0 \sin\left(kx_0 - \omega_{p0}t\right) \tag{3.9.3}$$

Inserting Eq. (3.9.3) to Eq. (3.6.3) and taking integration gives:

$$E = \frac{V_0}{\omega_{p0}} \cos\left(kx_0 - \omega_{p0}t\right) \tag{3.9.4}$$

The density profile can be solved with use of the continuity relation:

3.9 Large Amplitude Electron Plasma Waves

$$n_e dx = n_0 dx_0 \tag{3.9.5}$$

With (3.6.2), (3.9.5) reduces to:

$$\frac{\partial x}{\partial x_0} = 1 + V_0 \frac{k}{\omega_{p0}} \sin(kx_0 - \omega_{p0}t) \tag{3.9.6}$$

The density profile is obtained by inserting (3.9.6) to (3.9.5):

$$n = \frac{n_0}{1 + \Delta \sin(kx_0 - \omega_{p0}t)}$$
$$\Delta = V_0 \frac{k}{\omega_{p0}} \tag{3.9.7}$$

(3.9.7) is valid as far as $\Delta < 1$, namely, the density has negative part for $\Delta > 1$. At the time of $\Delta = 1$, the **wavebreaking** starts to happen to be described soon. It is important to note that the condition $\Delta = 1$ means that the oscillation velocity is equal to the phase velocity of the wave:

$$V_0 = \frac{\omega_{p0}}{k} \tag{3.9.8}$$

This means that the electrons at the maximum velocity point are trapped by the same phase of the electric field of the wave and will be continuously feel the electric force in the same direction.

These nonlinear wave profiles are given in Fig. 3.30. The parameter in Fig. 3.30 is chosen so that the amplitude just before the wave- reaking happens. If the normalized amplitude Δ exceeds unity in (3.9.7), there is no solution physically acceptable like the case the overtaking of the steepened shock front without viscosity. It is, however, noted that the plasma is collisionless, and the wave-breaking is physically acceptable. For $\Delta > 1$, the density is not given by (3.9.5) where single value of the $V(x,t)$ is assumed along the axis of x. In such large amplitude, the solution of (3.9.2) is known to provide the multi-value function of $V(x,t)$ in x-space, and the particle trapping is seen. This allows the acceleration of the trapped electrons to high energy like a DC acceleration. This is one of the origins of the generation of hot electrons.

3.9.1 Wave-Breaking

It should be noted that (3.9.1) is an exact equation to the electron plasma oscillation for the cold electron fluid. (3.6.5) is in the form of a forced oscillation of the electron fluid in an inhomogeneous plasma density profile. The coordinate x appears only through ω_{pe0}, and x is the Lagrange coordinate x_0 defined in (3.9.2), and it is independent of time t in (3.6.5).

Assume the initial condition

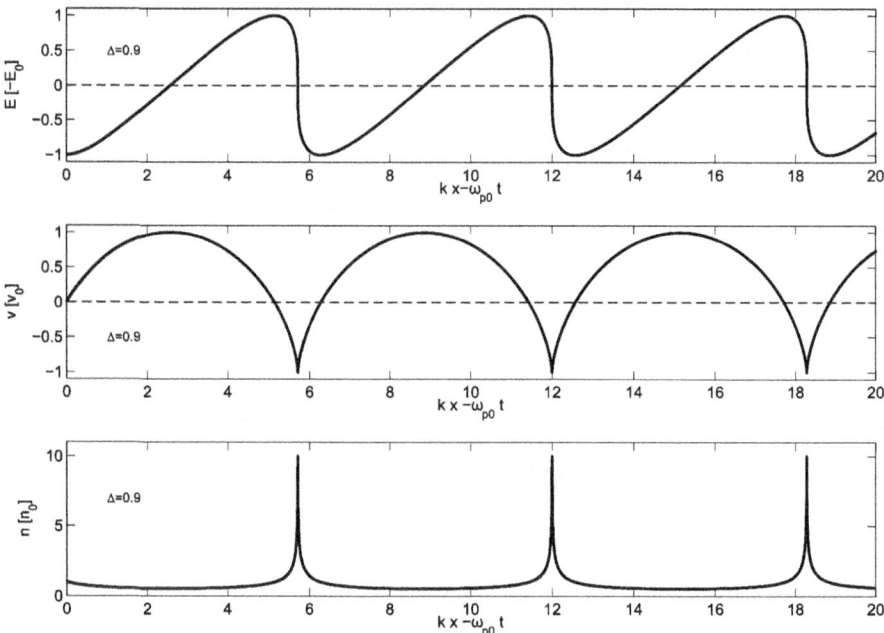

Fig. 3.30 The snapshots of a nonlinear wave propagating to the right direction. From the top, the minus of electric field, velocity, and density profile. The wave-breaking will happen near the density spike if the amplitude is a bit stronger

$$V = 0, \quad \frac{d}{dt}V = 0 \quad (t = 0)$$

After an initial phase, the asymptotic solution of (3.6.5) for the liner density profile

$$n_e(x) = (1 + x/L)n_{cr} \quad (-L < x) \quad (3.9.9)$$

is obtained in the form:

$$V(t, x) = \frac{V_d}{2}\left(\frac{x}{2L}\right)^{-1} \sin\left(\frac{x}{2L}\omega t\right) \cos\left[\left(1 + \frac{x}{2L}\right)\omega t\right] \quad (3.9.10)$$

Time evolution of this solution is plotted in Fig. 3.31. It is seen in Fig. 3.31 that the wave propagates from high density- to low-density region (right to left), and the oscillation amplitude increases monotonically with time. In addition, effective wavelength decreases near the critical point $x = 0$. This solution (3.9.10) gives the following amplitude of the oscillation velocity in time at $x = 0$:

3.9 Large Amplitude Electron Plasma Waves

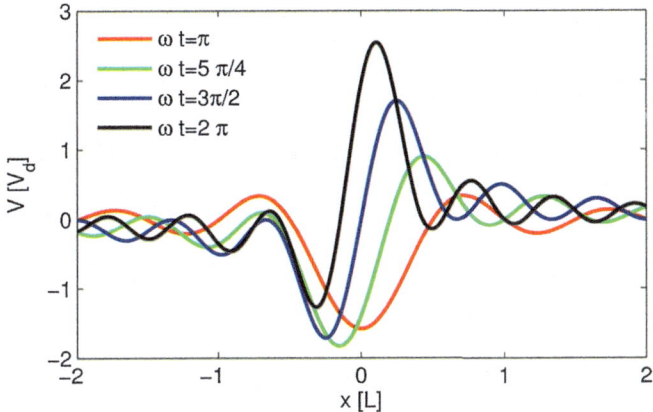

Fig. 3.31 The analytical solution of the resonantly excited plasma oscillation near the critical density. The oscillation profiles at four times shown near the wave will break. It is seen that the wave is localized near the critical point and propagating toward the lower-density region

$$|V| = \frac{V_d}{2}\omega t \tag{3.9.11}$$

The effective wavenumber and phase velocity of this wave are evaluated from the last term of (3.9.10):

$$k_{eff} = \frac{\omega t}{2L}, \quad V_{ph} = \frac{\omega}{k_{eff}} = -\frac{2L}{t} \tag{3.9.12}$$

(3.9.12) means the effective wavelength becomes shorter, and the phase velocity decreases with time. When the amplitude grows to satisfy the condition

$$\left|\frac{\partial}{\partial t}\right| \approx \left|V\frac{\partial}{\partial x}\right| \tag{3.9.13}$$

the wave-breaking condition given in (3.9.8) is subject to satisfy. The wave-breaking is taken place around the time when the oscillation velocity increases as follows:

$$\omega t = \left(\frac{2L\omega}{V_d}\right)^{1/2}, \quad V = (L\omega V_d)^{1/2} \tag{3.9.14}$$

Namely,

$$|V_{ph}| \approx |V(t,0)| \tag{3.9.15}$$

If this is the case of conventional fluid, the fluid viscosity becomes important, and a shock wave structure will appear to inhibit the wave-breaking. In the present

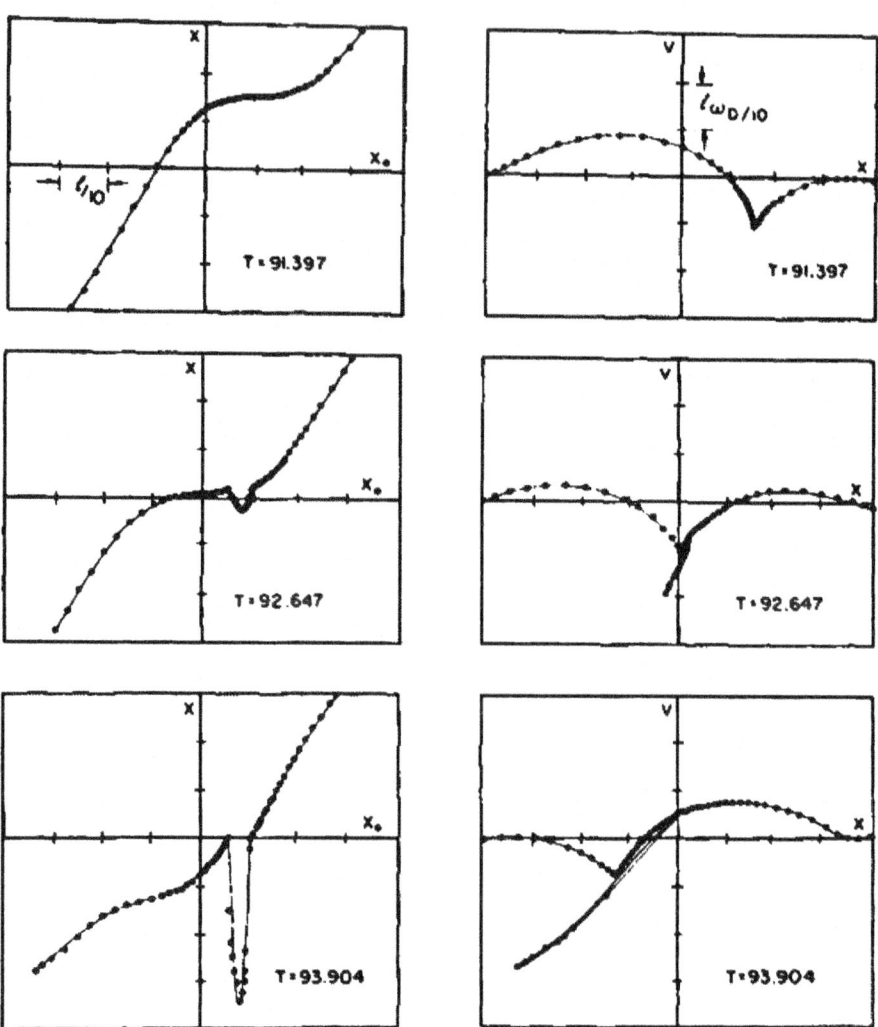

Fig. 3.32 A numerical solution of the exact equation of plasma oscillation (3.6.5). The left figures are the real position x as function of the initial points of electron elements x_0. The right figures are the oscillation velocity as function of the real position of each particle. This easily demonstrates the behavior of wave-breaking

electron fluid, however, the Coulomb collision can be neglected, and fluid elements can catch up and overtake with each other.

When the catch up and overtaking of the wave components occur, the phenomenon is called **wave-breaking**. In order to see the physics image before and after the wave-breaking, the result of numerical solution is shown in Fig. 3.32, where x_0 is the initial coordinate, x is the fluid element position at time t, and V is the oscillation velocity [20]. On the left of Fig. 3.32, the displacement of cold fluid elements is shown for three times when the wave-breaking happens. It is seen that many fluid elements are oscillated at the same point in the top of the left figures, and the next time, the small fraction of the elements are trapped by the wave phase in the next neighbor to be continuously accelerated selectively. The trapped electrons are ejected with higher velocity from the region of resonant point. They are called **high-energy electrons** or **hot electrons**.

From Fig. 3.24, in order to obtain substantial absorption in finite angle θ, the density scale length L should be of the order of the laser wavelength (kLsinθ ~ 1). It looks like not the case of laser plasma where hydrodynamic expansion makes the density scale length longer as the time goes. It will be clear, however, in the later session that the laser photon pressure will modify the density profile near the critical density and makes it of the order of unity.

As shown in Fig. 2.42, a small fraction of electrons deviates from its initial phase and escapes from the plasma oscillation region with high energy. This is one example of high-energy electron production process, and the absorbed laser energy is converted not to the bulk electron heating but is given to selected electrons in plasma. All collisionless heating processes are due to such energy deposition physics, and laser energy is once transferred to the selected electrons and relaxes to the thermal energy of the bulk electrons through classical collisional process.

The energy of the high-energy electron is roughly estimated in this case like:

$$E_{he} = mV^2 = eLE_d \tag{3.9.16}$$

This is the work done by laser electric field over the distance L. This is intuitively reasonable evaluation. Roughly evaluate the energy of high-energy electrons accelerated in such scheme. Using the value of laser field $E = 6 \times 10^8$ V/m at $I_L = 10^{15}$ W/cm^2, the energy is about 6 keV for L = 10 μm.

3.10 Vacuum Heating

When an ultra-short laser pulse is irradiated on solid density target, it is already discussed about the classical absorption in Sect. 3.2. If the laser intensity is high enough and the collisional absorption can be neglected, what kind of collisionless absorption process may happen? It is proposed [21] that the following physical process happens in a very short time, during which the ions have no time to move, and the electrons in the solid target are affected by a strong laser field penetrated into the solid surface due to the plasma skin effect, which is the same as the tunneling effect of the laser field into the higher-density region. This is the case of p-polarization as shown in Fig. 3.33, where the oscillating electric field penetrates into solid plasmas over the skin depth. This electric field drives the following electrons motions in the solid and vacuum regions.

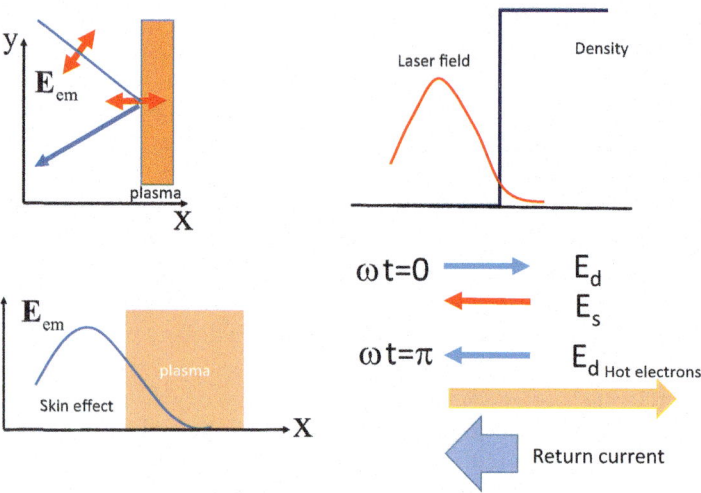

Fig. 3.33 When laser electric field (E_d) is initially inward, the electrons near the surface are pulled to the vacuum, inducing electrostatic field by charge separation so that it cancels the driver field. After a half cycle of laser field, the electrons pulled out are pushed back to the solid and penetrate into the solid with high energy. Then, the bulk electrons move to the surface by the electric field to repeat the same process above

In such case, it is appropriate to assume that the electron density is a step function-like structure, and the charge neutrality is satisfied with the immobile ions before laser comes. Since the energy-binding electrons and ions are, however, very small compared to the laser oscillation energy of the electrons, it is possible to assume that the electrons move freely under both of the external laser field and the electrostatic field generated by charge separation due to the electron motion by laser field. As the result, the electrons obtain substantial amount of energy from laser. Such collisionless absorption is called **vacuum heating** or **Brunel heating** [21, 22].

The motion of electron fluid layer in plane geometry is considered in the x-coordinate. The electron density profiles are initially a constant n_0 for $x > 0$ and vacuum for $x < 0$. For mathematical convenience, each electron displacement ξ from its initial coordinate (**Lagrange coordinate**) is introduced:

$$\xi = x - x_0 \tag{3.10.1}$$

The electrostatic field due to electron motion is obtained by integrating Eq. (3.6.2):

$$E = \frac{e}{\varepsilon_0} \int_{-\infty}^{x} (n_i - n_e) dx \tag{3.10.2}$$

3.10 Vacuum Heating

As long as the wave-breaking doesn't take place in the aligned electron fluid elements, analytical solution of (3.10.2) exists. In case without the wave-breaking, Eq. (3.10.2) can be solved to obtain:

$$E(x,t) = \frac{en_0}{\varepsilon_0} x_0 \quad (x < 0) \tag{3.10.3}$$

$$E(x,t) = \frac{en_0}{\varepsilon_0} \xi \quad (x > 0) \tag{3.10.4}$$

Inserting (3.10.3) and (3.10.4) into (3.6.5), it is found that the equation of motion for the all the electron elements initially aligned is:

$$\frac{d^2}{dt^2}\xi + \omega_{p0}^2 x_0 = -\frac{e}{m}E_d \quad (x < 0) \tag{3.10.5}$$

$$\frac{d^2}{dt^2}\xi + \omega_{p0}^2 \xi = -\frac{e}{m}E_d \quad (x > 0) \tag{3.10.6}$$

These equations can be solved when the laser electric field normal to the solid surface is give:

$$E_d = E_0 \sin(\omega t), \tag{3.10.7}$$

(3.10.6) has the solution for $x > 0$ region which satisfies the initial condition in the form:

$$\xi_+ = \frac{\xi_{d0}}{\omega_{p0}^2/\omega^2 - 1}\left[\frac{\omega}{\omega_{p0}}\sin(\omega_{p0} t) - \sin(\omega t)\right] \quad (x > 0) \tag{3.10.8}$$

where

$$\xi_{d0} = \frac{e}{m\omega^2} E_0 \tag{3.10.9}$$

The electrons escaping from the surface are found to the initial position satisfying the condition:

$$x_0 < \frac{\omega^2}{\omega_{p0}^2}\left(1 - \frac{\omega^2}{\omega_{p0}^2}\right)^{-1} \xi_{d0} \tag{3.10.10}$$

An electron initially located at the position x_0 escapes the surface at the time t_0 satisfying the relation:

$$x_0 + \xi_+(t_0, x_0) = 0 \qquad (3.10.11)$$

At this time, the electron goes into the vacuum region, and its motion should be governed by (3.10.5) then. The general solution of (3.10.5) is obtained:

$$\xi_- = \frac{\omega_{p0}^2 x_0}{2} t^2 + At + B + \xi_{d0}\sin(\omega t) \qquad (x < 0) \qquad (3.10.12)$$

where A and B are integral constants and are determined so as to satisfy the following condition. At the time t_0 given in (3.10.11), the position and velocity should be continued in both of (3.10.8) and (3.10.12):

$$\begin{aligned}\xi_+(t_0, x_0) &= \xi_-(t_0, x_0) \\ \frac{\partial \xi_+}{\partial t} &= \frac{\partial \xi_-}{\partial t} \quad \text{for } t_0, x_0\end{aligned} \qquad (3.10.13)$$

It should be noted that the first term in (3.10.12) is a constant acceleration term into the positive x-direction, and constant acceleration is kept in the vacuum region. This is the reason why it is called **vacuum heating**. It is intuitively clear that this acceleration occurs by the electric field due to the charge separation. This electrostatic force is induced because the electrons near the surface are pulled out by laser field.

It is easily understood that after certain time, the wave-breaking may happen at certain point, and the above analytic solutions cannot be applicable. Therefore, the longtime evolution of the orbits of electron layers can be obtained by numerically solving (3.6.3) and (3.10.2) self-consistently. This is the same as one-dimensional particle simulation.

In Fig. 3.34, a computational result is plotted in x and t space for the case of laser irradiation of 60° from the normal, the density $n_0 = 10 n_{cr}$, and the laser intensity $I\lambda^2 = 10^{19}$ W/cm²µm². The time is normalized by one cycle, and space is normalized by laser wavelength. We can see the target oscillation with the fundamental laser frequency as shown in (3.10.8) and (3.10.12). The particles are accelerated into the front side; most of them inject into the solid region and escape from the rear side as high-energy electrons. The analytical relations given in (3.10.8) and (3.10.12) can be applicable only before the wave-breaking. It is noted that most of the data shown in Fig. 3.34 are the electron orbits numerically obtained after the wave-breaking.

3.10.1 Physical Image and Absorption Rate

It is important to grasp the intuitive image of the vacuum heating process. It is easily understood by considering the motion of electrons and generation of electrostatic field over one cycle of laser field. During the time when the laser field vector is in the

3.10 Vacuum Heating

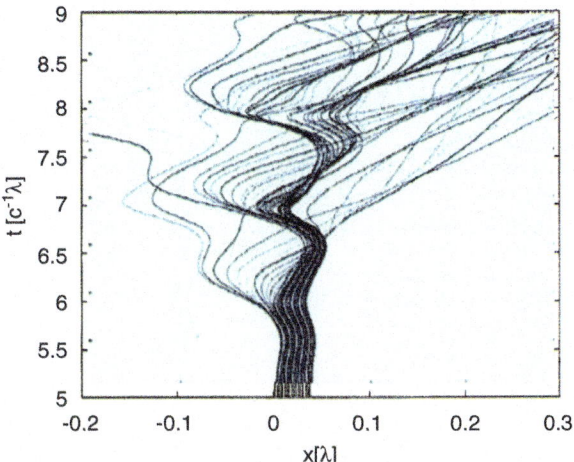

Fig. 3.34 The trajectories of electron elements initially aligned in the solid. They are accelerated by the vacuum heating, and high-energy electrons are produced. This computation is equivalent to 1-D PIC simulation

direction to the right in Fig. 3.33, the electrons near the solid surface are accelerated to the vacuum. Then, strong electrostatic field due to the charge separation appears in the direction so that the electrons are pulled back toward the inside of the solid. When the sigh of the laser field becomes opposite after the half cycle, the laser and electrostatic field both work as strong force so that the electrons are accelerated toward the inward of the solid. Then, these electrons escape from the interaction region near the surface to the solid inside with accelerated high energy. Of course, a part of the electrons accelerated to the vacuum can continue to accelerate over the next cycles of laser as seen in Fig. 3.34.

From such intuitive view, it is possible to evaluate the fraction of laser absorption roughly. Assume that E_d is the value of laser field near the surface in the skip depth. It is assume that the skin depth is much longer than the following charge separation distance Δx at the surface. Then, the condition that the electrostatic field E_s increases to the value of laser electric field is given by the relation:

$$-\frac{en_e}{\varepsilon_0} \Delta x \sim E_s \sim E_d \tag{3.10.14}$$

It is reasonable to assume that the electrons initially located in the depth Δx return toward the solid region and escape from the skin depth region as high-energy electrons with average energy of the oscillation velocity by laser field, v_{os}. The total energy of the escaping electrons over one cycle is regarded by the absorbed energy from the laser field, namely, absorption power per unit surface P_{abs} is:

$$P_{abs} \sim n_e \Delta x \frac{1}{2} m v_{os}^2 \frac{\omega}{2\pi} \tag{3.10.15}$$

Since the laser intensity with incident angle θ, P_L is

$$P_L = \frac{c}{2}\varepsilon_0 E_0^2 \cos\theta \tag{3.10.16}$$

where E_0 is the laser electric field in the vacuum and E_d is given:

$$E_d = E_0 \sin\theta \tag{3.10.17}$$

Assuming these relations, the absorption rate is approximately give to be:

$$\eta_{abs} = \frac{P_{abs}}{P_L} \sim\sim \frac{4}{\pi} a_0 \frac{\sin^3\theta}{\cos\theta} \tag{3.10.18}$$

As clear in the above description, the electros obtain their energies in the acceleration in the vacuum by both forces due to laser and electrostatic fields. These energies are carried into the solid by electrons, and as a result, the laser energy is absorbed. This is the reason why this mechanism is called **vacuum heating**.

It is clear from (3.10.18) that the absorption rate is enhanced in proportion to the laser strength a_0 (v_{os}/c). It is suggested that the vacuum heating becomes important in the laser intensity in the relativistic regime. It is straightforward to extent the above model to the relativistic case with $a_0 > 1$. The physics of the interaction of relativistic laser and plasmas is described in Chaps. 5, 6, 7, and 8, and here it is enough to show the relativistic form of (3.10.18):

$$\eta_{abs} \sim \frac{8}{\pi}\left(\sqrt{1+a_0^2}-1\right)\frac{\sin^2\theta}{\cos\theta} \tag{3.10.19}$$

(3.10.18) also shows that the laser absorption rate increases as the incident angle increases against the normal direction. This is because the driver field E_d increases for the oblique incident. However, (3.10.18) diverges at the shallow incidence because of invalid assumption to the driver field in the skin depth region in the solid as seen below.

It is found below in the discussion on the skin depth that the driving force inside the solid in (3.10.14) should be reduced by a factor k_x/K, where k_x is the wavenumber in the normal direction and $k_x = k_0 \cos\theta$, and K is the inverse of the penetration depth determined the density ratio of the solid to the critical point. Inserting this $\cos\theta$ dependence in (3.10.14), the divergence of the absorption rate is eliminated.

The above analysis is not valid when the skin depth is shorter than Δx in (3.10.14). In order to avoid such a divergence of the absorption rate, the pump depletion model used for resonance absorption in Sect. 3.8 can be used. For example, the driver strength is reduced by the pump depletion as:

$$E_d \rightarrow (1-\eta_{abs})E_d \tag{3.10.20}$$

Then, an effective absorption rate less than unity is given as:

$$\frac{\eta_{abs}}{(1-\eta_{abs})^3}\frac{\tilde{8}}{\pi}\left(\sqrt{1+a_0^2}-1\right)\frac{\sin^2\theta}{\cos\theta} \qquad (3.10.21)$$

3.10.2 Skin Depth at Sharp Boundary

It is useful to calculate the structure of the skin depth in the oblique incident of lasers. This is given by solving the Helmholtz wave equation of (3.1.1) for the present condition of the solid surface. Since the solid plasma frequency is higher than the critical density, it is required to connect the following solutions at the solid surface:

$$\begin{aligned} E &= \exp(-ik_x x) + \exp(+ik_x x + i\delta) \quad (x < 0) \\ E &= A\exp(-Kx) \quad (x > 0) \end{aligned} \qquad (3.10.22)$$

where

$$\begin{aligned} k_x &= k_0 \cos\theta \\ K &= k_0 \sqrt{\frac{n_e}{n_{cr}} - \cos^2\theta} \end{aligned} \qquad (3.10.23)$$

The **skin depth** λ_s is defined by:

$$\lambda_s = 1/K \qquad (3.10.24)$$

In (3.10.22), δ is a phase shift by the reflection at the surface, and A is the amplitude of the penetration field at the surface. Both of δ and A are unknown to be determined with the connecting condition at the solid surface $x = 0$. Requiring the value and derivative should be continued at the solid surface; they are easily calculated, and A is given:

$$A = \frac{2}{1 + iK/k_x} \qquad (3.10.25)$$

It is clear that $A \propto \cos\theta$ near the shallow incidence disappears.

Finally, it is noted that the neglected term by $\mathbf{v} \times \mathbf{B}$ force becomes important when the laser intensity approaches the relativistic one, namely, the field strength a_0 increases near and over the unity. The force by this nonlinear motion oscillates with the frequency of 2ω, and the above discussion should be done for the duration of a half period of the fundamental oscillation period. This becomes essential in the relativistic regime.

References

1. B. Rethfeld et al., J. Phys. D. Appl. Phys. **50**, 193001 (2017)
2. D.F. Price et al., Phys. Rev. Lett. **75**, 252 (1995)
3. I.B. Foldes et al., Web Conf. **167**, 04001 (2018)
4. H.M. Milchberg, R.R. Freeman, J. Opt. Soc. Am. B **6**, 1351 (1989)
5. K. Eidmann, J. Meyer-ter-Vehn, T. Schlegel, Phys. Rev. E **62**, 1202 (2000)
6. S. P. Marsh (ed.), *LASL Shock Hugoniot Data* (University of California Press, Berkeley, 1980)
7. R.L. More et al., Phys. Fluids **31**, 3059 (1988)
8. R. Fedsejevs et al., Phys. Rev. Lett. **64**, 1250 (1990)
9. H.M. Milchberg, R.R. Freeman, S.C. Davey, R.M. More, Phys. Rev. Lett. **61**, 2364 (1988)
10. D. Semkat, R. Redmer, T. Bornath, Phys. Rev. E **73**, 066406 (2006)
11. C.D. Decker et al., Phys. Plasmas **1**, 4043 (1994)
12. V.P. Silin, Sov. Phys. JETP **20**, 1510 (1965)
13. T. Bornath et al., Phys. Rev. E **64**, 026414 (2001)
14. A. Bratov et al., Phys. Plasmas **10**, 3385 (2003)
15. S.-M. Weng, Z.M. Sheng, J. Zhang, Phys. Rev. E **80**, 056406 (2009)
16. N. David et al., Phys. Rev. E **70**, 056411 (2004)
17. V.L. Ginzburg, *The Propagation of Electromagnetic Waves in Plasmas* (Pergamon, New York, 1964)
18. D.E. Hinkel-Lipsker, B.D. Fried, G.J. Morales, Phys. Fluids **B4**, 559 (1992)
19. D.W. Forslund et al., Phys. Rev. A **11**, 679 (1975)
20. J. Albritton, P. Koch, Phys. Fluids **18**, 1136 (1975)
21. F. Beunel, Phys. Rev. Lett. **59**, 52 (1987); F. Brunel, Phys. Fluids 31, 2714 (1988)
22. P. Mulser, D. Bauer, *High-Power Laser-Matter Interaction* (Springer, New York, 2010)

Chapter 4
Nonlinear Physics of Laser-Plasma Interaction

4.1 Ponderomotive (PM) Force

Consider the time-averaged force to electrons by the laser force. This time-averaged force stems from the nonuniform profile of the electromagnetic energy density. Intuitively speaking, the electrons or any charged particles are affected by nonzero force after the time averaging the electromagnetic force, when the electromagnetic energy is localized spatially. In other words, the electrons obtain an effective pressure stemming from the quivering motions, and this pressure gradient provides a new force to the electrons. Since the electron energy density of the quivering motion is proportional to the energy density of the electromagnetic fields, the time-averaged force to the electrons is regarded as the repelling force to the electrons from localized field energy density. In order to confirm this explanation, derive mathematically the form of this force. Note that this force is called **ponderomotive force** and is written as **PM force** in short.

Consider the case of non-relativistic electron motion governed by (2.3.1) in an electromagnetic field. In order to use Taylor expansion, it is assumed that the spatial variation length of the laser amplitude is much longer than the electron quivering distance ξ. The spatial dependence of the electric field structure can be approximated with Taylor expansion form:

$$\mathbf{F} = -e\mathbf{E} - e(\xi \cdot \nabla)\mathbf{E} - e\mathbf{v} \times \mathbf{B} \qquad (4.1.1)$$

It is clear that the second term of RHS in (4.1.1) remains for the case of the electric field is not constant in space. Since the non-relativistic case is considered, the third term in RHS of (4.1.1) is small enough. Taking the time average of (4.1.1) yields:

$$\langle \mathbf{F} \rangle = -e\langle (\xi \cdot \nabla)\mathbf{E} \rangle - e\langle \mathbf{v} \times \mathbf{B} \rangle, \qquad (4.1.2)$$

where < > means taking time average over the electron oscillation.

In the non-relativistic limit ($v \ll c$), the following relations can be used:

$$\mathbf{v} = -i\frac{e}{m\omega}\mathbf{E}, \quad \xi = \frac{e}{m\omega^2}\mathbf{E}, \quad \mathbf{B} = -i\frac{1}{\omega}\nabla \times \mathbf{E} \qquad (4.1.3)$$

Inserting (4.1.3) into (4.1.2), the time-averaged force is obtained:

$$\begin{aligned}\mathbf{f}_{PM} &= -\frac{e^2}{m\omega^2}\langle(\mathbf{E}\cdot\nabla)\mathbf{E} + \mathbf{E}\times(\nabla\times\mathbf{E})\rangle \\ &= -\frac{e^2}{m\omega^2}\frac{1}{2}\nabla\langle\mathbf{E}^2\rangle = -\nabla\left(\frac{m}{2}\langle\mathbf{v}^2\rangle\right),\end{aligned} \qquad (4.1.4)$$

where the mathematical formula:

$$(\mathbf{E}\cdot\nabla)\mathbf{E} + \mathbf{E}\times(\nabla\times\mathbf{E}) = \frac{1}{2}\mathbf{E}^2 \qquad (4.1.5)$$

is used.

This is called **ponderomotive force** for an electron in the non-relativistic case. It is noted that this force is shown as a potential force:

$$\mathbf{f}_{PM} = -\nabla\phi_{PM}, \quad \phi_{PM} = \left\langle\frac{m}{2}v^2\right\rangle \qquad (4.1.6)$$

This potential ϕ_{PM} is called **ponderomotive potential**. The force working to the unit volume is easily obtained:

$$\mathbf{F}_{PM} = -\frac{\omega_{pe}^2}{\omega^2}\nabla W, \quad W = \frac{\varepsilon_0}{2}\langle|\mathbf{E}|^2\rangle \qquad (4.1.7)$$

Note that the W in (4.1.7) is the energy density of electric field as seen in (2.2.7).

Now, consider also the case of electrostatic waves, for example, electron plasma waves. It is straightforward to show that the ponderomotive force is in the same form as (4.1.7) even in the electrostatic field case. For the electrostatic oscillations, there is no contribution by the magnetic field, and the third term in (4.1.1) is neglected. Although $\nabla \times \mathbf{E}$ in (4.1.3) vanishes, however, the resultant form of the ponderomotive force is the same as (4.1.4) because (4.1.5) is universal relation.

The derivation of the ponderomotive force in general from also covering the relativistic motion of electron is given in Chap. 6. It is informative to show the resultant form for convenience:

4.2 Nonlinear Schrodinger Equation

$$\mathbf{f}_{PM} = -\nabla U_p$$
$$U_p = \langle \gamma - 1 \rangle mc^2 \qquad (4.1.8)$$

It is noted that (4.1.8) is general form and tends to (4.1.6) in the non-relativistic limit. In addition, the ponderomotive potential is the kinetic component of the total energy of oscillating electron.

In the present book, the cases with long pulse non-relativistic of ns order and short pulse relativistic of sub-picoseconds are considered. It is important to keep in mind that in the long pulse phenomena, ion has enough time to be moved by ambipolar field, and the electron and ion density are almost the same for the phenomena longer than the Debye length. On the other hand, ultra-short phenomena happen during the time the ion cannot move and better to assume that the PM force acts only with electrons, so strong charge separation is induced.

4.2 Nonlinear Schrodinger Equation

When the intensity of laser is strong, we cannot neglect the density modification by the ponderomotive force by the laser. Consider the case where the laser propagates with depression of the density by the ponderomotive force and time scale is relatively long so that charge neutrality is assumed. The laser propagation equation in plasma of (2.5.2) can be written for the case without external current:

$$\frac{\partial^2}{\partial t^2}\mathbf{E} - c^2 \nabla^2 \mathbf{E} + \omega_{pe}^2 \mathbf{E} = 0 \qquad (4.2.1)$$

We assume that the laser propagates in the x-direction with its electric filed in the perpendicular direction, say, y in the form:

$$\mathbf{E} = \mathbf{i}_y E(x,t) e^{i(kx - \omega t)} \qquad (4.2.2)$$

Inserting (4.2.2) into (4.2.1) and taking account of the dispersion relation (2.5.3) (4.2.1) can be modified in the following form after neglecting the second derivative in time, where the time variation of the complex amplitude $E(x,t)$ in (4.2.2) is assumed:

$$\left(2i\omega \frac{\partial}{\partial t} + 2ikc^2 \frac{\partial}{\partial x} - c^2 \frac{\partial^2}{\partial x^2} + \frac{e^2}{m\varepsilon_0} n_{e1} \right) E = 0 \qquad (4.2.3)$$

In (4.2.3), n_{e1} is the electron density perturbation by the ponderomotive force. The first two terms govern the propagation of wave with group velocity:

$$v_g = \frac{c^2 k}{\omega} \tag{4.2.4}$$

Divide (4.2.3) by $2\omega^2$ and introduce the following new variables:

$$\tau = \omega t, \quad \xi = \sqrt{2}k(x - v_g t) \tag{4.2.5}$$

Then, the equation in the frame moving with the group velocity is obtained in the form:

$$i\frac{\partial}{\partial \tau}E = -\frac{\partial^2}{\partial \xi^2}E + \frac{1}{2}\frac{n_{e1}}{n_{cr}}E \tag{4.2.6}$$

If the density perturbation is given as a function of space, (4.2.6) has the following form of equation same as Schrodinger equation in a given potential U(x) by regarding E the wave function Φ:

$$i\hbar\frac{\partial}{\partial t}\Phi = -\frac{\hbar^2}{2m}\frac{\partial^2}{\partial x^2}\Phi + U(x)\Phi \tag{4.2.7}$$

Now assuming that the solution of (4.2.6) is slow enough, the ion motion follows to keep the change neutrality with the electrons. Then, the force by the electron pressure is assumed to balance with the ponderomotive force to satisfy the relation:

$$\frac{T_e}{n_e}\nabla n_e = -\frac{1}{n_{cr}}\nabla W, \tag{4.2.8}$$

where W is given in (4.1.6), and it is assumed that the electron temperature T_e is constant. The relation (4.2.8) can be easily solved:

$$n_e = n_0 \exp\left(-\frac{W}{n_{cr} T_e}\right) \tag{4.2.9}$$

Namely, the density perturbation is only the function of the ponderomotive potential.

In the limit where W is smaller than the electron pressure, (4.2.9) can be approximated:

$$n_{e1} \approx -\frac{\varepsilon_0}{2T_e}\frac{\omega_{p0}^2}{\omega^2}|E|^2 \tag{4.2.10}$$

Inserting (4.2.10) into (4.2.6) and defining a new variable Ψ

4.2 Nonlinear Schrodinger Equation

$$\Psi(\tau,\xi) = CE(\tau,\xi), \quad C = \frac{1}{2n_{cr}}\sqrt{\frac{\varepsilon_0 n_0}{T_e}}, \quad (4.2.11)$$

the following equation is obtained:

$$i\frac{\partial}{\partial \tau}\Psi = -\frac{\partial^2}{\partial \xi^2}\Psi - |\Psi|^2\Psi \quad (4.2.12)$$

This is called **nonlinear Schrodinger equation**.

It is clear that the potential is negative where the wave intensity is high, and the wave may be trapped by this self-potential. It is well-known that (4.2.12) has an exact solution. Assume the solution in a form with a frequency shift Ω:

$$\Psi = e^{i\Omega\tau}f(\xi) \quad (4.2.13)$$

Inserting (4.2.13) into (4.2.12), the following solution is obtained:

$$f(\xi) = \sqrt{2\Omega}\,\text{sech}\left(\sqrt{\Omega}\xi\right) \quad (4.2.14)$$

This spatial structure is a solitary wave, and such wave is called **soliton**. Note that the soliton propagates stationary in a homogeneous plasma with the group velocity. It is informative to note that the concept of soliton by self-force confinement is one of the standard models for giving stable elementary particles and widely used in the particle physics.

It is clear from (4.2.14) that for large amplitude wave, the density dip becomes deeper as:

$$\frac{n_{e1}(\xi)}{n_0} = -2\Omega\,\text{sech}^2\left(\sqrt{\Omega}\xi\right) \quad (4.2.15)$$

The density and wave amplitude are plotted in Fig. 4.1. The solution Ψ is the envelope of the soliton, and there are many oscillations satisfying (2.5.3) of the propagation wave. Such soliton is called **envelope solitons**. Since the soliton accompanies the density dip, it is sometimes called **cavitons**. It should be noted that the soliton is stable in 1 dimension but not clear whether it is stable in 2 and 3 dimensions.

It is easy to drive the same nonlinear Schrodinger equation for the case of the electron plasma wave. This is because the dispersion relation of electromagnetic wave and the plasma wave has the same dependence on the density.

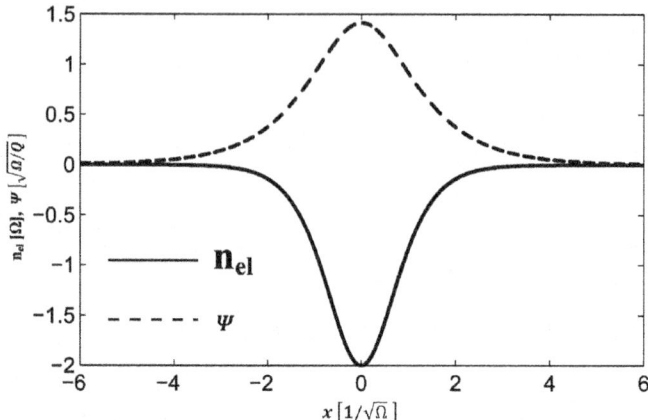

Fig. 4.1 The structure of an envelope soliton Ψ and the density profile to balance the ponderomotive force

4.3 Filament Instability of Lasers

The soliton solution was obtained with the assumption of one-dimensional geometry. It is natural question if the soliton propagates stationary in the three-dimensional geometry. This is essentially the same question about the stability of intense laser beam propagation in plasmas. In what follows, consider the physics for the case where the intense laser propagates in x-direction with a beam size of radius R in the perpendicular direction. It is assumed that R is much longer the wavelength of the laser. If this is the case of laser propagation in the vacuum, the laser light cannot propagate as plane wave for a long distance because of diffraction effect, and it diverges with an angle of roughly $\theta \sim k_0 R$, where k_0 is the wavenumber of the laser wave.

The effect of the ponderomotive force, however, is an attractive force effectively to keep the high-intensity region by depressing the electron density. It is clear that if the ponderomotive effect to laser intensity change is stronger than the diffraction effect, the laser intensity at the beam central may increase, and the initial structure becomes unstable. If the laser intensity is not so high to result the enough nonlinear force, on the other hand, the diffraction is dominant, and the beam intensity decreases as a function of time or space. The precise calculation on this **filamentation** and related **modulation instabilities** is intensively studied from the start of the research on laser-plasma interaction physics, and their precise mathematical treatment is given, for example, in Ref. [1].

In the present study, it is shown intuitively why the laser propagation is unstable and the instability condition of the filamentation instability without precise calculations. Start with the equation of laser propagation in three-dimensional space, where the effect of the density modification by the ponderomotive force is included with the use of the relation (4.2.10):

4.3 Filament Instability of Lasers

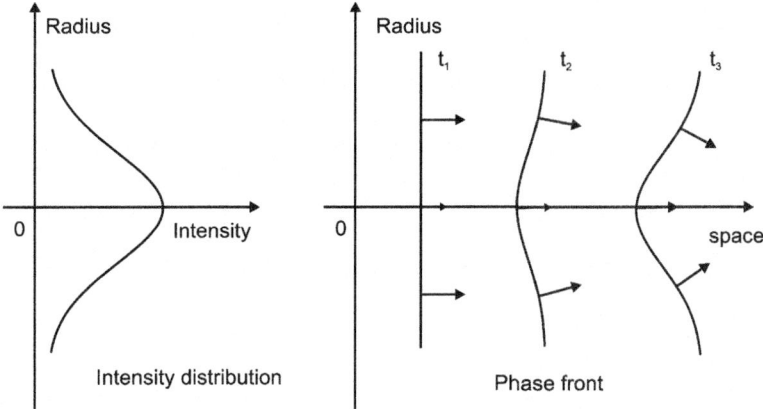

Fig. 4.2 The schematics showing the principle of filament instability by self-focusing. The phase velocity is lower at the center of laser beam where the intensity is highest. As the result, the wave front goes to bend toward the center to induce the self-focusing

$$\frac{\partial^2}{\partial t^2}\mathbf{E} - c^2\nabla^2\mathbf{E} + \omega_{p0}^2\left(1 - \alpha|\mathbf{E}|^2\right)\mathbf{E}, \quad (4.3.1)$$

where from (4.2.10) and (4.1.6),

$$\alpha|\mathbf{E}|^2 = \frac{1}{2}\left(\frac{v_{os}}{v_e}\right)^2 \quad (4.3.2)$$

For the validation of Taylor expansion, $v_{os}/v_e \ll 1$ is assumed.

For the case of uniform laser beams propagating in the x-direction with the wavenumber k in plane geometry is governed the nonlinear dispersion relation:

$$\omega^2 = c^2k^2 + \omega_{p0}^2\left(1 - \alpha|\mathbf{E}|^2\right) \quad (4.3.3)$$

This relation suggests that for a small perturbation of the laser intensity in the perpendicular direction, the region of relatively high-intensity increases the wavenumber k, and as a result, the phase velocity ω/k decreases. This is schematically shown in Fig. 4.2 for the case when the laser intensity of the beam central region increases. As the result of such perturbation, the wave front is deformed as shown in Fig. 4.2, and the laser beam is focused toward the high-intensity region. This is an intuitive image of the laser filamentation due to the ponderomotive force in plasmas.

In the above evaluation, the diffraction effect is not taken into account to restrict the condition of filamentation instability. For this purpose, take into account the finiteness of the beam size in the radial direction to the beam propagation direction.

Assume that the laser beam size is about R and $Rk_0 \gg 1$ is satisfied. (4.3.1) can be reduced to the following approximate dispersion relation:

$$\omega^2 = c^2k^2 + \frac{c^2}{R^2} + \omega_{p0}^2\left(1 - \alpha|E|^2\right) \quad (4.3.4)$$

It is noted that if the laser electric field is given by the 0th order Bessel function, (4.3.4) is the exact solution of (4.3.1) for the case of radially symmetric beam. With addition of the diffraction, the wave front deformation is found to be controlled by the sign of the η:

$$\eta = \frac{c^2}{R^2} - \omega_{p0}^2\alpha|E|^2 \quad (4.3.5)$$

The beam is stable for $\eta > 0$ but unstable for $\eta < 0$. At a glance, the beam is unstable for filamentation for large R but stops focusing at the certain radius.

It is not true in the case of three-dimensional system. The reason is simple because the laser power proportional to R^2E^2 is conserved. It is noted that in two-dimensional simulation in the Cartesian coordinates, the power conservation relation is RE^2, and the first term in (4.3.5) becomes dominant when the beam shrinks and the filamentation is stabilized a certain radius satisfying $\eta = 0$.

Evaluate the critical power for the filamentation instability. Since the laser power $P_L \approx \varepsilon_0|E|^2R^2$, the following critical power P_{cr} for the filamentation instability is obtained:

$$P_{cr} \approx \frac{\omega^2}{\omega_{p0}^2}n_0T_ec\lambda_s^2, \quad \lambda_s = \frac{c}{\omega_{p0}}, \quad (4.3.6)$$

where λ_s is the **plasma skin depth**.

The critical power is very low for low-temperature plasmas, since the relation (4.2.10) has been used to relate the ponderomotive force to the electron density change. It is noted that (4.2.8) is good approximation when the ambipolar field is weak enough so that the phenomena is very slow, and the ions can follow the change of electron density to keep charge neutrality. If the electrostatic field by charge separation is not neglected, the following fluid equations have to be solved for linear density perturbation:

From (3.5.6), (3.5.7), the equation of electron fluid motion with addition of the ponderomotive force is derived as:

$$m_e\frac{\partial}{\partial t}\mathbf{u}_{e1} = -\frac{T_e}{n_{e0}}\nabla n_{e1} - e\mathbf{E}_1 - \nabla\phi_{PM}, \quad (4.3.7)$$

where the motion perpendicular to the laser beam is only taken into account. Taking time derivative of (3.5.6) to insert in (4.3.7), the following equation is obtained:

$$\left(\gamma^2 + \omega_{p0}^2 - \frac{T_e}{m_e}\nabla^2\right)n_{e1} = \frac{\varepsilon_0}{e^2}\omega_{p0}^2\nabla^2\phi_{PM}, \tag{4.3.8}$$

where γ is the **growth rate** of **filamentation instability**. It is clear that the condition that the pressure force is larger than the force by the charge separation is satisfied only when:

$$\omega_{p0}^2 < \frac{T_e}{m_e R^2} \quad \Rightarrow \quad R < \lambda_{De} \tag{4.3.9}$$

This is not realistic. For the case of the charge separation force is dominant, the electron density perturbation is given to be:

$$\frac{n_{e1}}{n_0} = \left(\frac{\lambda_{De}}{R}\right)^2\left(\frac{v_{os}}{v_e}\right)^2 \tag{4.3.10}$$

This density depletion is much smaller than that obtained with pressure balance relation (4.2.10).

In general, depending on the time scale of the filamentation instability, namely, the laser intensity and the beam radius R, it is also plausible to consider that the ion motion also couples in determining the strength of ambipolar field, $\mathbf{E}_{es}$ in (4.3.7). Such precise calculation in order to obtain the growth rate of the filamentation instability has been carried out in [1], and the growth rate is found to proportional to the ion plasma frequency ω_{pi} as:

$$\gamma = \frac{\omega_{pi}}{\sqrt{2}}\frac{v_{os}}{c} \tag{4.3.11}$$

The experimental data of the filament is shown in Fig. 4.3 [2].

The same formulation can be done in the case of solitons by the ponderomotive force. It is, therefore, concluded that the solitary waves are unstable to the filamentary instability and collapse toward the center of the propagation axis when the laser power is larger than the critical values. This is the collapse of **3D soliton**.

4.4 Ion Fluid and Ion Acoustic Waves

In the laser-plasma interaction for relatively long pulse intense lasers (ns pulses), the ion motion is important to the electron density distribution. The ambipolar field generated by the charge separation is very strong as seen above, and the ion motion is induced by this electrostatic field. It is well-known that the density scale length is longer than the electron Debye length; it is reasonable to assume that both density profiles are same to keep the charge neutrality.

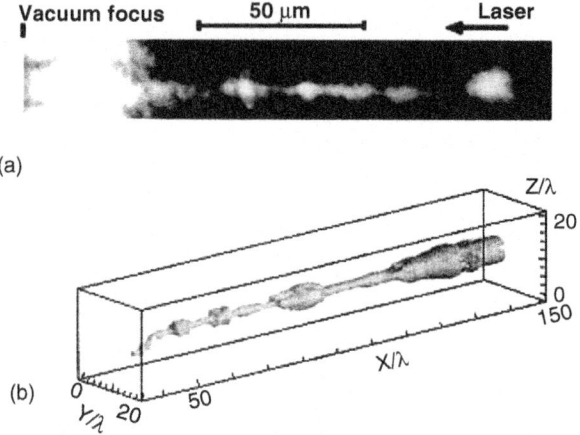

Fig. 4.3 (a) the features of the channel structure are clearly visible in an experiment, where the self-emission from focused laser has been detected. The diameter of the channel is about 5 μm and its length about 130 μm. Closer observation reveals that the channel changes in size periodically over distances of 15–20 μm with the transverse dimension varying within a few microns. (b) Results of 3D PIC simulations based on the density profile observed experimentally. A perspective snapshot of the self-focusing pulse after 150 laser cycles (i.e., about 0.5 ps); the plotted surface corresponds to 67% of the cycle-averaged maximum intensity. [Ref. 2]

4.4.1 Charge Neutral Plasma Fluids

In general, the laser wavelength is much longer than the Debye length as seen in Fig. 3.28, and neutral plasma condition is good approximation for discussing the density profile near the critical density, which is very sensitive to the resonance absorption. The fluid equations to the ion fluid are given with the following continuity equation and equation of motion to be described in detail in Vol. 2:

$$\frac{\partial}{\partial t} n_i + \nabla (n_i \mathbf{u}_i) = 0$$
$$m_i \frac{d\mathbf{u}_i}{dt} = e\mathbf{E} - \nabla P_i$$
(4.4.1)

Since the phenomena are slow, the electric field **E** in (4.4.1) is replaced with **E** by electron pressure force after neglecting the inertial term of the electron fluid. Assuming the charge neutral condition ($n_i = n_e = n_0$) and stationary state in a frame moving with the density profile, the following relations are obtained in one-dimensional plane system:

4.4 Ion Fluid and Ion Acoustic Waves

$$n_0 u_0 = J_0 : \text{const.}$$
$$m_i n_0 u_0^2 + P_e + P_i = \text{const.} \quad (4.4.2)$$

In (4.4.2), the first relation shows the density flux J_0 constant; the second one shows the total pressure including **dynamic pressure** balance in the system.

For simplicity, neglect the ion pressure term in (4.4.2). The ratio of the dynamic pressure to the static electron pressure is found to have the relation:

$$\frac{m_i n_0 u_0^2}{P_e} = \frac{u_0^2}{C_s^2} \equiv M^2$$
$$C_s = \sqrt{\frac{T_e}{m_i}}, \quad (4.4.3)$$

where C_s is the **plasma sound velocity** defined for constant temperature, and M is **Mach number**. When the plasmas are heated by lasers and expand into the vacuum, the plasma flow velocity changes from subsonic (M < 1) to supersonic (M > 1), and the dynamic pressure is important to determine the density profile of the expanding plasmas.

4.4.2 Ion Sound Waves

Consider the waves governed mainly by the ion motion in a uniform plasma. This is a fundamental wave of plasma corresponding to the **acoustic waves** or **sound waves** in neutral gas. In the case of plasma, however, the acoustic wave oscillating with heavier particle ions can also be affected by electron through the electric field due to charge separation. As a result, the dispersion relation is different from the sound waves with a constant phase velocity. It is seen below that the phase velocity is a function of the wavenumber of the ion acoustic waves.

Since the ion acoustic waves are electrostatic waves, the magnetic field is neglected. The perturbations of the waves from (4.4.1) are the following four physical quantities: the linear perturbations of ion density, ion flow velocity, electron density, and electric field:

$$n_{i1}, \quad \mathbf{u}_{11}, \quad n_{e1}, \quad \mathbf{E}_1 \quad (4.4.4)$$

Inserting these perturbations to (4.4.1), (3.5.7), and Poisson equation, and then neglecting the electron inertial, the following coupled linear equations are obtained:

$$\frac{\partial}{\partial t} n_{i1} + n_{i0} \nabla \mathbf{u}_{i1} = 0 \quad (4.4.5)$$

$$m_i \frac{\partial}{\partial t} \mathbf{u}_{i1} = e\mathbf{E}_1 - \frac{1}{n_{i0}} \nabla P_{i1} \tag{4.4.6}$$

$$-\frac{1}{n_{e0}} \nabla P_{e1} - e\mathbf{E}_1 = 0 \tag{4.4.7}$$

$$\nabla \mathbf{E}_1 = \frac{e}{\varepsilon_0}(n_{i1} - n_{e1}) \tag{4.4.8}$$

If we assume that the ion and electron temperatures are constant and the pressures are only functions of the density, (4.4.5), 4.4.6, 4.4.7, and (4.4.8) are closed relation and can be solved.

After **Fourier-Laplace transformation** of (4.4.5), 4.4.6, 4.4.7, and (4.4.8) and inserting the electric field of (4.4.7) into (4.4.8), we can obtain the relation to the ion and electron density perturbations:

$$n_{i1} = \left(1 + k^2 \lambda_{De}^2\right) n_{e1}, \tag{4.4.9}$$

where we assumed that the electron pressure is proportional to the density and electron temperature is constant. The relation (4.4.9) indicates that for the case with wavelength long enough compared to the Debye length, the charge neutral condition is satisfied. When the wavelength becomes short and near $k\lambda_{De} = 1$, the electron density cannot follow the density change of ion fluid. This is because the attractive force to an electron by ion Coulomb force is relatively smaller compared to the repulsive force due to thermal motion of electrons. Then, the electrons cannot follow the ion motion because Debye length is much longer than the wavelength, and the electron density becomes uniform in space.

The algebraic relations after Fourier-Laplace transformation, the dispersion relation of the **ion acoustic waves** are derived:

$$\omega^2 = \frac{\gamma_i T_i}{m_i} + \frac{\gamma_e T_e}{m_i} \frac{1}{\left(1 + k^2 \lambda_{De}^2\right)} \tag{4.4.10}$$

The adiabatic constants γ_e and γ_i in (4.4.10) are evaluated by assuming that the electron temperature is uniform owing to good thermal conduction, while the ions behave as adiabatic fluid, namely,

$$\gamma_i = \frac{5}{3}, \quad \gamma_e = 1 \tag{4.4.11}$$

Usually the plasma satisfies the condition $T_i \ll T_e$; therefore, the following condition is satisfied:

4.4 Ion Fluid and Ion Acoustic Waves

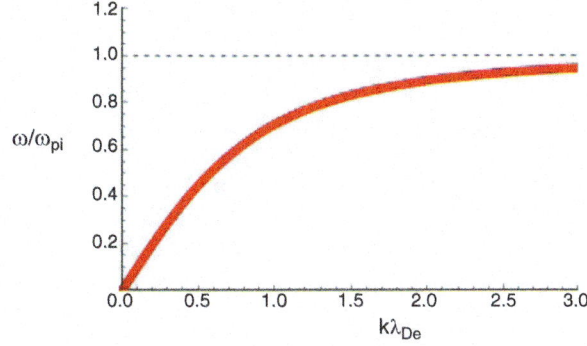

Fig. 4.4 Dispersion relation of the ion acoustic waves. The ion plasma frequency ω_{pi} is $(m_e/m_i)^{1/2}$ times lower than the electron plasma frequency

$$v_i < \frac{\omega}{k} \ll v_e \tag{4.4.12}$$

It should be noted that when the ion temperature is near the electron temperature, the interaction between the waves and ion particles causes the damping of the ion acoustic waves. This is called **Landau damping** and to be explained in Vol. 3. Assuming $T_e \gg T_i$, we can obtain the following dispersion relation:

$$\omega^2 = C_s^2 k^2 \frac{1}{1 + k^2 \lambda_{De}^2}, \tag{4.4.13}$$

where C_s is the sound velocity of the ion acoustic wave and defined to be:

$$C_s = \sqrt{\frac{\gamma_e T_e}{m_i}} \tag{4.4.14}$$

It is clearly understood that the ion acoustic waves feel the repulsive force via electric field due to the electron fluid which balances the electron thermal pressure force, while the ions bear the mass of the oscillating motion, and the combination of electron temperature and ion mass appears in (4.4.14).

The dispersion relation of (4.4.13) is shown in Fig. 4.4. The phase velocity near $k = 0$ is the ion sound velocity given in (4.4.14). Clearly different from the sound waves in neutral gas, the phase velocity is a function of the wavelength. Such property of waves is called **dispersive wave**. It is because an initially localized wave will disperse in space with propagation.

The dispersive property stems from the Debye shielding of the ions by electrons. The longer wavelength component propagates faster than the shorter wavelength one. This suggests that if such dispersive wave is locally steepened by some reason, the wave structure becomes gentle and spreads. Such property of the dispersion leads to a formation of **solitons** balancing the dispersion term with the nonlinear convection term.

4.5 Density Profile Modification

So far, the plasma density profile is assumed to be given as freely expanding ion flow, and the density scale length is of the order of (sound speed) times (laser pulse duration). In the ultra-short laser irradiation, the density scale length of the order of laser wavelength was assumed. When the laser ponderomotive pressure is comparable to the plasma pressure given in (4.4.2), it is more realistic taken into account the plasma profile near the cutoff density determined by the balance between the ponderomotive force and total plasma fluid pressure. If the laser pulse is long enough such as ns-pulse, it is appropriate to assume that a stationary density profile is formed due to the balance of these two pressures.

The basic equation for laser waves propagating normally to the expanding plasma is given in (2.7.1) and is shown for a given stationary density profile as:

$$k^2 \frac{d^2}{dx^2} E + \left(1 - \frac{n_0}{n_{cr}}\right) E = 0, \qquad (4.5.1)$$

where n_0 is the electron density profile varying in space x, and n_{cr} is the critical density. Plasma flow is assumed to be governed by (4.4.1) with ponderomotive force. Then, the basic equations to be solved for stationary solution are the equation of continuity in (4.4.1) and the equation of motion

$$\frac{d}{dx}\left(\frac{u_0^2}{2}\right) = -\frac{C_s^2}{n_0}\frac{dn_0}{dx} - \frac{m_e}{2m_i}\frac{d}{dx}\langle V_{os}^2 \rangle \qquad (4.5.2)$$

In solving (4.5.1) and (4.5.2), the local Mach number M defined in (4.4.3) is used instead of the fluid flow velocity.

Then, M = 1 point is a singular point mathematically. The solution is selected to change from subsonic to supersonic from higher density to lower density smoothly crossing the sonic flow point M = 1. Such solution can be obtained for the normal incident of laser [3]. The resultant profiles of the density and E^2 are shown in Fig. 4.5 (a). It is noted that in this case, the laser wave is assumed to be a standing wave, and the electric field in (4.5.1) is enough to be assumed real value, since no dissipation is taken into account.

The above model can be extended to the case of the mode conversion from the driver electric field to the plasma waves near the critical point [4]. Although the wave-breaking becomes dominant for the cold plasmas, the stationary solution of the mode conversion and the plasma wave propagation has been seen in Fig. 3.28. Assume that the electron temperature is high enough and no wave-breaking is taken place. Then, the wave equation to the plasma waves near the critical density is written for a given electron density in the form:

4.5 Density Profile Modification

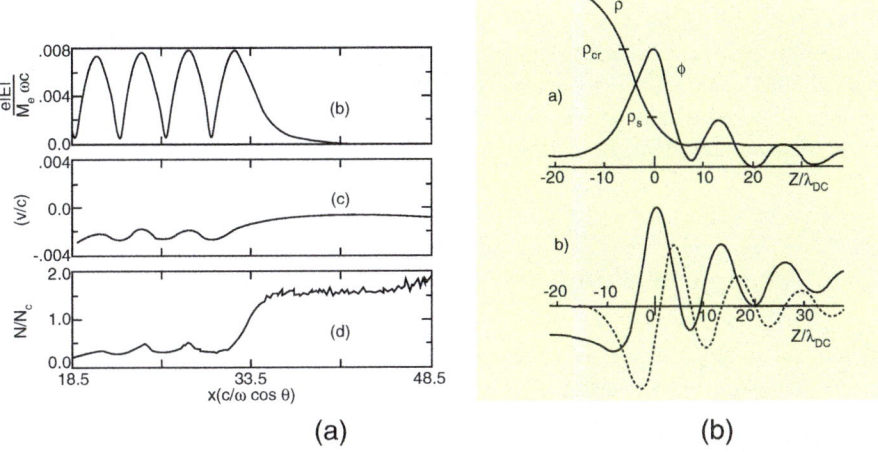

Fig. 4.5 (a) a stationary solution of the balance between plasma flow and laser standing wave obtained with a hybrid code, where electrons are fluid and ions are treated as particles. The figures show from the top to bottom the field structure, velocity structure, and density profile, respectively. (b) A stationary structure of density and ponderomotive potential (above) and the real and imaginary parts of the induced plasma wave propagating to the right (low density direction). [Refs. 3, 4]

$$\frac{3k_D^2}{1+i\nu/\omega}\frac{d^2}{dx^2}E + \left(1 - \frac{n_0}{n_{cr}}\right)E = E_d \qquad (4.5.3)$$

The wave Eq. (4.5.3) should be solved consistently with (4.5.2) for the complex value of the electric field E. In (4.5.3), an effective collisionless damping rate ν is included so that the plasma wave amplitude is reduced along the propagation to the downstream region. Since the solution E becomes complex, the problem is regarded as an eigenvalue problem in two-dimensional space in the complex quantities of E. A typical solution is shown in Fig. 4.5(b) [4].

Since the typical scale length of the plasma wave is of the order of Debye length, the density becomes much steeper than the case of PM force by laser field. If the density scale length is kept short, the absorption fraction by the resonance absorption is expected higher. This is confirmed by 2D computation in Ref. [5]. However, in the case of monochromatic laser irradiation, such steady state is found to be disturbed by the parametric decay instability, which will be discussed in Sect. 4.8. The ion density ripple is induced in the transverse direction by the decay instability, and the ripple enhances the resonance absorption. From the view point of reduction of the hot electron generation for better hydrodynamics, this enhancement is not welcomed. In addition, the parametric decay instability also produces hot electrons. In order to avoid such collective absorption process, broadband lasers have been widely used for hydrodynamics purpose experiments.

It is concluded in Ref. [5] that by the use of a broadband laser with $\Delta\omega/\omega_0$ a few percent, the absorption fraction by the resonance absorption is reduced by 20–40%.

Since the absorption is linear process, the absorption rate is not altered even for the broadband case, if the density scale is given. However, due to the delocalization of resonance, point $\Delta x_{cr} \sim \Delta\omega/\omega_0 L$ changes the density scale length longer than that for the coherent case, reducing the absorption efficiency.

It is noted that such broadband lasers are now standard for the purpose to drive high pressure and idealistic hydrodynamic phenomena with intense lasers.

4.6 Principle of Parametric Instabilities

When an intense laser is irradiated into plasmas, the laser propagates as electromagnetic waves in the plasmas by following the basic equation in (2.5.2). Most of the parametric instabilities are induced by the nonlinear external current $\mathbf{j}_{ext}$. When the laser propagates in plasmas, it may couple with another two waves fluctuating in the plasmas as thermal noise. In the case the laser field has its frequency and wavenumber, $(\omega_0, \mathbf{k}_0)$, consider coupling with the two waves characterized with $(\omega_1, \mathbf{k}_1)$ and $(\omega_2, \mathbf{k}_2)$. Although the amplitude of the latter two waves in plasmas is very small, their amplitudes also increase due to the parametric instabilities, and the energy conversion from the laser to these waves happens in plasmas. Very compact and precise explanation on such parametric instabilities in laser plasmas is given in a textbook by Kruer (Chap. 1, Ref. [2]), and the readers are recommended this book to know more details. In the present book, intuitive explanation is described below for readers to be familiar with the parametric instabilities.

The nonlinear external current in (2.5.2) is expressed as:

$$\mathbf{j}_{NL} = -e\delta n_e \delta \mathbf{v}_e \qquad (4.6.1)$$

Assume, for example, that the velocity perturbation is due to electromagnetic fluctuation and the density perturbation is due to electrostatic fluctuation. If the three waves satisfy the following condition, the nonlinear current becomes a resonant force to the third wave:

$$\begin{aligned} \omega_0 &= \omega_1 + \omega_2 \\ \mathbf{k}_0 &= \mathbf{k}_1 + \mathbf{k}_2 \end{aligned} \qquad (4.6.2)$$

This is called **matching condition** of three waves, and the condition of the wavenumber is plotted in Fig. 4.6. Different from the forced oscillation discussed in Sect. 3.7, the laser plays as energy source to the force term and contributes to the amplification of the nonlinear force. This is parametric instability, and the wave equations for the two fluctuating waves have the external source term as nonlinear force to amplify them.

4.6 Principle of Parametric Instabilities

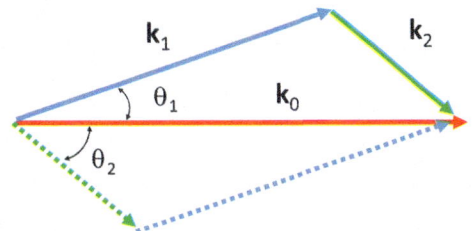

Fig. 4.6 The matching condition of wavenumbers in parametric instability

When the matching condition (4.6.2) is satisfied and the electric field of the linearly polarized laser propagating in the x-direction is assumed in the y-direction, (2.5.2) is written as:

$$
\begin{aligned}
\mathbf{j}_{NL}|_y &= -e\delta n_e \delta v_e|_y \\
&= -e\delta n_2 \delta v_1 \cos\theta_{01} \exp\left[-i(\omega_1+\omega_2)t + i(\mathbf{k}_1+\mathbf{k}_2)\mathbf{x}\right] \\
&= -e\delta n_2 \delta v_1 \cos\theta_{01} \exp\left(-i\omega_0 t + i k_0 x\right)
\end{aligned}
\quad (4.6.3)
$$

Then, resonant coupling of three waves is clear.

The equation to the laser with $(\omega_0, \mathbf{k}_0)$ and electromagnetic fluctuation with $(\omega_1, \mathbf{k}_1)$ are both given from (2.5.2) in the form:

$$
\left(\frac{\partial^2}{\partial t^2} - c^2 \frac{\partial^2}{\partial x_0^2} + \omega_{p0}^2\right) V_0 = -\omega_{p0}^2 \cos\theta_1 \frac{\delta n_2}{n_0} V_1 \quad (4.6.4)
$$

$$
\left(\frac{\partial^2}{\partial t^2} - c^2 \frac{\partial^2}{\partial x_1^2} + \omega_{p0}^2\right) V_1 = -\omega_{p0}^2 \cos\theta_1 \frac{\delta n_2}{n_0} V_0 \quad (4.6.5)
$$

In (4.6.4) and (4.6.5), x_0 and x_1 are the liner coordinate along which the waves 1 and 2 propagate, respectively, and both are assumed plane waves along these coordinates. In order to close the relation, we need the equation to the electrostatic wave $(\omega_2, \mathbf{k}_2)$.

Assume that the density perturbation is due to a component of the electron plasma wave fluctuations satisfying the matching condition (4.6.2). In order to take into account the nonlinear terms to (4.6.4) and (4.6.5) due to the ponderomotive force by the beat waves of two above waves, the following forced oscillation should be included. Returning to the derivation of the ponderomotive force in Sect. 4.1 due to beat waves, the term oscillating with $(\omega_2, \mathbf{k}_2)$ in the ponderomotive force is easily derived. The linearized equation of plasma wave (4.3.7) newly includes the following term.

$$
\left.\frac{\partial \mathbf{u}_{e1}}{\partial t}\right|_{PM} = -\frac{e^2}{m^2} \frac{\omega_{p0}^2}{\omega_0 \omega_1} \nabla(\mathbf{E}_0 \cdot \mathbf{E}_1) = -\omega_{p0}^2 \nabla(\mathbf{V}_0 \cdot \mathbf{V}_1), \quad (4.6.6)
$$

where the oscillation velocities by two electromagnetic waves are introduced.

Fig. 4.7 Graphic depiction of parametric instabilities. [6]

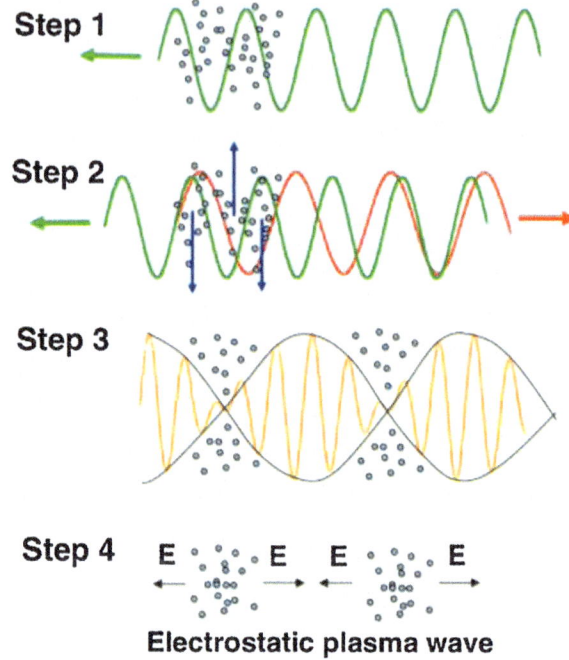

$$\mathbf{V}_0 = \frac{e\mathbf{E}_0}{m\omega_0}, \qquad \mathbf{V}_1 = \frac{e\mathbf{E}_1}{m\omega_1} \tag{4.6.7}$$

Taking time derivative of (3.5.6) and inserting the new velocity perturbation of (4.6.6), the following equation to the plasma waves is obtained:

$$\left(\frac{\partial^2}{\partial t^2} - 3v_e^2 \frac{\partial^2}{\partial x_2^2} + \omega_{p0}^2\right) \frac{\delta n_2}{n_0} = -\cos\theta_1 k_2^2 V_0 V_1 \tag{4.6.8}$$

And for the case of the ion waves, it is easy to obtain the equation:

$$\left(\frac{\partial^2}{\partial t^2} - C_s^2 \frac{\partial^2}{\partial x_2^2}\right) \frac{\delta n_2}{n_0} = -\cos\theta_1 \frac{m_e}{m_i} k_2^2 V_0 V_1, \tag{4.6.9}$$

where the third coordinate x_2 is introduced for the electrostatic waves to propagate in this direction. It is noted that three coordinates have the geometrical relation in Fig. 4.6, and the angle θ_1 is defined there.

It is useful to grasp the image of the parametric instability with the work by the ponderomotive force in (4.6.6). Graphic depiction of a parametric instability such as SRS or SBS is given in Fig. 4.7 [6]. Assume that $\theta_1 = \pi$ and the laser is incident from the right in (4.6.8) or (4.6.9). In step 1, the laser propagates from the right to left in a plasma. In step 2, electrons oscillating in the laser electric field radiate light via Thomson scattering that is Doppler shifted due to the collective electron motion. In step 3, the beating of the laser wave and scattered light wave creates a ponderomotive force, pushing the plasma particles into the troughs of the beat-wave envelope, causing the particles to bunch into a wave. In step 4, if the wave matches an electrostatic mode in the plasma, the three waves are then resonant and grow exponentially.

4.6.1 Coupled Oscillator Model-1

Before solving three coupled equations, consider the property of such coupled equations. Make the problem simpler and assume that the system is uniform so that the amplitudes of three waves are uniform in space. Then, (4.6.4), (4.6.5), and (4.6.8) or (4.6.9) can be reduced to the following coupled three oscillators:

$$\frac{d^2 A_0}{dt^2} + \omega_0^2 A_0 = -\alpha_1 A_1 A_2$$
$$\frac{d^2 A_1}{dt^2} + \omega_1^2 A_1 = -\alpha_2 A_0 A_2 \qquad (4.6.10)$$
$$\frac{d^2 A_2}{dt^2} + \omega_2^2 A_2 = -\alpha_3 A_0 A_1,$$

where three frequencies are derived from the linear dispersion relations and satisfy the matching condition (4.6.2). The three coupling constants are shown as α_1, α_2, and α_3.

In the case where the initial amplitude of A_0 is much larger than A_1 and A_2, global solution of time evolution is found to be intuitively given like Fig. 4.8, where the initial value with a small A_1 and A_2 (=0) are assumed. As the amplitude of A_1 and A_2 grow in time, the energy of the wave A_0 is converted to the other two waves. The A_2 is also induced to increase the amplitude, and at the time t_0, all energy of the wave A_0 is converted to the other two waves. This mechanical model is easy to understand what the parametric instability is in three-mode coupling. However, it is essential to consider the initial amplitude level of the thermal noise in plasmas. Especially, the electrostatic wave A_2 easily couples with individual electron motions via wave-particle interaction process. In what follows, it is reasonable to assume the initial amplitudes of A_1 and A_2 are very small, and our interest is to know the growth rate of the very beginning of this mode growth in time. In what follows, the liner growth rate in very early time evolution of the three-wave coupling will be mainly discussed.

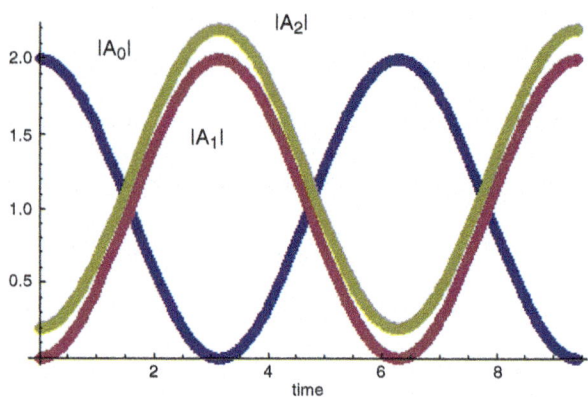

Fig. 4.8 Schematics of time evolution of the envelope of three oscillators interacting with a nonlinear coupling

4.6.2 Thermal Noise of Electrostatic Fluctuations

It is known that any thermal system has noise of physical quantities. It is derived, for example, in Ref. [7] that the fluctuation level of the electric field in one component plasma in thermodynamically equilibrium state is given with **the dynamic form factor** $S(k,\omega)$ defined for time correlation of the density fluctuation Fourier components, namely,

$$S(\mathbf{k}, \omega) = \frac{1}{2\pi} \int dt \langle \delta n_{\mathbf{k}}(t'+t) \delta n_{-\mathbf{k}}(t') \rangle \exp(i\omega t) \qquad (4.6.11)$$

It is given in Eq. (9.29) of Ref. [7] in the form:

$$S(\mathbf{k}, \omega) = \frac{n_e}{\pi \omega} \frac{k^2}{k_{De}^2} \operatorname{Im}\left\{\frac{1}{\varepsilon(\mathbf{k}, \omega)}\right\}, \qquad (4.6.12)$$

where ε is the dielectric constant. For a complex function $\varepsilon = \varepsilon_r + i\varepsilon_i$,

$$\operatorname{Im}\left\{\frac{1}{\varepsilon(\mathbf{k}, \omega)}\right\} = \frac{\varepsilon_i}{\varepsilon_r^2 + \varepsilon_i^2} \qquad (4.6.13)$$

The imaginary part in (4.6.12) has maximum for the fluctuation satisfying the dispersion relation, $\varepsilon_r(\mathbf{k}, \omega) = 0$.

It is seen that the fluctuation spectrum is high near the resonance points, because the resonant modes are always spontaneously excided in plasmas and decays due to some dissipation process. It is well-known that, roughly saying, the field energy due to the thermal noise ΔE_{TN} is of the order of:

$$\frac{\Delta E_{TN}}{n_e T} \sim \frac{1}{n\lambda_{De}^3} \approx \exp(-\ln\Lambda), \tag{4.6.14}$$

where $\ln\Lambda$ is the Coulomb log defined in Sect. 2.4. It varies weakly with the density and temperature and is about the value around 10. Therefore, if the growth exponent defined by (interaction time)×(the growth rate) of the parametric instability is about 10, the fluctuation waves A_1 and A_2 can grow substantially, of the order of the incident laser amplitude.

4.6.3 Coupled Oscillator Model-2

Derive the growth rate of the parametric instability from the model equation of (4.6.10). When an intense laser induces the other two modes, it is valid to assume that E_0 (A_0) is constant and given as an external parameter. We assume that both waves A_1 and A_2 propagate almost satisfying each linear dispersion relation, but due to the nonlinear coupling, their frequencies shift by a small amount. Setting this new frequency for the A_2 as ω, it is reasonable to assume that the wave A_1 oscillates with the frequency $\omega_0 - \omega$ from the matching condition. Then, it is found that the coupled oscillation of waves A_1 and A_2 should satisfy the following relation:

$$\begin{pmatrix} \omega_1^2 - (\omega_0 - \omega)^2 & \alpha_2 A_0 \\ \alpha_1 A_0 & \omega_2^2 - \omega^2 \end{pmatrix} \begin{pmatrix} A_1 \\ A_2 \end{pmatrix} = 0 \tag{4.6.15}$$

A new dispersion relation to this coupled system is given by requiring the determinant vanishes:

$$\left[\omega_1^2 - (\omega_0 - \omega)^2\right](\omega_2^2 - \omega^2) = \alpha_1 \alpha_2 A_0^2 \tag{4.6.16}$$

It is clear that LHS of (4.6.16) is the dispersion relations for the two waves without coupling, while the frequency is shifted by the coupling term on RHS.

In order to solve (4.6.16), assume that $\omega = \omega_2 + \delta\omega$ ($\omega_2 > > \delta\omega$). After Taylor expansion of (4.6.16), the following frequency shift is obtained:

$$(\delta\omega)^2 = -\frac{\alpha_1\alpha_2 A_0^2}{4\omega_2(\omega_0 - \omega_2)} \quad (4.6.17)$$

It is clear that $\delta\omega$ is pure imaginary, and the waves are unstable with the growth rate γ:

$$\gamma = \frac{\sqrt{\alpha_1\alpha_2}}{2\sqrt{\omega_2(\omega_0 - \omega_2)}} A_0 \quad (4.6.18)$$

In general the waves also have some damping in the plasmas as we see before. So, inserting the damping terms with damping rate $2\nu_1$ and $2\nu_2$ in (4.6.10) and carrying out the same calculating as above, the following new dispersion relation for the growth rate with damping γ' is obtained:

$$(\gamma' - \nu_1)(\gamma' - \nu_2) = \gamma^2 \quad (4.6.19)$$

where γ is the growth rate of (4.6.10). It is found from (4.6.19) that there is a **threshold of the parametric instability,** and the instability condition is given with:

$$\gamma^2 > \nu_1\nu_2 \quad (4.6.20)$$

It is noted about the up-shift conversion in above discussion. It is also found that the wave A_1 seems to satisfy the following matching condition also can resonate with the nonlinear current by the other two waves:

$$\begin{aligned}\omega_1 &= \omega_0 + \omega_2 \\ \mathbf{k}_1 &= \mathbf{k}_0 + \mathbf{k}_2\end{aligned} \quad (4.6.21)$$

With this condition, however, it is easily found that the sign of RHS in (4.6.17) becomes negative, and the system becomes stable. In general, when the modes satisfy the matching condition (4.6.2) and the scattered wave has lower frequency than the original light, the emission signal in spectrum of the scattered light is called **Stokes lines**. On the other hand, higher-frequency scattering satisfying (4.6.21) is called **anti-Stokes lines**. As seen below, the anti-Stokes line is used for collective Thomson scattering diagnostics.

4.7 Stimulated Raman and Brillouin Scattering

As clear in the discussion above, there are two cases where the incident lasers are scattered by the parametric instability. One is the case where the wave A_2 is the electron plasma waves already derived in (4.6.18), and the other is the case where the

4.7 Stimulated Raman and Brillouin Scattering

wave A_2 is the ion acoustic waves. The former is called **stimulated Raman scattering (SRS)**, and the latter is called **stimulated Brillouin scattering (SBS)**. Raman and Brillouin have found the light scattering phenomena with frequency shifts, when the irradiating light is scattered from molecules, liquids, and solids. In the case of Raman scatter, the frequency shift is due to the coupling with the optical mode by electrons, while Brillouin scattering is due to the phonon mode in solid. However, both are just scattering and no instability, so they are called just scattering, Raman scattering, and Brillouin scattering. They are due to the coupling of incident light with the waves in materials with enough amplitude. They can be evaluated with given amplitude on RHS in (4.6.5) and they are not instability.

The reason for "stimulated" in the names in case of plasmas is due to the enhancement of scattering by the parametric instability. Since the detailed theoretical analysis of SRS and SBS is given in the text by Kruer (Chap. 1, Ref. [2]), it is not necessary to repeat the same in the present book. Let us obtain the growth rates for uniform plasma, based on the model equation of the three coupled oscillators. The growth rate of the SRS was almost calculated in the previous section, and it is easy to obtain the following:

$$\gamma_{SRS} = \frac{\omega_{p0}|\cos\theta_1|}{2\sqrt{\omega_{ek}(\omega_0 - \omega_{ek})}} kV_{os}, \qquad (4.7.1)$$

where ω_{ek} is the frequency of the plasma wave.

From (4.7.1), it is clear that SRS grows predominantly to the forward ($\theta_1 = 0$) or backward ($\theta_1 = \pi$) directions. As clear in (4.6.6), there is no ponderomotive force to induce the electrostatic waves for the perpendicular scattering (side scattering). The growth time τ ($=1/\gamma_{SRS}$) in (4.7.1) is roughly equal to kV_{os}, and k is roughly the wavenumber of laser; therefore, the growth time is approximately:

$$\tau \sim \frac{1}{k_0 V_{os}} \qquad (4.7.2)$$

In the assumption of uniform plasmas, it is required that the plasma length L should be much longer than the following length:

$$L \sim c\tau \approx \frac{\lambda_L}{2\pi a_0} = \frac{0.16}{a_0} \ [\mu m], \qquad (4.7.3)$$

where the last relation is for laser wavelength $\lambda_L = 1$ μm. This is a very small number in the under-dense plasmas expanding to the vacuum, for example, L = 16 μm for the laser intensity of 10^{14} W/cm^2. It is natural to assume that the plasma scale length increases as the laser intensity and pulse duration increases. In (2.5.23), the absorption rate is increase in proportion to the pulse duration, but SRS prevents the penetration of laser near the turning point where the dominant classical absorption is taken place.

In addition, the SRS threshold is given by assuming that the scattered electromagnetic wave damps due to the electron-ion collision frequency ν_{ei}, given in (2.6.14), for example, and the plasma wave are mainly due to collective phenomena such as Landau damping with ν_{ek}, in the form:

$$a_0^2 \equiv \left(\frac{V_{os}}{c}\right)^2 > \left(\frac{\omega_{p0}}{\omega_0}\right)^2 \frac{\nu_{ei}\nu_{ek}}{\omega_0 \omega_{p0}} \qquad (4.7.4)$$

In the case of wave damping in collective process, damping rate changes in time along the evolution of the wave amplitude. This will be discussed in a later section relating to nonlinear evolution of SRS.

In the case of SBS, the maximum growth rate from is obtained as:

$$\gamma_{SBS} = \frac{1}{2\sqrt{2}} \frac{\omega_{pi}|\cos\theta_1|}{\sqrt{\omega_0 k_0 C_s}} k V_{os} \qquad (4.7.5)$$

In (4.7.5), the relation $\omega_0 > > \omega_{pi} > > k_0 C_s$ is in general satisfied; therefore, almost the same evaluation of the growth time (4.7.3) can be applicable to SRS. SBS is also easily occurs in long pulse intense laser irradiations.

4.8 Decay-Type Parametric Instabilities

In Fig. 4.6, waves A_1 and A_2 can both be electrostatic waves. Such parametric instability is called **parametric decay instabilities**. Two cases are possible. One is the case where the waves A_1 and A_2 are electron plasma waves and ion waves, respectively. This is called simply **decay instability**. On the other hand, both of waves A_1 and A_2 can be electron plasma waves. This is called **two-plasmon decay instability**. Since these instability helps to converge the laser energy to the energy in the plasmas, it looks beneficial from the view of absorption increase. However, such wave energy finally goes to a small fraction of electrons to generate **high-energy electrons** (**hot electrons**), and the bulk heating is not expected. It is not preferable for the use of laser energy to generating hydrodynamic phenomena by laser-driven high pressure.

The formulation to the decay instability is almost the same as SRS and SBS, but one thing to be noted is that the direction of the electric field is parallel to the propagation direction. Start with the basic equations to electron plasma waves and ion waves in the form (2.2.11) and insert the liner operators from the dispersion relations (3.5.10) and (4.4.13), respectively, the following equations are obtained:

4.8 Decay-Type Parametric Instabilities

$$\left(\frac{\partial^2}{\partial t^2} - 3v_e^2 \nabla^2 + \omega_{p0}^2\right) \mathbf{E} = -\frac{1}{\varepsilon_0} \frac{\partial \mathbf{j}_{ext}}{\partial t} \tag{4.8.1}$$

$$\left(\frac{\partial^2}{\partial t^2} - C_s^2 \nabla^2\right) \mathbf{E} = -\frac{1}{\varepsilon_0} \frac{\partial \mathbf{j}_{ext}}{\partial t} \tag{4.8.2}$$

The difference is only the linear operators and the evaluation of the nonlinear current are done with the same process for both electrostatic waves. According to (4.6.3), the nonlinear currents are only due to the density perturbation by the electrostatic waves, and the oscillation velocity is due to the laser field. For the density perturbation δn_2, the nonlinear current to the vector $\mathbf{k}_1$ is:

$$\mathbf{j}_{NL}|_{\mathbf{k}_1} = -e\delta n_2 \sin\theta_1 V_{os} = -\frac{\omega_1 k_2}{4\pi} \sin\theta_1 V_{os} E_2 \tag{4.8.3}$$

In deriving (4.8.3), we have used Poisson relation to the density perturbation δn_2. It is noted that the angle dependence is *sin* not *cos* like in (4.6.5), because the direction of the electric fields is different as shown in Fig. 2.1. Following the same process as deriving (4.6.16), the dispersion relation is obtained:

$$\left[(\omega - \omega_0)^2 - \omega_1^2\right](\omega^2 - \omega_2^2) = \frac{\omega_1 \omega_2 k_1 k_2}{(4\pi)^2} \sin\theta_1 \sin\theta_2 V_{os}^2 \tag{4.8.4}$$

It is clear that (4.8.4) has instability solution same as the case of SRS. Inserting the same dispersion relation (3.5.10) to ω_1 and ω_2, the growth rate of two-plasmon decay instability is obtained. On the other hand, inserting the dispersion relation of the ion wave (4.4.13) to ω_2, the growth rate of the decay instability is obtained from (4.8.4).

The dependence to the angle in (4.8.4) indicates that forward and backward decays are forbidden. Side scattering has large growth rate. Consider the easiest case of two-plasmon decay for $\theta_1 = \theta_2$. It is clear that this condition is satisfied near the density of $n_e = n_{cr}/4$, a quarter critical density. The growth rate of this instability is obtained from (4.8.4):

$$\gamma_{decay} = \frac{1}{8\sqrt{2\pi}} k_0 V_{os} \tag{4.8.5}$$

It is informative to point out that since such decay instability is taken place near the quarter critical density region, strong plasma waves oscillating with the frequency $\omega_0/2$ are generated in the plasmas, and another nonlinear coupling with the incident laser subsequently appears. Then, the nonlinear current with frequency $1/2\, \omega_0$ and $3/2\, \omega_0$ (another harmonics) is induced to work as new external source of current in (2.5.2) to generate the electromagnetic fields with these frequencies. The scattering from the plasmas with $1/2\, \omega_0$ and $3/2\, \omega_0$ is the indirect evidence of the two-plasmon decay instability near the quarter critical density.

4.9 Experimental Data for SRS

Precise experiment has been carried out with Trident laser in LANL [8]. In order to produce uniform plasmas, a line focusing system is used. Then, an intense laser is irradiated to induce the parametric instability in the uniform plasma. For diagnosing the electron and temperatures and electron density of the plasma, another relatively week intensity lasers are irradiated to the interaction region. **Collective Thomson scattering** (CTS) diagnostics are used to determine the plasma parameters from the diagnostic lasers scattered from the plasma. It is informative to explain briefly the principle of CTS methods, since it is related to SRS and SBS physics.

As already explained, the incident lasers are scattered by the electrostatic waves in plasma. Irradiating laser beam for diagnostics to the laser-plasma interaction region, it is scattered by the ion acoustic wave and electron plasma waves induced by the parametric instabilities. The scattering is not due to the parametric instability. Assume that E_1 in (4.6.5) is induced by the electron density fluctuation δn_2 and incident laser for diagnostics E_0. In this case, the electron density fluctuation is driven by the parametric instability by the main pulse. Measuring the scattered light, we can obtain the information of the electron density wave. The incident laser for diagnostics is scattered by the electrons in plasma like Thomson scattering in Sect. 2.3, while CTS is due to the scattering by the collective motions of electrons in the plasma.

The intensity, angle, and spectra of the scattered laser by CTS have the information of the amplitude and dispersion relation of these waves. Since they depend on the electron and ion temperatures and and electron density, the precise diagnostics can be done for determination of the physical parameters of the plasmas in the interaction region.

For example, CTS spectrum from the ion waves is plotted in Fig. 4.9, where the solid line is the experimental data. The ion waves are induced by the parametric decay instability. The frequency shift is due to the scattering by forward and backward propagating ion waves. This shift reflects the dispersion relation

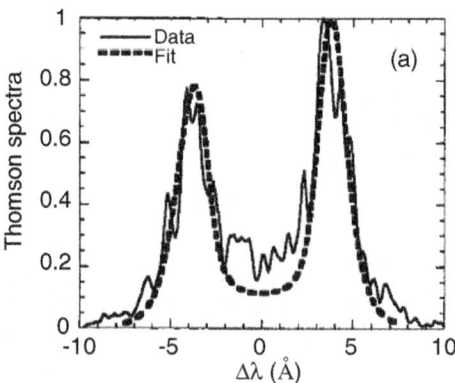

Fig. 4.9 Experimental and calculated spectra of collective Thomson scattering from the laser-plasma interaction region. [8]

4.10 Physics of Saturation of SRS Instability

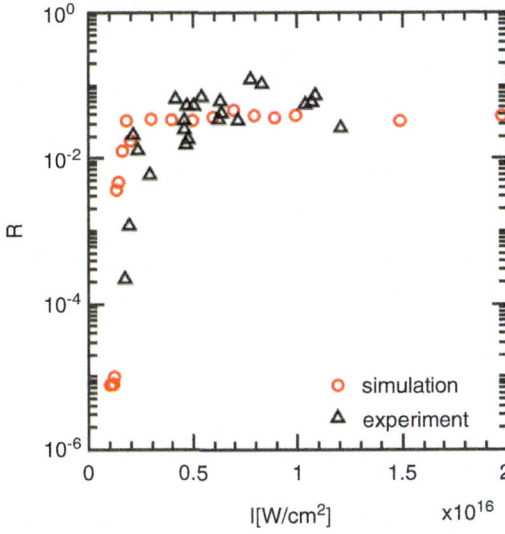

Fig. 4.10 The time-averaged reflectivity vs. incident laser intensity. The black triangles are from experiment, and red circles are from simulation [Fig. 1 in Ref. 9]

(4.4.13). The spreads of the two peaks can be related to property of Landau damping which is mainly determined by the ion temperature. The electron and ion temperatures are identified by comparing to the theoretical curve plotted with the dashed line in Fig. 4.9. In Ref. [8], it is reported that Te~500 ± 20 ev, Ti~100 ± 30 ev from this comparison. In addition, the dispersion relations yield n_e/n_{cr} ~ 0.3, and the wavenumber of the plasma wave in SRS is $k\lambda_{De} = 0.33 - 0.35$.

The reflectivity of the incident laser by a backward SRS is observed experimentally in [8]. It is plotted as a function laser intensity in Fig. 4.10 with triangle (black). The threshold is clearly seen around laser intensity 2×10^{15} W/cm^2, and the reflectivity abruptly increases around 5–10% at higher intensity above the threshold. It is clear that the parametric instability SRS saturates over the intensity 5×10^{15} W/cm^2, and the reflectivity remains almost constant about 10% in the higher intensity region. Let us study theoretically what physics determines such values of reflectivity at higher intensity.

4.10 Physics of Saturation of SRS Instability

At first, consider the intensity dependence from the instability threshold. In the experiment, the plasma length of uniform density region for the parametric instability is observed, L ~ 1 mm. The fluctuation level is theoretically calculated for the parameters as shown in Fig. 4.11 [8, 9]. In plotting the figure, the experimental data for electron plasma waves are used. The imaginary part in (4.6.12) is obtained from the linear curve (dotted line) in Fig. 4.11. It is found that

Fig. 4.11 The property of the dielectric constant for plasma parameters observed in experiment. The dotted line is from the linear analysis, and solid line is after including the effect of trapped electrons. [Fig. 11 in Ref. 8]

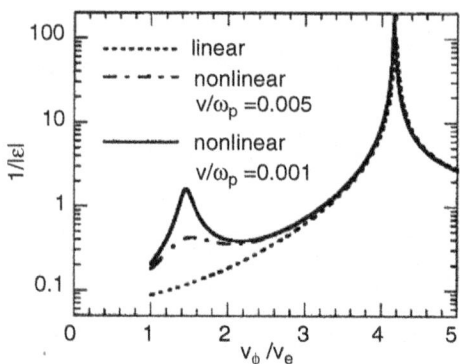

the plasma waves have a strong resonance for the phase velocity, $v_\phi/v_e = 4.2$, and Debye length, $k\lambda_{De} = 0.27$, in the experiment.

In order to do analysis of the experimental result, the finiteness of the parametric instability matching region of the relation (4.6.2) is important. For analysis of the localized instability, it is convenient to use **Rosenbluth-Nishikawa coupled equation** [10]. Following Ref. [9], consider the effect of density inhomogeneity and nonlinear evolution after the liner growth phase of the backward SRS scattering. In deriving the coupled equations, the waves 1 and 2 in (4.6.5) and (4.6.8) are assumed in the form:

$$E_1 \propto A_1(x,t)\exp(-i\omega_1 t - k_1 x)$$
$$\frac{\delta n_2}{n_0} \propto A_2(x,t)\exp(-i\omega_2 t + k_2 x) \qquad (4.10.1)$$

Inserting (4.10.1) into (4.6.5) and (4.6.8), and neglecting the slowly varying components proportional to the second derivative to A_1 and A_2, the following simple model equations are obtained [10]:

$$\left(\frac{\partial}{\partial t} - V_{g1}\frac{\partial}{\partial x}\right)A_1 = \gamma_0 A_2 - \nu_1 A_1$$
$$\left(\frac{\partial}{\partial t} + V_{g2}\frac{\partial}{\partial x}\right)A_2 = \gamma_0 A_1 - \nu_2 A_2, \qquad (4.10.2)$$

where V_{g1} and V_{g2} are the group velocities of scattered light and plasma waves:

$$V_{g1} = \frac{c^2}{\omega_1/k_1}, \qquad V_{g2} = \frac{3v_e^2}{\omega_2/k_2} \qquad (4.10.3)$$

In (4.10.2), γ_0 is the growth rate of SRS for uniform plasma given in (4.7.1), and ν_1 and ν_2 are the damping rate used in (4.6.19).

4.10.1 Effect of Inhomogeneity

Before solving (4.10.2), it's intuitively understood that the density inhomogeneity reduces the growth rate compared to that in the uniform plasma, since the decayed plasma waves mainly localize near the matching region. Their spatial structures may be determined by an eigen value problem like a localized wave function of Schrodinger equation. In addition, the convective loss of the waves also reduces the growth rate as additional loss process in the parametric instability growth. Note that such additional loss process increases the threshold intensity of the instability. This is the same image as saturation amplitude of the plasma wave in the linear mode conversion as seen in Figs. 3.28 and 4.5(b), where the amplitude of the generated plasma wave decreases as its group velocity increases.

Assume that the three-wave matching condition in Fig. 4.6 is satisfied for the backward SRS with $\theta_1 = \theta_2 = 0$. Then, the absolute instability has the form:

$$\gamma_0 = \frac{\omega_{p0}}{4\sqrt{\omega_1 \omega_2}} k V_{os} \qquad (4.10.4)$$

Taking a Laplace transformation in time of (4.10.2) with the Laplace transformation variable p, neglecting the damping terms, and eliminating A_1 from (4.10.2), the following equation is obtained:

$$\frac{\partial^2}{\partial x^2} A_2 + K^2 A_2 = 0, \qquad (4.10.5)$$

where

$$K^2 = \frac{\gamma_0^2}{V_{g1} V_{g2}} - \frac{p^2}{4} \left(\frac{1}{V_{g1}} + \frac{1}{V_{g2}} \right)^2 \qquad (4.10.6)$$

For given p and the boundary condition that $A_2 = 0$ at $x = 0$ and L, (4.10.5) is easily solved to have the solution:

$$A_2 = \sin\left(\frac{\pi n}{L} x\right), \qquad \left(\frac{\pi n}{L}\right)^2 = K^2, \qquad (4.10.7)$$

where n is an integer. It is clear that for $p^2 > 0$, the solution is unstable. Including the damping term, instability condition is obtained:

$$\frac{\gamma_0^2}{V_{g1} V_{g2}} > \left(\frac{\pi}{L}\right)^2 + \frac{1}{4} \left(\frac{\nu_1}{V_{g1}} + \frac{\nu_2}{V_{g2}} \right)^2 \qquad (4.10.8)$$

It is clear that the matching condition in an inhomogeneous plasmas increases the threshold intensity. In obtaining the growth rate, we have to solve (4.10.5) as an eigen value problem for the given boundary condition. This is the same as in solving Schrodinger equation, and the K^2 in (4.10.7) is regarded to the kinetic energy, {E-U (x)}. In (4.10.6), the eigen value is p^2. With this analogy, it is easy to know the property of (4.10.2) for more general case.

4.10.2 Nonlinear Saturation of SRS

In analyzing the experimental data, the linear analysis is useful to identify the instability threshold. It is about 2×10^{15} W/cm^2 in experiment. With the phase velocity shown in Fig. 4.11, the damping rate is evaluated. This damping is mainly due to wave-particle interaction known as Landau damping to be explained in Vol. 3.

However, in order to explain the rapid growth and saturation of SRS reflectivity shown in Fig. 4.10, a computer simulation is required. In Fig. 4.10, the simulation result is plotted with red circles. The simulation could well explain the experimental data. The nonlinear physics seen in the simulation is as follows.

In Ref. [9], it is insisted that Landau damping is stabilized in higher intensity by electron trapping in the plasma wave potential. The electrons with velocities near the phase velocity of the plasma waves directly interact the electrostatic field, and some are accelerated, and some are decelerated. Landau damping is due to net energy conversion from the waves to the electrons, because the number of the accelerating electrons is larger than the decelerated ones in Boltzmann velocity distribution. Once the wave amplitude becomes large, however, such trapped electrons take bounce motions in the wave potential, and finally no net energy conversion happens. In Fig. 4.11, a nonlinear dielectric function is plotted with a solid line by taking into account the bounce motion by the trapped particle, where a new resonance appears around the phase velocity 1.5.

For further increase of the laser intensity, it is expected that the wave amplitude becomes higher, even though the particle trapping is taken place. As seen in the large amplitude plasma wave, however, the **wave-breaking** is induced to prevent the further growth of wave amplitude. It is reasonable to understand that at the intensity much higher than the threshold one, the particle trapping and wave-breaking are simultaneously seen, and as a result, a part of the laser energy is converted to the high-energy electron production. Namely, significant amount of SRS scattering suggests a production of hot electrons in plasmas.

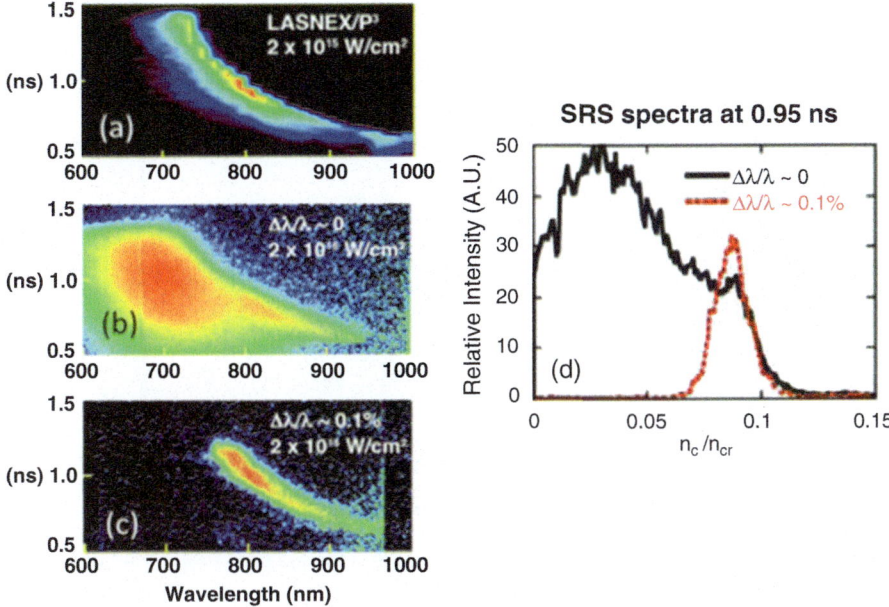

Fig. 4.12 Time-dependent SRS spectra from (**a**) simulation (growth rate), (**b**) coherent laser irradiation, and (**c**) broadband laser are irradiated. (**d**) is the emission intensity at 0.95 ns after evaluation of density of emitting region from the frequency shift. [Fig. 2 in the Ref. 8]

4.11 Broadband Effect

As already mentioned in Sec. 4.6, the broadband lasers are widely used to suppress the collective phenomena in laser-plasma interaction physics. As the plasma scale increases with the increase of incident laser energy, the parametric instabilities become more important, and laser energy is reflected by SRS and SBS processes. In addition, the large amplitude electron waves and ion waves are generated in plasmas, and their energy is not used for hydrodynamics but goes to hot electrons which heats over the long rage in plasmas. It is important to show how the broadband laser can suppress such parametric instabilities.

In the same experiment as in Fig. 4.10, the backscattered SRS spectra were observed as a function of time for the case of laser intensity 2×10^{15} W/cm^2 [6]. The result is shown in Fig. 4.12. As guidance for the experiment, linear growth rate of expanding plasma is obtained from a hydrodynamic simulation in Fig. 4.12 (a). It is clear that since the density of the expanding plasma and the electron plasma frequency decreases with time, the SRS wavelength, $\lambda_1 \propto \omega^{-1} \approx (\omega_0 - \omega_{pe})^{-1}$, decreases in time as predicted in Fig. 4.12(a). In the experiment without broadband, however, the spectrum dramatically changed as in Fig. 4.12(b). The SRS are induced

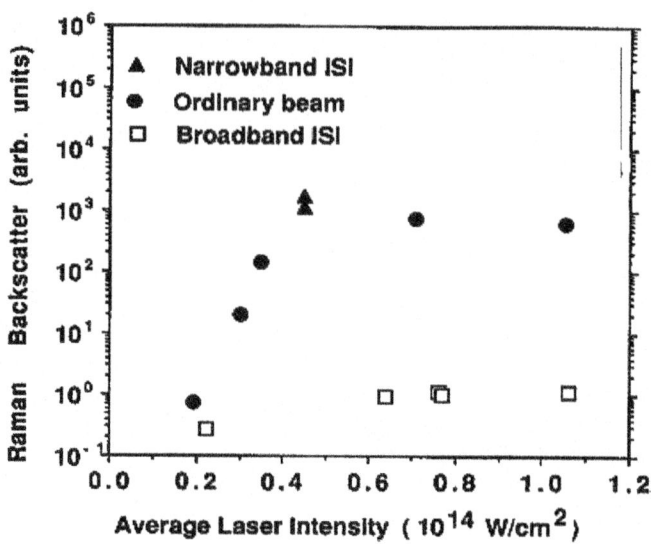

Fig. 4.13 The peak Raman backscatter in the band 1350–1750 nm for the case of broadband ISI, narrowband ISI, and an ordinary laser beam with narrow bandwidth. [Fig. 2 in Ref. 11]

in the wide range of the density and SRS stemming from relatively low density was observed in the early time. This is explained that without band; the density ripples are induced by the self-focusing and filamentation, which also produces electron plasma waves with wide range of frequency.

In order to suppress such nonlinear phenomena, a broadband laser with $\Delta\omega/\omega_0 = 0.1\%$ is irradiated. Then, the SRS spectrum is suppressed as shown in Fig. 4.12 (c). It is clear that only 0.1% of the band width is effective to kill the growth of self-focusing and filamentation instability at the intensity of 2×10^{15} W/cm^2. The SRS spectra at time 0.95 ns are compared for with and without broadband in Fig. 4.12(d). The density is calculated from the spectrum using the matching condition for SRS instability. It is clear that with self-focusing, electron density is almost 100% perturbed in the interaction region from the maximum density of about $0.1n_{cr}$. The sharp peak in red in Fig. 4.12(d) indicates that the band width of 0.1% is not enough to suppress SRS instability; wider band is preferable for killing the instability.

It should be noted that this intensity in Fig. 4.10 is a bit above the threshold intensity, and the linear growth rate gives us a good indication of the SRS backscatter. For higher intensity, however, there is no evidence that broadband can suppress the parametric instability significantly. It is, however, reported that the broadband laser can suppress the SRS dramatically over the wide range above the threshold intensity as shown in Fig. 4.13 [11]. The ordinary and narrowband ISI beams have no broadband, and the SRS signal rapidly increases above the threshold intensity and saturate as shown in Fig. 4.10. On the other hand, it is demonstrated that the broadband ISI beam with about 0.2% of the bandwidth can suppress SRS dramatically even at the intensity of nonlinear saturation region.

References

1. R. Bingham, C.N. Lasmore-Davis, Nucl. Fusion **16**, 1 (1976)
2. M. Borghesi et al., Phys. Rev. Lett. **78**, 879 (1997)
3. K.G. Estabrook, E.J. Valeo, W.L. Kruer, Phys. Fluids **18**, 1151 (1975)
4. H. Takabe, P. Mulser, Phys. Fluids **25**, 2304 (1982)
5. J.P. Palastro et al., Phys. Plasmas **25**, 123104 (2018)
6. D.S. Montgomery, Phys. Plasmas **23**, 055601 (2016)
7. S. Ichimaru, *Basic Principle of Plasma Physics* (Benjamin Inc., New York, 1972)
8. D.S. Mongomery et al., Phys. Plasmas **9**, 2311 (2002)
9. Y.X. Wang et al., Phys. Plasmas **25**, 100702 (2018)
10. M.N. Rosenbluth, R.W. White, C.S. Liu, Phys. Rev. Lett. **31**, 1190 (1973)
11. S.P. Obenschain et al., Phys. Rev. Lett. **62**, 768 (1989)

Part II
Relativistic Lasers

Chapter 5
Relativistic Laser-Electron Interactions

5.1 Introduction

For the case when the laser intensity approaches 10^{18} W/cm^2 for 1 μm wavelength lasers, relativistic effect should be taken into account to any of phenomena induced by laser-matter interaction. In order to obtain electron motion in such laser field, relativistic electron motion should be solved, and the radiation due to retardation effect, coming from the finiteness of the speed of light compared to the velocity of electron, cannot be neglected. Relativistic effect to mass correction of a single electron is itself nonlinear effect to laser dynamics interacting with an electron. When electron oscillation velocity approaches the speed of light, the relativistic mass correction changes the plasma frequency of the matters. Since the magnetic force to the electron becomes comparable to the electric one, in addition, laser field in matter can also induce the electrostatic wave through $\mathbf{v} \times \mathbf{B}$ force.

In this section, we briefly summarize the important relation of special relativity. When the relations of relativistic physics are shown without derivation or proof in what follows, readers are recommended to refer mainly the Chaps. 1, 2, and 3 in the textbook by Landau and Lifshitz [1] and the Chap. 11 in the book by J. D. Jackson [2].

5.2 Special Relativity for Electron Motion

Derive the equation of the momentum conservation of electromagnetic field. It is useful to use the momentum change by coupling with charged particles via Lorentz force. The momentum change per unit volume is given by

$$q(\mathbf{E} + \mathbf{v} \times \mathbf{B}) \Rightarrow \rho\mathbf{E} + \mathbf{j} \times \mathbf{B} \quad (5.2.1)$$

Carrying out the operation of scalar product of **E** to (1.3.3) and taking vector product of **B** from the left in (1.3.2), the term with the above coupling term appears after the sum of both. Then, this equation pulse the vector product of $\varepsilon_0\mathbf{E}$ to (1.3.1) leads to the following equation to the momentum density of the electromagnetic fields:

$$\frac{\partial}{\partial t}\mathbf{P}_{EM} = -(\rho\mathbf{E} + \mathbf{j} \times \mathbf{B}) + \nabla\overleftrightarrow{T}_{EM} \quad (5.2.2)$$

where the **momentum density** is introduced:

$$\mathbf{P}_{EM} = \varepsilon_0 \mathbf{E} \times \mathbf{B} = \frac{\mathbf{S}}{c^2} \quad (5.2.3)$$

The last term in (5.2.2) is given as

$$\nabla\overleftrightarrow{T}_{EM} = \varepsilon_0[(\nabla \cdot \mathbf{E})\mathbf{E} - \mathbf{E} \times (\nabla \times \mathbf{E})] - \frac{1}{\mu_0}\mathbf{B} \times (\nabla \times \mathbf{B}) \quad (5.2.4)$$

This tensor is **Maxwell stress tensor** consisting of electric and magnetic contributions:

$$\overleftrightarrow{T}_{EM} = \overleftrightarrow{T}_E + \overleftrightarrow{T}_M \quad (5.2.5)$$

The explicit form of the tensor is given in Ref. [1], but it is enough to know that (5.2.2) can be modified in the following form of transport in vacuum limit:

$$\left(\frac{\partial}{\partial t} + \mathbf{c} \cdot \nabla\right)\mathbf{P}_{EM} = -(\rho\mathbf{E} + \mathbf{j} \times \mathbf{B}) \quad (5.2.6)$$

In order to calculate the relativistic coupling of electrons with ultra-intense laser field, it is convenient to use the vector potentials to Maxwell equations. Then, the electric and magnetic fields are given with the **vector potential, A**, and **scalar potential, ϕ**:

$$\mathbf{E} = -\frac{\partial \mathbf{A}}{\partial t} - \nabla\phi$$
$$\mathbf{B} = \nabla \times \mathbf{A} \quad (5.2.7)$$

Inserting (5.2.7) to Maxwell equations, (1.3.1) and (1.3.4) are automatically satisfied, and (1.3.2) reduces to

5.2 Special Relativity for Electron Motion

$$\frac{\partial^2}{\partial t^2}\mathbf{A} - c^2 \nabla^2 \mathbf{A} = \frac{1}{\varepsilon_0}\mathbf{j} \qquad (5.2.8)$$

And (5.2.3) reduces to

$$\frac{\partial^2}{c^2 \partial t^2}\phi - \nabla^2 \phi = \frac{1}{\varepsilon_0}\rho \qquad (5.2.9)$$

In deriving (5.2.8) and (5.2.9), **Lorentz gauge** condition are required:

$$\frac{\partial}{c^2 \partial t}\phi + \nabla \cdot \mathbf{A} = 0 \qquad (5.2.10)$$

It is noted that the propagation of electromagnetic waves in plasmas are governed by (5.2.8) with the induced and external electron currents:

$$\mathbf{j} = \mathbf{j}_{\text{ind}} + \mathbf{j}_{\text{ext}}. \qquad (5.2.11)$$

On the other hand, the electrostatic waves such as plasma waves are governed by (5.2.9) due to the induced and external charge density ρ. It is easily shown that the charge conservation law is automatically satisfied by (5.2.8), (5.2.9) and (5.2.10):

$$\frac{\partial}{\partial t}\rho + \nabla \cdot \mathbf{j} = 0 \qquad (5.2.12)$$

In general, there are external and induced currents in (5.2.9). The "induced" means the ones generated by the $\mathbf{A}$ or ϕ in LHS of (5.2.8) and (5.2.9), respectively, while the "external" is the source or sink terms for the wave equations in (5.2.8) and (5.2.9). In order to obtain RHS in (5.2.8) and (5.2.9), we have to solve the electron motions in plasmas. Since the laser field is extremely high and electrons in plasmas should be described by relativistic dynamics, consider at first an electron motion in the laser field. It is noted that the ions can be regarded background charge at rest in most of cases, because we consider the phenomena driven by ultra-intense and ultra-short pulse. The ions have no time to respond to the laser and electron motion because of much higher mass, inertial.

5.2.1 Equation of Relativistic Electron Motion

The Lagrangian of a relativistic charged particle with charge q and mass m in electromagnetic field is given in the form:

$$L = -mc^2\sqrt{1 - v^2/c^2} + q\mathbf{A} \cdot \mathbf{v} - q\phi \qquad (5.2.13)$$

The principle of minimum action is used to derive the Lagrangian (5.2.13). In (5.2.13), the last two terms represent the contribution by field and charged particle interaction:

$$L_{int} = q\mathbf{A} \cdot \mathbf{v} - q\phi \qquad (5.2.14)$$

The equation of motion is derived from Euler-Lagrange equations with two independent coordinate **r** and **v** = d**r**/dt:

$$\frac{d}{dt}\left(\frac{\partial L}{\partial \mathbf{v}}\right) = \frac{\partial L}{\partial \mathbf{r}} \qquad (5.2.15)$$

From LHS of (5.2.15), the following **Canonical momentum** $\mathbf{P}^c$ is defined:

$$\frac{\partial L}{\partial \mathbf{v}} = \mathbf{p} + q\mathbf{A}$$
$$\mathbf{P}^c = \mathbf{p} + q\mathbf{A} \qquad (5.2.16)$$

where particle momentum **p** and **Lorentz factor** γ are

$$\mathbf{p} = \gamma m\mathbf{v} \qquad (5.2.17)$$

$$\gamma = \frac{1}{\sqrt{1 - v^2/c^2}} \qquad (5.2.18)$$

Note that (**r**, $\mathbf{P}^c$) is the pair of **generalized coordinate and momentum**. Then, the equation of motion is

$$\frac{d}{dt}(\mathbf{p} + q\mathbf{A}) = q[\mathbf{v} \cdot \leftrightarrow \nabla \mathbf{A}] - q\nabla\phi \qquad (5.2.19)$$

where $\leftrightarrow \nabla \mathbf{A}$ is matrix and (I, j) component is

$$\leftrightarrow \nabla \mathbf{A} = \frac{\partial A_j}{\partial x_i} \qquad (5.2.20)$$

Using the mathematical formulae

$$[\mathbf{v} \cdot \leftrightarrow \nabla \mathbf{A}] = \mathbf{v} \times (\nabla \times \mathbf{A}) + (\mathbf{v} \cdot \nabla)\mathbf{A} \qquad (5.2.21)$$

The equation of motion of (5.2.19) becomes

5.2 Special Relativity for Electron Motion

$$\frac{d}{dt}(\mathbf{p}+q\mathbf{A}) = -q\nabla\phi + q\mathbf{v}\times(\nabla\times\mathbf{A}) + q(\mathbf{v}\cdot\nabla)\mathbf{A} \tag{5.2.22}$$

By use of the relation

$$\frac{d}{dt}\mathbf{A} = \left(\frac{\partial}{\partial t} + \mathbf{v}\cdot\nabla\right)\mathbf{A}$$

(5.2.22) is reduced to well-known equation of motion in Lorentz force:

$$\frac{d}{dt}\mathbf{p} = -q\left(\frac{\partial\mathbf{A}}{\partial t} + \nabla\phi\right) + q\mathbf{v}\times(\nabla\times\mathbf{A})$$
$$\Rightarrow \frac{d}{dt}\mathbf{p} = q(\mathbf{E} + \mathbf{v}\times\mathbf{B}) \tag{5.2.23}$$

Using the relation of Hamiltonian and Lagrangian given as

$$H = \mathbf{v}\cdot\frac{\partial L}{\partial \mathbf{v}} - L$$

The Hamiltonian is derived:

$$H = \frac{mc^2}{\sqrt{1-v^2/c^2}} + q\phi = \gamma mc^2 + q\phi \tag{5.2.24}$$

However, the Hamiltonian should be given as a function of general coordinate and momentum, and it is finally

$$H = \sqrt{m^2c^4 + c^2(\mathbf{P}^c - q\mathbf{A})^2} + q\phi \tag{5.2.25}$$

The kinetic energy of the particle is the first term of RHS in (5.2.24) and

$$E_{kin} = mc^2\left[1 + (p/mc)^2\right]^{1/2}$$
$$\Rightarrow E_{kin} = \gamma mc^2 \tag{5.2.26}$$
$$\Rightarrow E_{tot} = \gamma mc^2 + q\phi$$

The change of the particle kinetic energy is the work done per unit time by the external force, and it is easily obtained

$$\frac{d}{dt}E_{kin} = \mathbf{v} \cdot \frac{d\mathbf{p}}{dt} = q\mathbf{v} \cdot \mathbf{E} \qquad (5.2.27)$$

Taking the sum of all electrons in a unit volume of (5.2.27), we find $\mathbf{j} \cdot \mathbf{E}$ in (1.3.5) in the form:

$$\sum_i - e\delta(\mathbf{r} - \mathbf{r}_i(t))\mathbf{v}_i \cdot \mathbf{E} = \mathbf{j} \cdot \mathbf{E}$$

It is clear that the energy change only due to the force by electric field and magnetic field doesn't change particle energy. This is consistent with (1.3.5). It is obvious that the magnetic field only works to change particle momentum, namely, the direction of particle orbit.

It is convenient to summarize the relations to be used frequently later:

$$\begin{aligned} \mathbf{p} &= \gamma m\mathbf{v} = \gamma m \frac{d\mathbf{r}}{dt} \\ \varepsilon &= \gamma mc^2, \qquad \varepsilon^2 = c^2 p^2 + m^2 c^4 \\ \gamma &= \frac{1}{\sqrt{1-\beta^2}}, \qquad \beta = \frac{v}{c} \\ \gamma^2 &= 1 + \beta^2 \gamma^2 \end{aligned} \qquad (5.2.28)$$

5.2.2 Lorentz Transformation of Time and Space

H. A. Lorentz found the transformation of time and space so that Maxwell equations do not change. He thought that Maxwell equations should be invalid in any coordinates because the experimental result by Michelson-Morley denied the existence of any absolute coordinate. He proposed the Lorentz transformation in 1899. Strikingly at the same time, he also pointed out the relativistic mass increases and the Lorentz contraction.

Let us summarize the **Lorentz transformation**. Define the coordinate at rest with $S = (t, x, y, z)$ and the coordinate moving in the x-direction with a constant velocity $\mathbf{V}_0$ as $S' = (t', x', y', z')$. This is shown in Fig. 5.1. Then, it is well-known that the following relations are satisfied for physical quantities between the two coordinates: $S \rightarrow S'$. Since the perpendicular y and z to the direction of the velocity $\mathbf{V}_0$ are kept the same, only x and t change in Lorentz transformation. It is given more general form:

5.2 Special Relativity for Electron Motion

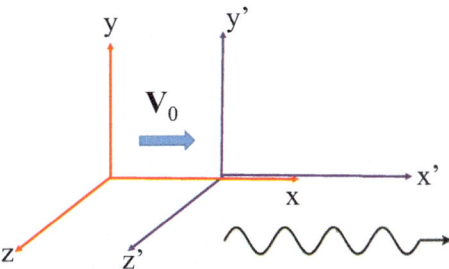

Fig. 5.1 The laboratory frame $S = (x, y, z)$ and the frame moving with velocity $\mathbf{V}_0$ in the x-direction $S' = (x', y', z')$. Lorentz transformation shown in (5.2.29) relates physical quantities in the two frames. In most of cases in the text, ultra-intense lasers are assumed to propagate from –x- to +x-direction as being shown

$$\begin{aligned} t &= \gamma_0(t' + \mathbf{V}_0 \cdot \mathbf{x}'/c^2) \\ \mathbf{x} &= \mathbf{x}_\perp' + \gamma_0(\mathbf{x}_\parallel' + \mathbf{V}_0 t) \end{aligned} \quad (5.2.29)$$

where γ_0 is Lorentz factor defined by $\mathbf{V}_0$ and $\mathbf{x}_\perp$ and $\mathbf{x}_\parallel$ means the coordinate perpendicular and parallel to the moving frame direction, respectively:

$$\gamma_0 = \frac{1}{\sqrt{1-\beta_0^2}}, \quad \beta_0 = \frac{V_0}{c} \quad (5.2.30)$$

Solving (5.2.29) for $\mathbf{x}'$ and t' provided the following relation equivalent to (5.2.29).

$$\begin{aligned} t' &= \gamma_0(t - \mathbf{V}_0 \cdot \mathbf{x}/c^2) \\ \mathbf{x}' &= \mathbf{x}_\perp + \gamma_0(\mathbf{x}_\parallel - \mathbf{V}_0 t) \end{aligned} \quad (5.2.31)$$

It is noted that Lorentz transformation (5.2.31) can be obtained just by changing the velocity $\mathbf{V}_0$ to $-\mathbf{V}_0$ and regarding S' frame to S frame.

The Lorentz transformation is obtained from two assumptions that:

1. The speed of light is always constant in any inertial frame.
2. The length of four dimensional vector (-ct, x, y, z) is kept constant in the transformation:

$$ds^2 = c^2 dt^2 - d\mathbf{x}^2 = c^2 dt'^2 - d\mathbf{x}'^2 \quad (5.2.32)$$

It is well-known that (5.2.29) indicates that the time becomes slow and x-space contracts both by $1/\gamma_0$ (<1) in moving frame S'.

The elongation of the time in a moving frame is experimentally observed in cosmic ray research. High-energy proton cosmic rays collide nuclei in the atmosphere and muons with short life time are produced. The life time of the muon is

1.5 μs at the rest frame, but they are observed by detectors on the grand. Although the length by the product of the life time and the speed of light are only 450 m, they can arrive at the grand after traveling about 20 km of the atmosphere. This is the evidence of the delay of the time in the frame traveling with the speed of the muon.

This phenomenon is also understood by use of the contraction of the space. Since the velocity of the muon is almost the speed of light, the thickness of the atmosphere is roughly less than hundreds of meter in the muon frame. The space and time contractions are consistent like this example.

5.2.3 Lorentz Transformation of Velocities

It is easy to obtain the relation of Lorentz transformation of velocities:

$$\mathbf{v} = \frac{d\mathbf{x}}{dt}, \quad \mathbf{v}' = \frac{d\mathbf{x}'}{dt'} \tag{5.2.33}$$

Using (5.2.29) it is easy to obtain the transformation of the velocity in the parallel direction at first:

$$v_\parallel = \frac{dx_\parallel}{dt} = \frac{dx_\parallel'/dt' + V_0}{1 + V_0/c^2 \left(dx_\parallel'/dt'\right)}$$

Then, the perpendicular velocity vector is also obtained, and they are written in the form:

$$\begin{aligned} v_\parallel &= \frac{v_\parallel' + V_0}{1 + V_0 v_\parallel'/c^2} \\ \mathbf{v}_\perp &= \frac{\mathbf{v}_\perp'}{\gamma_0 \left(1 + V_0 v_\parallel'/c^2\right)} \end{aligned} \tag{5.2.34}$$

Note that the velocity transformation in the parallel direction has no Lorentz factor dependence, while the perpendicular velocity reduced by a factor of Lorentz contraction.

(5.2.34) can also be applicable to light, and consider the light is irradiated isotropic in $\mathbf{S}'$ frame. For a light emitted with an angle θ' in $\mathbf{S}'$ frame, it is observed to propagate in θ direction in $\mathbf{S}$ frame with the following relation:

$$\tan\theta = \frac{1}{\gamma_0} \frac{\sin\theta'}{(\beta_0 + \cos\theta')} \tag{5.2.34a}$$

This is shown in Fig. 5.2 schematically. The uniform beam emission is elongated in the direction of the motion of the light source as seen in Fig. 5.2. This means that

5.2 Special Relativity for Electron Motion

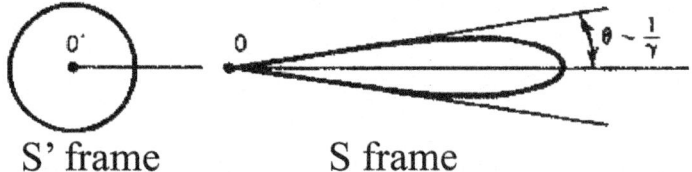

Fig. 5.2 The light emitted uniformly to all angle in the moving frame is observed in the laboratory frame as shown in the right, when Lorentz factor γ given by the velocity of the moving frame is much larger than unity, namely, the velocity V_0 is almost speed of light

something is ejected isotropic in space from a space ship with velocity V_0; it looks like spreading like a beam to the forward direction with angle of $1/\gamma_0$. This is called **relativistic beaming**. In considering the radiation property emitted by electrons moving with a velocity near the speed of light, this beaming effect and the relativistic Doppler effects should be correctly taken into account as seen in the following sections.

This is the reason why the synchrotron radiation is dominantly emitted to the forward direction for large Lorentz factor beam bending. This relativistic beaming effect is also important to relate short time observation of gamma-rays from the **gamma-ray burst** objects in far distant universe.

5.2.4 Lorentz Transformation of Fields

It is obvious that the equation of motion should be the same in the frame moving with a constant velocity V_0. Since the charge and mass of particle are Lorentz invariant and the equation in the S′ frame is

$$\frac{d}{dt}\mathbf{p}' = q(\mathbf{E}' + \mathbf{v}' \times \mathbf{B}')$$
$$\mathbf{p}' = \gamma' m \mathbf{v}' \qquad (5.2.35)$$

In order to solve (5.2.35), the field and velocity vectors in the S′ frame should be known with the values in the S frame. The velocity in the S′ frame is defined in (5.2.31). It is clear that the velocity $\mathbf{v}'$ in the S′ frame is calculated from (5.2.34):

$$\mathbf{v}' = \frac{d\mathbf{x}'}{dt'} = \frac{\mathbf{v}_\perp/\gamma_0 + \mathbf{v}_\| - \mathbf{V}_0}{1 - \mathbf{V}_0 \cdot \mathbf{v}/c^2} \qquad (5.2.36)$$

There is a relation between the velocities in the both frames as

$$1 - \beta'^2 = \frac{1 - \beta^2}{\gamma_0^2(1 - \mathbf{V}_0 \cdot \mathbf{v}/c^2)} \quad (5.2.37)$$

where β and β' are v/c and v'/c, respectively. In addition, (5.2.31) and (5.2.37) give the relation:

$$\frac{dt'}{dt} = \gamma_0(1 - \mathbf{V}_0 \cdot \mathbf{v}/c^2) = \sqrt{\frac{1 - \beta^2}{1 - \beta'^2}} \quad (5.2.38)$$

This results the important relation for the **proper time** τ, being also a Lorentz invariant and convenient to use in mathematics:

$$d\tau = dt\sqrt{1 - \beta^2} = dt'\sqrt{1 - \beta'^2} \quad (5.2.39)$$

It is noted that in the any inertial frame, the proper time of moving particles is the same:

$$d\tau = \frac{dt}{\gamma(t)} \quad (5.2.40)$$

After mathematics, the field values are found to be converted in the form:

$$\begin{aligned} \mathbf{E}' &= \mathbf{E}_\parallel + \gamma_0(\mathbf{E}_\perp + \mathbf{V}_0 \times \mathbf{B}_\perp) \\ \mathbf{B}' &= \mathbf{B}_\parallel + \gamma_0(\mathbf{B}_\perp - \mathbf{V}_0 \times \mathbf{E}_\perp/c^2) \end{aligned} \quad (5.2.41)$$

The electric and magnetic fields depend on which coordinate we observe. For example, a point charge induces magnetic field when it moves, but no magnetic field exists when it is at rest. This is clear from (5.2.41). It is informative to show a relation for electromagnetic field in the moving frame. When S' is moving in the same direction as the wave propagates, the electric field of the wave is

$$\mathbf{E}_\perp' = \gamma_0(1 - \beta_0)\mathbf{E}_\perp = \left(\frac{1 - \beta_0}{1 + \beta_0}\right)^{1/2} \mathbf{E}_\perp \quad (5.2.41a)$$

Thus, the field strength becomes weaker in the frame moving with the propagation direction, while of course it becomes stronger in the frame moving in the opposite direction of wave propagation ($\beta_0 \rightarrow -\beta_0$).

In the case where there is no electric field but external magnetic field in the laboratory frame, the electric field of $\mathbf{E}' = \mathbf{V}_0 \times \mathbf{B}$ will appear in the moving frame S' as shown in (5.2.41). This electric filed is called **motional electric filed**. Considering a charge particle trapped in the wave potential moving with a velocity $\mathbf{V}_0$, it will be accelerated by this motional electric field toward the perpendicular direction in the

5.2 Special Relativity for Electron Motion

moving frame. Such particle acceleration is called the **surfing acceleration**. How is the case with no magnetic field but finite electric field? It is clear from (5.2.41) that the Lorentz correction is negligible in the limit of $V_0 \ll c$, but this correction will be important when V_0 approaches the speed of light.

It is also useful to know the Lorentz transformation of the vector potential **A** and scalar one ϕ. Four vectors made of $A^i = (\phi, \mathbf{A})$ have the same structure as (5.2.29) and

$$\begin{aligned} \phi &= \gamma_0 \left(\phi' + \mathbf{V}_0 \cdot \mathbf{A}'/c^2 \right) \\ \mathbf{A} &= \mathbf{A}_\perp' + \gamma_0 \left(\mathbf{A}_\parallel' + \mathbf{V}_0 \phi \right) \end{aligned} \quad (5.2.42)$$

Consider the **relativistic Doppler shift** to light propagating in the x-direction. Assume V_0 is in the x-direction, for simplicity. By use of the fact that the phase of the wave is an Lorentz invariant

$$\varphi(x, t) = kx - \omega t = k'x' - \omega't' \quad (5.2.43)$$

Inserting (5.2.29 to 5.2.43), the Doppler shift relation in relativistic velocity is obtained:

$$\begin{aligned} [\omega', k'] &= \gamma_0 (1 - \beta_0)[\omega, k] \\ &= \sqrt{\frac{1 - \beta_0}{1 + \beta_0}} [\omega, k] \end{aligned} \quad (5.2.44)$$

This indicates that the phase velocity of waves is also Lorentz invariant.

5.2.5 Plane Electromagnetic Waves in Vacuum

Assume that the laser field is plane electromagnetic wave propagating to x-direction and the vector potential is given in the form:

$$\mathbf{A} = \mathbf{A}_0 \cos(kx - \omega t) \quad (5.2.45)$$

From the Canonical momentum in (5.2.19), **p** and e**A** have the same dimension to the electrons. The following normalized vector potential can be defined:

$$\mathbf{a} = \frac{e\mathbf{A}}{mc} \quad (5.2.46)$$

$$\mathbf{a} \equiv \frac{e\mathbf{A}}{mc}, \quad a_0 = \frac{eE_0}{mc\omega} \left(= \frac{v_{os}}{c} \right)$$

For electromagnetic field, the perpendicular component of **A** in (5.2.42) is Lorentz invariant, and the normalized **a** in (5.2.24) is also independent of any inertial frame. Let us derive this fact from the equation of motion in the both frames. In the moving frame with velocity $\mathbf{V}_0$ along the laser propagation, the equation of motion with laser electric field is

$$\frac{d\mathbf{p}'}{dt'} = -e\mathbf{E}'$$
$$-i\omega' m\gamma' \mathbf{v}'_\perp = e\mathbf{E}'_\perp \quad (5.2.47)$$

Inserting (5.2.41a) and (5.2.44) into (5.2.47), the laser strength parameter **a** is proved to be a Lorentz invariant as follows:

$$\mathbf{a}' = \frac{\mathbf{v}'_\perp}{c} = \frac{e}{m'c}\frac{\mathbf{E}'_\perp}{\omega'} = \frac{e}{mc}\frac{\mathbf{E}_\perp}{\omega} = \mathbf{a} \quad (5.2.48)$$

This is because the perpendicular component of **A** in (5.4.18) is also a Lorentz invariant.

The normalized amplitude a_0 has the following relation with oscillation velocity v_{os} defined in the non-relativistic limit as shown in (1.3.11):

$$a_0 = \frac{v_{os}}{c} = k\xi_{os}, \quad \xi_{os} = \frac{v_{os}}{\omega} \quad (5.2.49)$$

In (5.2.49), ξ_{os} is the quivering distance of the electron motion. The quivering distance becomes comparable to the laser wavelength for a_0 approaches unity.

In the plane wave limit in the vacuum, the following relations is obtained:

$$\frac{\omega}{k} = c, \quad B_0 = \frac{E_0}{c}$$
$$\langle W \rangle = \frac{\varepsilon_0}{2} E_0^2 \quad (5.2.50)$$
$$\langle P_{EM} \rangle = \frac{\langle W \rangle}{c} = \frac{\langle W \rangle}{\hbar \omega} \hbar k$$

Given laser intensity and wavelength, we can call a_0 as **laser strength parameter**. Inserting the parameters, we obtain

$$a_0 = 0.85 \sqrt{I_{18} \lambda_{\mu m}^2} \quad (5.2.51)$$

where I_{18} is the laser intensity divided by 10^{18} W/cm^2 and $\lambda_{\mu m}$ is the laser wavelength in μm unit.

5.3 Electron Motion in a Relativistic Strong Field

In the case of laser intensity near or more than $I_L = 10^{18}$ W/cm^2 ($a_0 = 1$), relativistic analysis is essential to electron motions in laser field. Then, it seems that complex mathematics is probably required. However, very simple conservation laws are obtained as seen below by assuming plane wave of laser field. Before proceeding to the laser-plasma interaction, derive several important relations of an electron in relativistic laser field. The electron motions in relativistic fields are well developed in textbook, for example, [3, 4].

Let us assume that the laser field is monochromatic plane wave given with the vector potential **A** and $\phi = 0$ in (5.2.7), and therefore

$$\mathbf{E} = -\frac{\partial \mathbf{A}}{\partial t}, \quad \mathbf{B} = \nabla \times \mathbf{A} \quad (5.3.1)$$

At first, derive the time evolution of the electron energy given intuitively in (5.2.28). Operate the vector **v** product to (5.2.23) to find the following relation:

$$\mathbf{v} \cdot \frac{d\mathbf{p}}{dt} = -e\mathbf{v} \cdot \mathbf{E} \quad (5.3.2)$$

The LHS of (5.3.2) can be reduced to

$$\frac{\mathbf{v}}{mc^2} \cdot \frac{d\mathbf{p}}{dt} = \beta \frac{d}{dt}\left(\frac{\beta}{\sqrt{1-\beta^2}}\right) = \beta\gamma^3 \frac{d\beta}{dt}$$

On the other hand, it is easy to show that γ and β have the following relation:

$$\frac{d\gamma}{dt} = \beta\gamma^3 \frac{d\beta}{dt}$$

Therefore, (5.3.2) can be written in the energy equation for an electron same as in (5.2.28):

$$mc^2 \frac{d\gamma}{dt} = -e\mathbf{v} \cdot \mathbf{E} \quad (5.3.3)$$

It is noted that the energy exchange between the laser and an electron is done through the $\mathbf{j} \cdot \mathbf{E}$ term.

The equation of the momentum is written from (5.2.23) with the vector potential **A** in the form:

$$\frac{d}{dt}\mathbf{p} = +e\left(\frac{\partial \mathbf{A}}{\partial t} - \mathbf{v} \times (\nabla \times \mathbf{A})\right) \tag{5.3.4}$$

(5.2.19) is written in the present case as

$$\frac{d}{dt}(\mathbf{p} - e\mathbf{A}) = -e[\mathbf{v} \cdot \leftrightarrow \nabla \mathbf{A}] \tag{5.3.5}$$

where

$$\mathbf{v} \cdot \leftrightarrow \nabla \mathbf{A}|_i = v_j \frac{\partial A_j}{\partial x_i} \tag{5.3.6}$$

Since the vector potential is assumed perpendicular to the propagation direction, (5.3.5) yields a conservation relation to the perpendicular component:

$$\mathbf{p}_\perp - e\mathbf{A} = 0 \tag{5.3.7}$$

where the integration constant is taken null due to the following reason. (5.3.5) is the equation to an electron, and it has no momentum before the laser comes. We assume that the laser intensity increases gradually with the period of many frequencies. The quantity of (5.3.7) is conserved to the value before the laser come, where $\mathbf{p}_\perp = 0$, $\mathbf{A} = 0$ is satisfied.

5.3.1 Constant of Motion in Vacuum

The time evolution of the parallel momentum $\mathbf{p}_\parallel$ in the x-direction should satisfy the relation from (5.3.5):

$$\frac{d}{dt}p_x = -e\left(v_y \frac{\partial}{\partial x}A_y + v_z \frac{\partial}{\partial x}A_z\right) \tag{5.3.8}$$

On the other hand, the energy Eq. (5.3.3) can be rewritten as

$$mc\frac{d}{dt}(\gamma) = \frac{e}{c}\left(v_y \frac{\partial}{\partial t}A_y + v_z \frac{\partial}{\partial t}A_z\right) \tag{5.3.9}$$

Taking the difference and sum of (5.3.8) and (5.3.9) leads to

5.3 Electron Motion in a Relativistic Strong Field

$$\frac{d}{dt}(p_x \mp mc\gamma) = -\frac{e}{c}\left[v_y\left(\frac{\partial}{\partial t}A_y \pm c\frac{\partial}{\partial x}A_y\right) + v_z\left(\frac{\partial}{\partial t}A_z \pm c\frac{\partial}{\partial x}A_z\right)\right] \quad (5.3.10)$$

For the wave propagating to +x-direction in vacuum, the relation $\mathbf{A} = \mathbf{A}_0(x - ct)$ is satisfied, and the sign up in (5.3.10) leads to the following conservation relation as long as the wave $\mathbf{A}$ propagates with the speed of light:

$$\frac{p_x}{mc} - \gamma = -\alpha \quad (5.3.11)$$

In (5.3.11), integration constant α is introduced to consider two cases with different α values in this section.

It is important to note that the two cases are $\alpha = 1$ and $\langle\gamma\rangle$, where $\langle\gamma\rangle$ is the time-averaged value of Lorentz factor. In the case of $\alpha = 1$, time average of the momentum in the x-direction $\langle p_x \rangle = \langle\gamma\rangle$, and an electron obtains a constant velocity as will be seen below. The force to accelerate to give this velocity is usually called **ponderomotive force**, and this force is explained in the later section in more detail. This is the case for electron beams or electrons in low-density plasmas. When the plasma density is high, this force induces charge separation from the ions at rest. In the dense plasma, it is better to assume that electrons in plasmas cannot move forward independently of the ion distribution. The electric field by the charge separation profibits the free motion of electrons. Therefore, ¼ 0 with α ¼ $\langle\gamma\rangle$ is better assumption for the electron motions in high-density plasmas.

5.3.2 Normalizations

For convenience of expressions, define normalization of physical quantities. The normalized variables are shown with the same letters but with a hut on it:

$$\widehat{t} = \omega_0 t, \quad \widehat{\mathbf{r}} = \frac{c}{\omega_0}\mathbf{r}, \quad \widehat{\mathbf{p}} = \frac{\mathbf{p}}{mc},$$
$$\widehat{\mathbf{A}} = \frac{e\mathbf{A}}{mc} \equiv \mathbf{a}, \quad \widehat{\mathbf{v}} = \frac{\mathbf{v}}{c} \equiv \boldsymbol{\beta}, \quad \widehat{E} = \frac{E}{mc^2} \equiv \gamma \quad (5.3.12)$$

Then, the constants of motions (5.3.7) and (5.3.11) are shown as

$$\widehat{\mathbf{p}}_\perp = \mathbf{a}$$
$$\widehat{p}_x = \gamma - \alpha \quad (5.3.13)$$

The kinetic energy of an electron in (5.2.27) is

$$\gamma = \sqrt{\widehat{p}_x^2 + \widehat{p}_\perp^2 + 1} \qquad (5.3.14)$$

Inserting (5.3.13) into (5.3.14), we obtain the following relation:

$$\gamma(t) = \frac{1}{2\alpha}\left[\mathbf{a}^2 + \alpha^2 + 1\right] \qquad (5.3.14a)$$

Eliminating γ from (5.3.13) and (5.3.14), the following simple relation is obtained:

$$\widehat{p}_x = \frac{1}{2\alpha}\left[1 - \alpha^2 + \widehat{p}_\perp^2\right] \qquad (5.3.15)$$

5.3.3 Electron Motion in Vacuum

Including any polarization of laser fields, the following expression to **A** is usually used:

$$\mathbf{A}(\phi) = \left[0,\ \delta A_0 \cos\phi,\ (1-\delta^2)^{1/2} A_0 \sin\phi\right] \qquad (5.3.16)$$

where ϕ is the phase of the laser:

$$\phi(t, x) = \omega t - kx \qquad (5.3.17)$$

The linear polarization to y or z direction is given for $\delta = 1, -1$ or 0. Right or left rotating circularly polarized lasers is given for $\delta = 1/\sqrt{2},\ -1/\sqrt{2}$.

It is useful to know the following relation in the frame of the moving electron $\mathbf{x}(t)$:

$$\frac{d}{dt}\phi[t, x(t)] = 1 - \frac{d\widehat{x}}{dt} = 1 - \frac{\widehat{p}_x}{\gamma} = \frac{\alpha}{\gamma} \qquad (5.3.18)$$

In deriving (5.3.18), (5.3.13) is used. Taking time average of (5.3.18), it is shown that the time-averaged frequency in the moving electron frame is given by

$$\omega' = \frac{\alpha}{\langle\gamma\rangle}\omega \qquad (5.3.19)$$

We consider the two cases with different α value. In the case of a free electron, $\alpha = 1$, and the other case is for $\alpha = \langle\gamma\rangle$. As clear from (5.3.19), an electron feels down-shift of frequency, and it keeps moving to the laser propagation direction. In the latter case, the average position of an electron has no move as clear from (5.3.19), and this case is appropriate for many electrons in plasmas as will be explained later.

5.3 Electron Motion in a Relativistic Strong Field

5.3.4 Free Electron Orbit

Let us consider the case of linearly polarized lasers to obtain the analytic solution of electron orbits. We start with the case of a free electron:

$$\alpha = 1.$$

Then, (5.3.13) becomes

$$\gamma = \widehat{p}_x + 1 \tag{5.3.20}$$

Assume that the laser is linearly polarized in y-direction ($\delta = 1$). The vector potential (5.3.16) is given as

$$a_y = a = a_0 \cos\left(\widehat{x} - \widehat{t}\right) \tag{5.3.21}$$

Then, the following relations are obtained:

$$\begin{aligned} \widehat{p}_y &= a \\ \widehat{p}_x &= \frac{1}{2}\widehat{p}_y^{\,2} = \frac{a^2}{2} \\ \gamma &= \frac{a^2}{2} + 1 \end{aligned} \tag{5.3.22}$$

From (5.3.18)

$$\begin{aligned} \widehat{p}_x &= \gamma \frac{d\widehat{x}}{d\widehat{t}} = \gamma \frac{d\phi}{d\widehat{t}}\frac{d\widehat{x}}{d\phi} = \frac{d\widehat{x}}{d\phi} \\ \widehat{p}_y &= \gamma \frac{d\widehat{y}}{d\widehat{t}} = \gamma \frac{d\phi}{d\widehat{t}}\frac{d\widehat{y}}{d\phi} = \frac{d\widehat{y}}{d\phi} \end{aligned} \tag{5.3.22a}$$

Therefore,

$$\begin{aligned} \frac{d\widehat{x}}{d\phi} &= \frac{1}{2}a^2 = \frac{a_0^2}{2}\cos^2\phi \\ \frac{d\widehat{y}}{d\phi} &= a(\phi) \end{aligned} \tag{5.3.22b}$$

Integrating (5.3.6), we obtain the orbit:

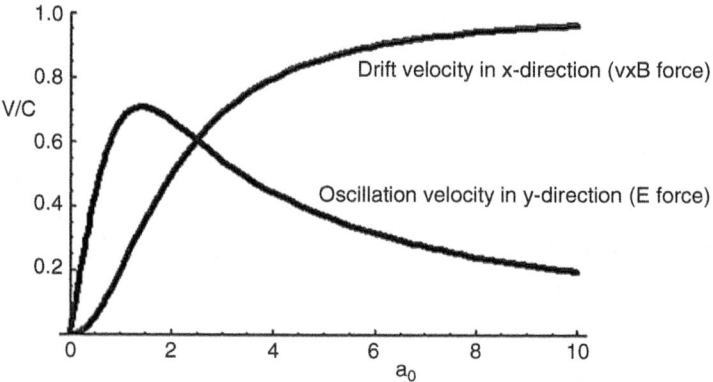

Fig. 5.3 Time-averaged drift velocity V_d of (5.3.24) (blue) and the oscillation amplitude in y-direction of (5.3.25) (red) as a function of the field strength a_0

$$\hat{x}(t) = \frac{a_0^2}{4}\left(\phi + \frac{1}{2}\sin 2\phi\right)$$
$$\hat{y}(t) = a_0 \sin(\phi) \tag{5.3.23}$$

Inserting the phase, we know that the electron is moving in the x-direction:

$$\hat{x}(t) = \frac{a_0^2}{4+a_0^2}\left(\hat{t} + \frac{1}{2}\sin 2\phi\right) \tag{5.3.23a}$$

which provides a constant drift velocity V_d in the x-direction with the form

$$\hat{V}_d = \frac{a_0^2}{a_0^2 + 4} \tag{5.3.24}$$

The electron velocities are easily obtained in the forms. (5.3.22b) can be rewritten with use of (5.3.18) to the form:

$$\hat{v}_x = \frac{a^2}{2\gamma} = \frac{\frac{1}{2}a^2}{\frac{1}{2}a^2 + 1}$$
$$\hat{v}_y = \frac{a}{\gamma} = \frac{a}{\frac{1}{2}a^2 + 1} \tag{5.3.25}$$

It is clear that the time average velocity in x-direction is equal to (5.3.24).

In Fig. 5.3, the time-averaged drift velocity V_d in the x-direction (red) and the amplitude of the quivering velocity in y-direction (blue) are plotted as functions of laser strength a_0. Typical orbits with a_0 much smaller or larger than the unity are

5.3 Electron Motion in a Relativistic Strong Field

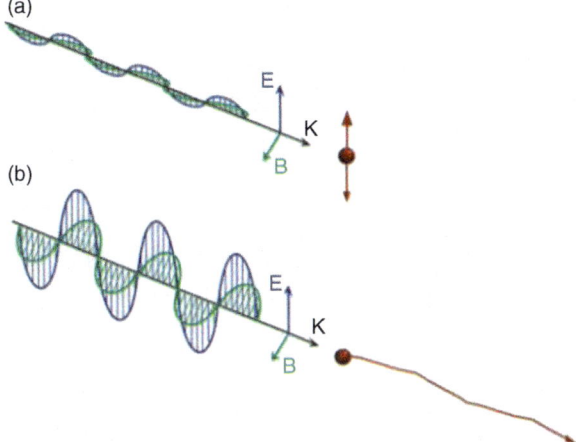

Fig. 5.4 Classical optics versus relativistic optics. (**a**) In classical optics, the amplitude of the light wave is small, electrons oscillate in the direction of the electric field at the light's frequency, and there is no displacement along the light's propagation direction. Note that only the E field acts on the electron and the electron-oscillation velocity is very small compared with the speed of light. (**b**) In relativistic optics, the amplitude of the light wave is very large, the light's magnetic field becomes important, and the combined action of the electric and magnetic fields pushes the electron forward. In this case, the electron velocity becomes close to the speed of light. [Fig. 3 in Ref. 5]

plotted in Fig. 5.4 [5]. In the non-relativistic limit, the oscillation is up and down by electric field, while its orbit becomes almost straight as imagined in Fig. 5.4.

Finally, it is informative to show the case of circularly polarized laser. The solutions are easily calculated to be

$$\widehat{x} = \frac{a_0^2}{a_0^2+4}\widehat{t} = \widehat{V_d}\widehat{t}$$
$$\widehat{y} = \frac{a_0}{\sqrt{2}}\sin\phi \qquad (5.3.26)$$
$$\widehat{z} = \pm\frac{a_0}{\sqrt{2}}\cos\phi$$

The electron takes a helical orbit, and the rotation frequency is reduced to ω/γ different from laser field rotation at rest frame ω. In fact, in the frame moving with x (t) in (5.3.26), the phase of laser field ϕ is

$$\phi = \widehat{t} - \widehat{V_d}\widehat{t} = \frac{1}{\langle\gamma\rangle}\widehat{t} \qquad (5.3.27)$$

The time-averaged γ in (5.3.27) is constant in the circularly polarized wave. It is informative to know that in the frame moving with the velocity in (5.3.26) in the x-direction, the orbit of the electron is only circular motion, the electric field force is

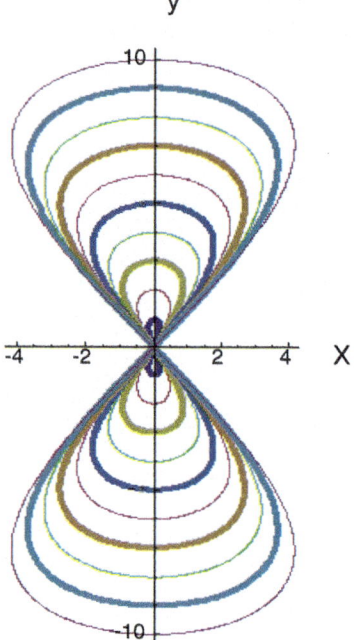

Fig. 5.5 Normalized orbit of the figure-of-eight motion in (x, y) plane for the case of $a_0 = 1$ (green), $a_0 = 5$ (blue), and $a_0 = 10$ (red)

always toward the center, the magnetic field is parallel with the velocity ($\mathbf{v} \times \mathbf{B} = 0$), and no force works to the electron by the magnetic field. This is shown in Fig. 5.5.

It is also very important to know the quantitative difference of the electron motion compared to the liner polarized case. The linear polarization has 2ω components in the x-direction, but it disappears in the circularly polarized case.

5.3.5 Electrons in Plasmas

In the plasmas, the forward drift of electron motion shown in (5.3.23) induces the charge separation from the background ions at rest. Soon after the laser penetrates to plasmas from free boundary, it is appropriate to consider that the accumulation of the electrons generates electrostatic field in plasmas and electrons cannot move stationary as seen in the cases of free electrons. We assume that time-averaged electron motion has no drift motion. This is the case with the following value of α in (5.3.11):

$$\langle \hat{p}_x \rangle = 0 \quad \Rightarrow \alpha = \langle \gamma \rangle \tag{5.3.28}$$

From (5.3.14a), we obtain

5.3 Electron Motion in a Relativistic Strong Field

$$1 - \alpha^2 + \langle a^2 \rangle = 0$$

namely,

$$\alpha = \sqrt{\frac{1}{2}a_0^2 + 1} \equiv \gamma_0 \qquad (5.3.29)$$

And from (5.3.14a), Lorentz factor is

$$\gamma = \frac{1}{2\gamma_0}\left[\mathbf{a}^2 + \gamma_0^2 + 1\right] \qquad (5.3.29a)$$

It is clear that taking time average of (5.3.29a), the relation $\gamma = \gamma_0$ is satisfied.

Lorentz factor of the free electron motion given in (5.3.24) is much larger than that in (5.3.29) because of relativistic motion V_d in the x-direction. It is very important to note that Lorentz factor for $\alpha = 1$ or above is

$$\langle \gamma_{\text{free}} \rangle = \langle \gamma_{\text{plasma}} \rangle^2 \qquad (5.3.30)$$

5.3.6 Linear Polarization

For linear polarization ($\delta = 1$), (5.3.7) and (5.3.11) become

$$\widehat{p}_y = a$$
$$\widehat{p}_x = \gamma - \gamma_0$$

Inserting the relation (5.3.29a) to the above relation and knowing that $\langle p_x \rangle = 0$, the momentums are obtained as

$$\begin{aligned} \widehat{p}_x &= \frac{a_0^2}{4\gamma_0}\sin 2\phi \\ \widehat{p}_y &= a_0 \cos\phi \\ \widehat{p}_z &= 0 \end{aligned} \qquad (5.3.31)$$

And the orbit is

$$\widehat{x} = \frac{a_0^2}{8\gamma_0^2} \cos 2\phi$$
$$\widehat{y} = \frac{a_0}{\gamma_0} \sin\phi \qquad (5.3.32)$$
$$\widehat{z} = 0$$

Eliminating the phase from (5.3.32), we obtain the famous **figure-of-eight** motion defined by the relation:

$$16\widehat{x}^2 = \widehat{y}^2\left(4q^2 - \widehat{y}^2\right), \quad q = \frac{a_0}{2\gamma_0} \qquad (5.3.33)$$

The orbit of the figure-of-eight motion is plotted for $a_0 = 1$ (green), $a_0 = 5$ (blue), and $a_0 = 10$ (Red) in Fig. 5.5.

5.3.7 Circular Polarization

For the case of circular polarization $\left(\delta = 1/\sqrt{2}\right)$, the orbit becomes a simple circle with radius $a_0/\sqrt{2}\gamma_0$:

$$\widehat{x} = 0$$
$$\widehat{y} = \frac{a_0}{\sqrt{2}\gamma_0} \sin\phi \qquad (5.3.34)$$
$$\widehat{z} = \frac{a_0}{\sqrt{2}\gamma_0} \cos\phi$$

It is noted that the circular electron motion is sustained by the balance of the centripetal force due to the rotating electric field and the centrifugal force of the electron, and magnetic field works no force to the electron because the velocity is always parallel to the rotating magnetic field.

5.4 Nonlinear Radiation Scattering

Let us consider the situation of Fig. 5.6 where an intense laser is coming from the left and an electron or electron beam is coming from the right. The scattering by oscillating electrons in laser field is called Thomson scattering, when the laser intensity is weak enough and irradiated on an electron at rest. Nonlinear effect becomes important for the case the laser intensity is extremely high. On the other hand, Compton scattering takes place even with the irradiation of optical laser, when an electron is impinging to the laser with relativistic kinetic energy as relativistic

5.4 Nonlinear Radiation Scattering

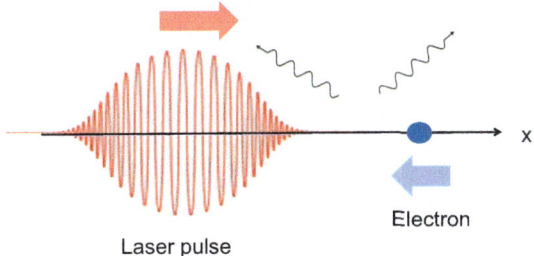

Fig. 5.6 Schematics of interaction of an electron with propagating laser field. Since the electron motion is nonlinear oscillation motion in strong laser field, it scatters radiation in nonlinear process generating higher harmonics of the laser frequency. When electron beams are injected as counter-propagation with the lasers, relativistic Doppler effects help to produce higher-energy photons

electron beam. This is called inverse Compton scattering. With increase of laser intensity to ultra-high intensity, the inverse Compton scattering is also affected nonlinear effect due to relativistic motion of electrons in the laser field. Let us explain these phenomena one by one.

5.4.1 Linear Thomson Scatterings

As shown in (5.2.8), the vector potential **A** can be generated, when an external current source is imposed. The simplest example is an antenna, where AC current flows in a tube to emit electromagnetic waves in space. In the non-relativistic laser limit, the induced current by an electron scatters the laser, and it is Thomson scattering. The directed photon energy is slightly scattered mainly to the perpendicular directions according to Larmor emission shown in (2.3.16). The classical Thomson scattering due to the retardation from oscillating electron is called **linear Thomson scattering**, and nonlinear one becomes important as seen below.

Thomson scattering is dominant for the case when the photon energy is much smaller than the electron one, $\hbar\omega < < mc^2$. When the photon has higher energy in colliding an electron, quantum effect becomes important, and Compton scattering is dominant with the shift of photon frequency. When a low-energy photon interacts with a relativistic electron, Compton scattering occurs in the frame moving with the electron. This is due to frequency up-shift of the photon by Doppler effects in the moving frame. We will see more details below.

In Chap. 2, we saw that electrons in a strong laser field induce the figure-of-eight relativistic motion of electrons in linear polarized laser and cycle motion in circularly polarized laser. In addition, the speed of oscillating electron is almost the speed of light so that we cannot neglect the finiteness of the light travel, namely, so-called retardation effect should be taken into account the evaluation of the radiation emission from such charged particle. The radiation emission from arbitrary motion of a charged particle is well explained in Chap. 14 in the book by Jackson [2]. The energy spectrum of the radiation emitted by a single electron in an arbitrary orbit **r**(t) and velocity **β** = v/c can be calculated from **Lienard-Wiechert potentials** [Eq. (14.67) in Ref. [2]]:

$$\frac{d^2 I}{d\omega d\Omega} = \frac{e^2 \omega^2}{4\pi^2 c} \left| \int_{-T/2}^{T/2} dt [\mathbf{n} \times (\mathbf{n} \times \boldsymbol{\beta})] \exp[i\omega(t - \mathbf{n} \cdot \mathbf{r}/c)] \right|^2 \quad (5.4.1)$$

where the definition of LHS is the energy per frequency ω per solid angle Ω during the interaction time T and **n** is a unit vector pointing to the scattering direction. (5.4.1) can be easily solved in the case of the linear Thomson scattering. Try to obtain the scattering to the angle θ to the laser propagation direction x with the polarization in y-direction. The direction θ is in the x-y plane. Since the oscillation motion in the y-direction is in the form of $\sin(\omega_0 t)$ for laser frequency ω_0, the following relation is obtained:

$$|\mathbf{n} \times (\mathbf{n} \times \boldsymbol{\beta})| = \beta(t)\sin\theta = a_0 \sin\theta \sin(\omega_0 t)$$
$$\omega \mathbf{n} \cdot \mathbf{r}/c = a_0 \omega/\omega_0 \cos\theta \sin(\omega_0 t) \quad (5.4.2)$$

Inserting (5.4.2) to (5.4.1), the exponential term in the integrand is found to be decomposed to a sum of higher harmonic oscillations by use of **Bessel function** identity:

$$\exp[i\alpha\sin(\omega_0 t)] = \sum_{n=-\infty}^{\infty} J_n(\alpha) \exp(in\omega_0 t) \quad (5.4.3)$$

$$\alpha = a_0 \omega / \omega_0 \cos\theta$$

It is clear that even for simple harmonic oscillation in y-direction, the scattering radiation has higher harmonics due to a finite amplitude due to the retardation effect. It is found from the following mathematical relation that only the higher harmonics of the fundamental frequency ω_0 are produced. That is

$$\int_{-\infty}^{\infty} \exp[-i(\omega t - n\omega_0 t)] \, dt \propto \delta(\omega - n\omega_0) \quad (5.4.4)$$

Then,

$$\omega = n\omega_0$$
$$\alpha = na_0 \cos\theta$$

In addition, knowing the property of Bessel function shown in Fig. 5.7 for a small argument that

5.4 Nonlinear Radiation Scattering

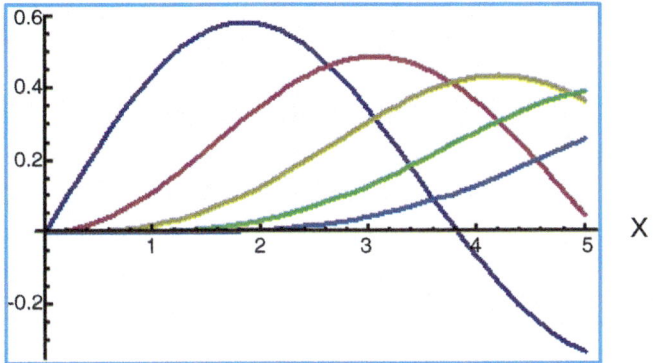

Fig. 5.7 Argument x and index n dependence of Bessel function $J_n(x)$. Near the $x = 0$, Bessel function starts to increase in proportion to x^n

$$J_n(\alpha) \propto \alpha^n \tag{5.4.5}$$

it is enough to assume that only the fundamental frequency is scattered in the linear Thomson scattering case.

5.4.2 Compton and Inverse Compton Scatterings

It is well-known that energy shift of the scattered photon is observed when the incoming photon energy is relativistic, $\hbar\omega \sim mc^2$. In the normal condition, the optical laser photon energy is about 1 eV ($\ll mc^2 = 500$ keV). It looks like intense laser application has no relation with Compton scattering. With use of relativistic electron beams, intense lasers are used to generate extremely high-energy photon via inverse Compton scattering.

The principle is very simple. Assume that a relativistic electron beam is traveling as shown in Fig. 5.8, where intense laser pulse collides the electron beam on the same axis. Then, the laser photons are scattered mainly backward direction at the beginning, or the laser intensity is weak enough so that the electrons can keep moving to the initial velocity direction. But if the beam kinetic energy is small compared to the photon pressure as shown in Fig. 5.8, the central orbit of oscillating motion will alter the direction. In such case, scattered radiations are Doppler shifted due to the time-dependent drift velocity. We assume that the electron beam energy is large enough under the not-so-strong laser strength and a constant drift velocity derived in (5.3.24) is enough to take into account the deceleration of the interacting electron beam. Under such condition, the mechanism of radiation emission is due to the linear Thomson scattering in the frame moving with electron beams, but the

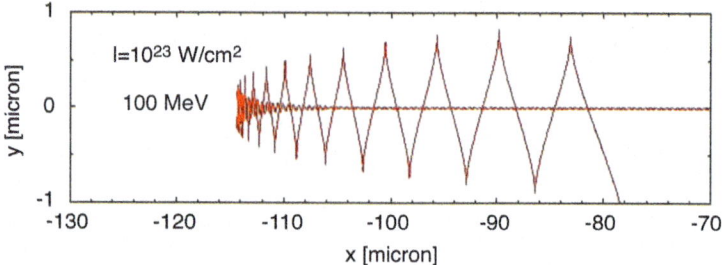

Fig. 5.8 An example of an electron orbit impinging from the right with 100 MeV kinetic energy to the ultra-intense laser with the peak intensity of 10^{23} W/cm². At the beginning the electron scattered the laser photon mainly to the backward direction to the laser (to left), but it is reflected back before the peak of laser intensity by the ponderomotive force to the right direction

incident and scattered photons are affected Doppler shift in (5.2.44) two times as shown below.

In the frame moving with the electron beam with its Lorentz factor β_e, the incoming laser frequency is upshifted as shown in (5.2.44):

$$\omega_1 = \left(\frac{1+\beta_e}{1-\beta_e}\right)^{1/2} \omega_0 \qquad (5.4.6)$$

It is reasonable to consider that the photon ω_1 is scattered by the Thomson scattering in this beam frame. This frequency is again upshifted in the laboratory frame by the Doppler shift to ω_2 in the form:

$$\omega_2 = \frac{1+\beta_e}{1-\beta_e}\omega_0 = \frac{(1+\beta_e)^2}{1-\beta_e^2}\omega_0 \approx 4\gamma_e^2\omega_0 \qquad (5.4.7)$$

It is intuitively understood that angular distribution of the photon flux near the frequency ω_2 has a strong peak at the direction to the electron beam because of relativistic beaming effect. It is informative to show the angular dependence of the radiation with frequency ω_2:

$$\omega_{peak}(\theta) = \frac{4\gamma_e^2\omega_0}{(1+\nu_0\gamma_e)}\left[\left(1+\frac{\gamma_e^2}{1+4\nu_0\gamma_e}\right)\theta^2\right]^{-1} \qquad (5.4.8)$$

$$\nu_0 = \frac{\hbar\omega_0}{mc^2}$$

It is seen in (5.4.8) that due to the **relativistic beaming effect** of radiation from an object moving near the speed of light, the scattered radiation is concentrated within an angle about

5.4 Nonlinear Radiation Scattering

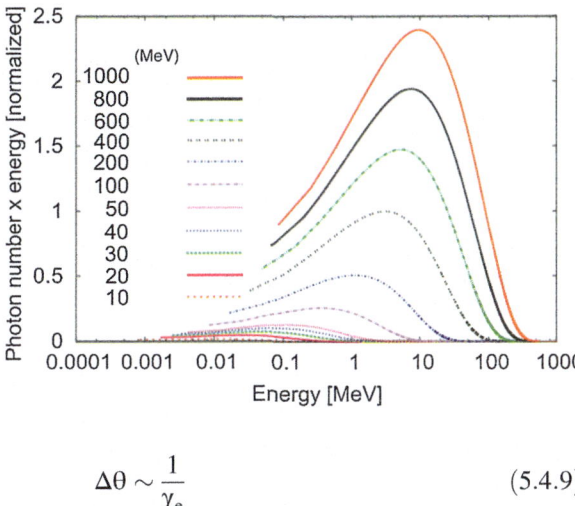

Fig. 5.9 Emission spectra of an electron with different initial energy interacting with 83 fs laser pulse of intensity 10^{21} W/cm^2. The peaks of the spectra reduce with the kinetic energy of electron beam. [Fig. 8 in Ref. 6]

$$\Delta\theta \sim \frac{1}{\gamma_e} \qquad (5.4.9)$$

It is noted that in the combination of the conventional accelerators and optical lasers, the condition $\nu_0 \gamma_e < \ < 1$ is valid at the present technology.

Using ultra-intense lasers and accelerated electron beams, we can design compact X-ray and γ-ray sources with high intensity. Radiation spectra computationally obtained including energy coupling with an electron and laser field are shown in Fig. 5.9 for different energy of electrons, where laser peak intensity is 10^{21} W/cm^2 and laser is Gaussian pulse with short duration 83 fs. In this case, beam energy reduces substantially over the initial 40 fs, and most of the energy is converted to the radiation energy [6]. Note that such energy change of electron makes continuum.

Such high-energy photons (γ-rays) have been used to study the inner structure of nuclei by installing laser in synchrotron facility. In Spring-8, Japan, an argon laser with the wavelength 351 nm is irradiated to the electron ring with 8 GeV ($\gamma_e = 16{,}000$). The initial photon energy from the laser 3.5 eV is converted to the photons with 2.4 GeV. The scattered angle is extremely small, and the photon intensity width at 100 m away from the scattering point is only 1.2 cm.

The inverse Compton scattering becomes very peculiar in the universe. One example is radiation spectra in X-ray and gamma-ray region from supernova remnants (SNR). The synchrotron emissions of cosmic rays are observed near the surface of SNR in X-ray region. Such X-rays are scattered by the inverse Compton by the cosmic rays with ultra-relativistic energy. In general two peaks are predicted and observed from many of SNRs as shown in Fig. 5.10 [7]. The lower-energy component is X-rays due to the synchrotron motion of cosmic ray electrons, while the higher-energy component is gamma-ray region due to the inverse Compton scattering of the X-ray by the cosmic rays.

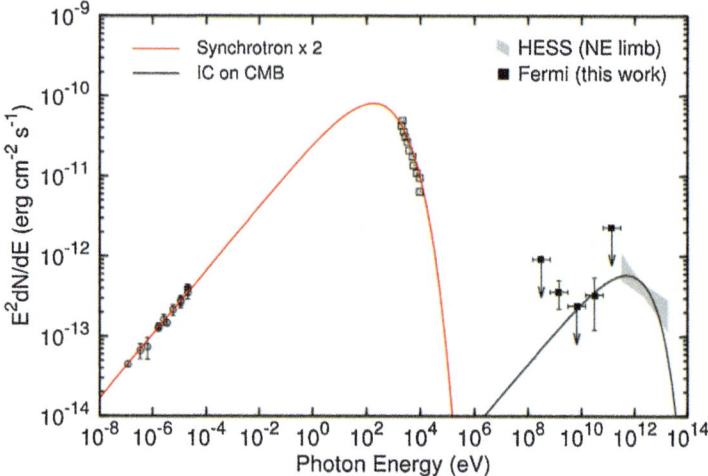

Fig. 5.10 Radiation spectrum from supernova remnant SN1006. The lower-energy photons are from radio observation, the middle are from X-rays, and the higher are from gamma-ray observation. The red line represents a theoretical emission spectrum from synchrotron of high-energy electron accelerated with cosmic ray protons. The black line at higher energy is calculated as inverse Compton emission of the synchrotron radiation by the high-energy relativistic electrons. Observation data points are also plotted. [Fig. 1 in Ref. 7]

5.4.3 Nonlinear Thomson Scattering

Nonlinear Thomson scattering provides **higher harmonic generation (HHG)**, and the maximum harmonic number n of HHG increases rapidly with the laser intensity. Efficient conversion of low-frequency laser photon energy to HHG of the order of n = 100 can provide the photon in the range of 100 keV. This is called **laser synchrotron source (LSS)** and may provide a practical method of generating tunable, near-monochromatic, well-collimated, short-pulse X-rays in a compact, relatively inexpensive source. As we see below, two examples of LSS configurations are possible theoretically: an electron beam LSS generating hard (30 keV, 0.4-A) X-rays and a plasma LSS generating soft (0.3 keV, 40 A) X-rays. The former one is based on the use of Doppler up-shift from an electron in strong field drifting with a relativistic velocity given in (5.3.24).

In the case where the normalized laser intensity a_0 increases toward unity or over unity, it is clear that the higher harmonic components in (5.4.3) dramatically increase. In the case of a simple harmonic oscillation assumed in deriving (5.4.3), we have all harmonic component of the laser frequency ω_0. It is found that an electron with the figure-of-eight motion emits only odd integer harmonics in the backward direction to the laser incidence. On the other hand, with the scattering toward the perpendicular direction along y-axis, all higher harmonics components appear. This fact is easily proven by the following mathematics.

5.4 Nonlinear Radiation Scattering

(a) Linear polarization case

For simplicity, we neglect the constant coefficients in the calculation and assume that an electron takes the figure-of-eight orbit:

$$x(t) = x_0 \sin(2\omega_0 t)$$
$$y(t) = y_0 \sin(\omega_0 t) \quad (5.4.10)$$

Consider the case of backscatter, $\theta = 0$, and **n** is in the x-direction. In (5.4.1), the first term and the argument of the exponent have such time dependence:

$$|\mathbf{n} \times (\mathbf{n} \times \boldsymbol{\beta})| = \beta_z(t) \propto \sin(\omega_0 t)$$
$$\omega \mathbf{n} \cdot \mathbf{r}/c \propto \sin(2\omega_0 t) \quad (5.4.11)$$

Using the relation

$$\sin(\omega_0 t) = \frac{1}{2i}[\exp(i\omega_0 t) - \exp(-i\omega_0 t)] \quad (5.4.12)$$

we obtain the following expression instead of (5.4.3):

$$\sin(\omega_0 t) \exp[i\alpha \sin(2\omega_0 t)]$$
$$\propto \sum_{n=-\infty}^{\infty} J_n(\alpha)\{\exp[i(2n+1)\omega_0 t] - \exp[i(2n-1)\omega_0 t]\}$$
$$= \sum_{n=-\infty}^{\infty} [J_n(\alpha) - J_{n+1}(\alpha)] \exp[i(2n+1)\omega_0 t] \quad (5.4.13)$$

Due to the property of the figure-of-eight motion, the even harmonics are accidentally cancelled to disappear. It is clear that after the Laplace transformation, only the odd integer harmonics are obtained.

On the other hand, the radiation to the polarization direction y has all higher harmonics, since in this case (5.4.11) has the form:

$$|\mathbf{n} \times (\mathbf{n} \times \boldsymbol{\beta})| \propto \cos(2\omega_0 t)$$
$$\omega \mathbf{n} \cdot \mathbf{r}/c \propto \sin(\omega_0 t) \quad (5.4.14)$$

It is easily understood that the exponential term provides all higher harmonic components.

The angular dependence of the radiation intensity of each higher harmonics are calculated and plotted in Fig. 5.11 [8] for the case of the linear polarized laser with the filed strength of $a_0 = 0.5$, 1.0, and 2.0 in Fig. 5.11 (a), (b), and (c), respectively. It is clear that in all cases, only the odd HH has finite value, and even HH also has finite value toward the y-direction ($\theta = \pi/2$). It is natural that the radiation intensity increases with the increase of laser intensity. For the case of $a_0 = 2$, the peak

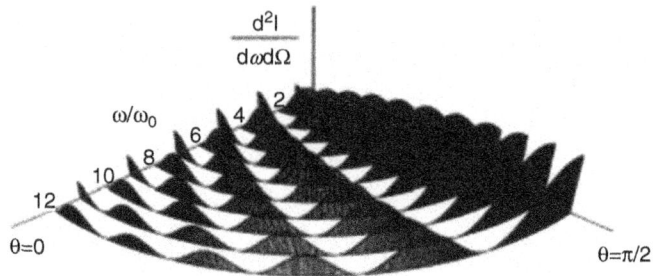

Fig. 5.11 . The normalized intensity of scattered radiation at the harmonic resonances frequency $\omega/\omega_0 = n$, as a function of angle θ in the $\phi = 0$ plane. This is the case of an intense laser scattering by a dense plasma electron, where the laser is linearly polarized. Figure shows the first 12 harmonics for the laser strength $a_0 = 2$. It is noted that only odd modes are scattered to the laser incident direction ($\theta = 0$), but all harmonics are observed to the perpendicular direction. [Fig. 3 (c) in Ref. 8]

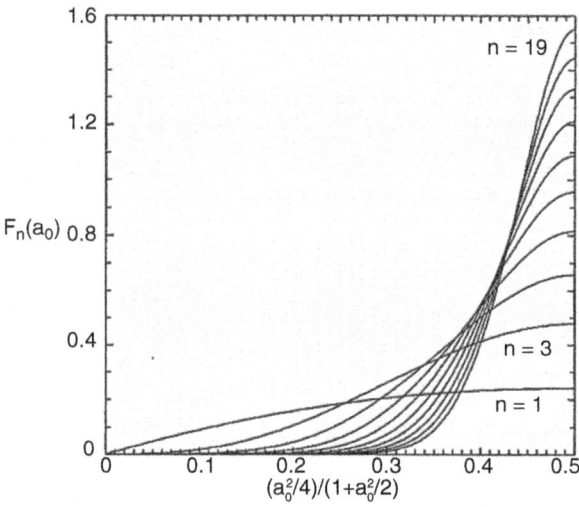

Fig. 5.12 The peak intensity of the scattered radiation to the laser incident direction ($\theta = 0$) as a function of laser strength a_0. As seen in Fig. 5.11, the harmonic number of the peak intensity changes as a function of laser strength. This can be imaged by the argument dependence of Bessel function shown in Fig. 5.7. [Fig. 4 in Ref. 8]

intensity is in the y-direction, and the maximum intensity is shifted to the higher harmonics components.

With increase of a_0, the harmonic number n of the peak intensity rapidly increases as shown in Fig. 5.12. It is noted that the horizontal axis becomes 0.5 for $a_0 \to \infty$. The number of the harmonics with the peak intensity among all HHs is shown in the figure as a function of a_0. The fact that the number of HH with the maximum intensity increases with the laser intensity is easily understood mathematically. Since the Bessel function is strongly depends on α^n as shown in (5.4.5), the amplitude of HHs increases rapidly with the increase of laser intensity.

(b) **Circular polarized case**

For the circularly polarized case at the oscillation center at rest, the radiation spectrum is the same as the synchrotron radiation source, because an electron orbit is

5.4 Nonlinear Radiation Scattering

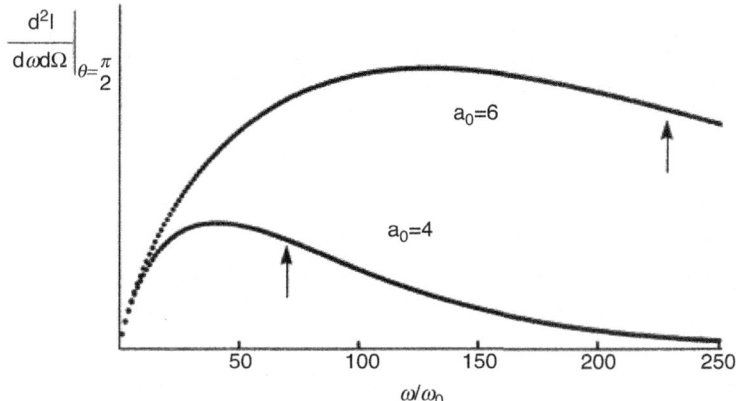

Fig. 5.13 The peak intensity of each harmonic in the transverse direction ($\theta = \pi/2$) versus normalized frequency $\omega/\omega_0 = n$ for a circularly polarized laser pulse scattering from a dense plasma electron. The cases $a_0 = 4$ and 6 are shown. The arrows indicate the approximate critical harmonic number $n = a_0^3$. [Fig. 8 in Ref. 8]

circular motion in the plane of y and z as obtained in (5.3.34). In this case, the emission spectrum is the same as the radiation from synchrotron facility and typical spectra for the cases of a0 = 4 and 6 are plotted in Fig. 5.13 [8]. The frequency is normalized by that of the incident laser.

It is useful to compare to the spectrum obtained theoretically for the radiation from a relativistic electron in a uniform magnetic field. Such synchrotron radiation spectrum F(x) is given in Ref. [9] as shown in Fig. 5.14. Normalized frequency x is defined as

$$x = \frac{\omega}{\omega_c}, \qquad \omega_c = \frac{3}{2}\gamma_0^3 \omega_{ce}^R, \qquad \omega_{ce}^R = \frac{eB}{\gamma_0 m} \qquad (5.4.15)$$

where the Lorentz factor is given in (5.3.29) and ω_{ce}^R is relativistic synchrotron frequency. F(x) is the continuum spectrum emitted from a relativistic electron accelerated by synchrotron motion in a constant magnetic field. A figure in the insert box of Fig. 5.14 is widely shown to indicate the performance of the output of synchrotron facility. Note that this log-log plot is the same as the main curve in Fig. 5.14 in linear scale.

5.4.4 Relativistic Beaming Effect and Doppler Shift

In the case of a free electron, its orbit drifts with a velocity near the speed of light, and the radiation spectra shown above are emitted in the frame moving with the drift velocity. We saw that the light emitted isotropic direction from a moving frame is

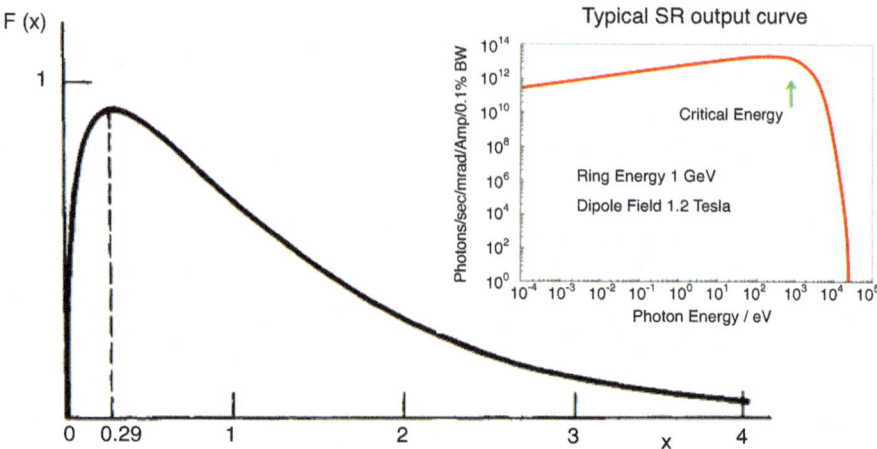

Fig. 5.14 F(x) is the normalized continuum spectrum emitted from a relativistic electron accelerated by synchrotron motion in a constant magnetic field, where x is defined in (5.4.15). A figure in the insert box indicates the performance of the output of synchrotron facility in log-log diagram. Note that both spectral shapes are same

observed in the laboratory as shown in (5.2.34a). The light is emitted predominantly toward the direction of relativistic motion. The angular dependence of the beaming is easily obtained from Lorentz transformation of velocity shown in Sect. 5.1.

The narrowing of the angle enhances the intensity. We obtain the relation of the enhancement factor $d\theta/d\theta'$ in the following form after taking derivative of (5.2.34a):

$$\frac{d\theta}{\cos^2\theta} = \frac{1}{\gamma_0} \frac{\beta_0 \cos\theta' + 1}{(\beta_0 + \cos\theta')^2} d\theta' \qquad (5.4.16)$$

Coupling (5.2.34a) and (5.4.16), we obtain the intensity conversion from the moving frame to the laboratory frame. (5.4.16) shows that all of the lights emitted with a finite angle θ' in the moving frame are observed only within the angle $\theta < 1/\gamma_0$ in the laboratory frame for relativistic motion, $\beta_0 \sim 1$. In Fig. 5.15, the angle dependence of the intensity in the moving frame and laboratory frame is plotted for two cases where the acceleration is on the axis (cf; dipole emission) and perpendicular to the axis (cf; synchrotron emission).

Consider Doppler shift of radiation emitted in the moving frame with the frequency ω'. Rewriting the relation (5.2.24) to the case with emission angle θ, we obtain

$$\omega = \frac{\omega'}{\gamma_0(1-\beta_0\cos\theta)} = \frac{[(1-\beta_0)(1+\beta_0)]^{1/2}}{(1-\beta_0\cos\theta)}\omega' \qquad (5.4.17)$$

5.4 Nonlinear Radiation Scattering

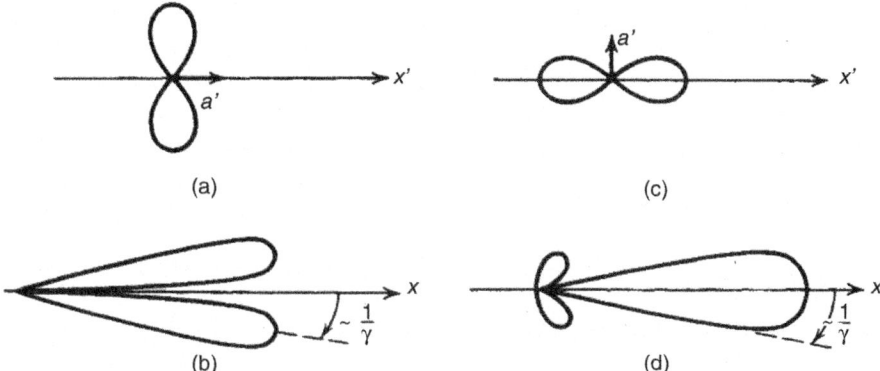

Fig. 5.15 (a) Radiation pattern from an electron accelerated (oscillated) in x′ direction with no drift motion. (b) Radiation pattern in laboratory frame (x) when an electron in (a) drifts with a relativistic velocity of Lorentz factor γ. (c) Radiation pattern from an electron accelerated (oscillated) in the perpendicular direction in the moving frame x′. (d) Its radiation pattern in the laboratory frame. [Fig. 4.11 in Ref. 9]

It is clear that Doppler up-shift of frequency occurs only near $\theta = 0$ and for $\beta_0 \sim 1$, while it is $\omega/\omega' = \sqrt{2(1-\beta_0)}$ for $\theta = \pi/2$, downshifted.

5.4.5 Nonlinear Compton Scattering

In the nonlinear Thomson scattering, the nonlinearity appeared due to the increase of the oscillation velocity for higher laser intensity. As seen above, the theory is purely classical, and no quantum mechanical effect is taken into account to explain it. In contrast, the **nonlinear Compton scattering (NCS)** is calculated with quantum electrodynamic effect. In quantum view, this corresponds to absorption of several laser photons accompanied by emission of a single photon of frequency ω:

$$e + n\omega_0 \rightarrow e' + \omega \quad (5.4.18)$$

As carefully explained in Ref. [10], it is not **multiple Compton scattering (MCS)**. The process MCS is used for the case where an electron is scattered more than one times during the interaction of counterstreaming laser photons. Therefore, the number of photons is conserved and the maximum energy of the scattered photon is limited by (5.4.7). On the other hand, in NCS the number of photons is not conserved, and n-photons are absorbed by an electron and a single higher-energy photon is emitted.

In NCS, not only the above quantum view but also the quantum electrodynamic effect is found to be important in calculating the scattered radiation spectrum and decelerated electron energy spectrum. An electron in a strong laser field has a different mass from the mass in the vacuum due to the strong coupling with laser field and behaves like the electron is dressed. It is called **dressed electron** in a strong field. The effective mass of the dressed electron is given in the form, for example, in [10]:

$$m_* = m\left(1 + |\mathbf{a}|^2\right)^{1/2} \qquad (5.4.19)$$

where **a** is the strength parameter of laser field defined in (5.2.46). The effective mass is obtained with **Volkov solution**. Volkov solution is the exact solution of Dirac equation in a plane electromagnetic wave with any amplitude. This solution provides many QED effects, and it is, for example, used to derive the **nonlinear QED** cross section of pair creation via multiphoton absorption in the strong laser and relativistic electron beam interaction scheme [11].

From the energy conservation relation for n-photon recoil, the minimum energy of the scattered electron to all angle is calculated for the incident angle of laser θ_0, for example from [10], in the form:

$$E_{min} = \frac{1}{1 + 2n\widehat{E}_0\widehat{\omega}_0(m/m_*)^2(1 + \cos\theta_0)} E_0 \qquad (5.4.20)$$

where E_0 is the initial electron beam energy and

$$\widehat{E}_0 = E_0/mc^2, \quad \widehat{\omega}_0 = \hbar\omega_0/mc^2$$

It is noted that taking $m_* = m$ and $n = 1$ reproduces the relation for the ordinary, linear Compton scattering (**inverse Compton scattering, ICS**).

For the case of $a_0 = 0.5$, $E_0 = 46.6$ GeV, and $\theta_0 = 17°$, the scattered electron energy spectra are plotted theoretically in Fig. 5.16 for the laser with 1.06 μm wavelength and the number of electron in a bunch of 5×10^9. It is noted that the photon energy in the beam frame is 211 keV upshifted from 1 eV and E_{min} is 25.6 GeV for $n = 1$. In order to calculate the scattering cross section beyond Klein-Nishina formula, Volkov solution is used to obtain the scattering matrix [11]. These parameters are taken from the corresponding experiment [10]. In Fig. 5.16, the dotted line below 25.6 eV is data only including the linear Compton scattering and MCS effect. The numerical values of NCS signal for $n = 1, 2, 3,$ and 4 in (5.4.20) are also spotted in Fig. 5.16.

In order to demonstrate NCS, the experiment was done with electron bean in SLAC [10]. The aim is to identify the electrons below 25.6 GeV to measure the excess electrons from the dashed line. It is reported that the above NCS theory is experimentally confirmed at the first time, and up to $n = 4$ events are experimentally confirmed as shown details of the experimental data and analysis [10].

The same type of experiment has been done with more intense laser and electron beam generated by laser wake field acceleration technique [12]. This is also an

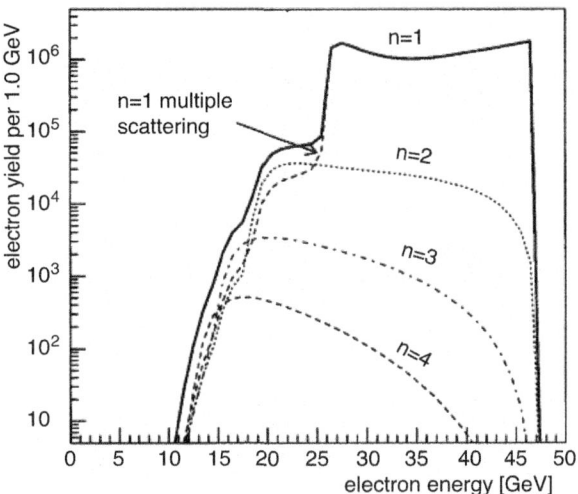

Fig. 5.16 Calculated yield of scattered electrons from the collision of 5×10^9 46.6 GeV electrons with a circularly polarized 1054 nm laser pulse of intensity strength $a_0 = 0.5$. [Fig. 1 in Ref. 10]

experiment of nonlinear Compton scattering, while the paper's title is multiphoton Thomson scattering. This is the same as the experiment at **SLAC**, and the difference is the laser and electron beam parameter. So, let us call it nonlinear Compton scattering (NCS) experiment. The laser photon energy is 1.55 eV and the normalized field strength $a_0 = 2 \sim 12$. The electrons in helium gas jet are accelerated to the energy of $200 \sim 300$ MeV. This relativistic electron beam is irradiated with counter directions by ultra-intense laser.

It is reported that for the case of $a_0 = 2$, HH photon up to n = 10 (10 MeV photon) was observed, and for $a_0 = 10$, HH photon of n = 500 (20 MeV photon) has been observed. Note that the laser photon is Doppler shifted downward by the drifting motion of (5.3.24). Therefore, the fundamental frequency of the incident laser is shifted to $\hbar\omega'' \sim 1$ MeV for $a_0 = 2$ case and $\hbar\omega'' = 40$ keV for $a_0 = 12$. The experimental data is compared to PIC simulation, and it is reported that the experimental spectra are well reproduced in the simulation [12].

References

1. L.D. Landau, E.M. Lifshitz, *The Classical Theory of Fields* (Nauka, Moscow, 1973)
2. J.D. Jackson, *Classical Electrodynamics*, 3rd edn. (John Wiley & Sons, New York, 1999), Chapter 11
3. A. Macchi, *A Super Intense Laser-Plasma Interaction Theory Primer* (Springer, Dordrecht, 2013)
4. P. Gibbon, *Short Pulse Laser Interaction with Matter: An Introduction* (Imperial College Press, London, 2005)
5. D. Umstadter, J. Phys. D. Appl. Phys. **36**, R151 (2003)
6. J.F. Ong et al., Phys. Plasmas **23**, 053117 (2016)
7. Y. Xing et al., Astrophys. J. **823**, 44 (2016)

8. E. Esary, S. Ride, P. Sprangle, Phys. Rev. E **48**, 3003 (1993)
9. G.B. Rybicki, A.P. Lightman, *Radiative Processes in Astrophysics* (Wiley-VCH, Weinheim, 2004), Chap. 4
10. C. Bula et al., Phys. Rev. Lett. **76**, 3116 (1996)
11. W. Griener, J. Reinhardt, *Quantum Electrodynamics*, 4th edn. (Springer, Berlin, 2009), p. 224
12. W. Yan et al., Nat. Photonics **11**, 524 (2017)

Chapter 6
Relativistic Laser Plasma Interactions

6.1 Charge Separation in Low-Density Plasma

In Sect. 5.3, we have solved two cases for $\alpha = 1$ and $\langle\gamma\rangle$. Study here the case when laser is impinging to low-density plasmas and the free electron model ($\alpha = 1$) is applicable. In order to make the mathematic simpler, consider the case of circular polarization. Then, $\gamma = a_0^2/2 + 1 (\equiv \gamma_0)$ is constant. Let us estimate under what condition the free electron model is validated by evaluating the accumulation of the charge separation from the plasma boundary.

6.1.1 Charge Separation by Photon Force

So far, an infinite plan wave is assumed to describe the electron motion in strong laser field. This is good assumption if the laser pulse is very long compared to the laser wavelength, and the focusing diameter is much larger than the laser wavelength. Based on the results above, however, let us consider what kind of physics is imagined intuitively for a given density of charge neutral plasma. This intuitive consideration may be important to plasma dynamics in the early time just after the laser irradiation, and the ions can be assumed not to move because of large mass. Consider the laser is impinged in the plasma whose electron density is n_0 and the electrons move forward due to the force in x-direction with the drift velocity (5.3.25). As shown in Fig. 6.1, it is reasonable to assume that the electrons move to the right until the time when the effective pressure due to coupling with laser field will balance with the pull-back force by the charge separation.

Consider one-dimensional system, and assume that the electrons are decelerated by the electric field E_{es} produced by the charge separation:

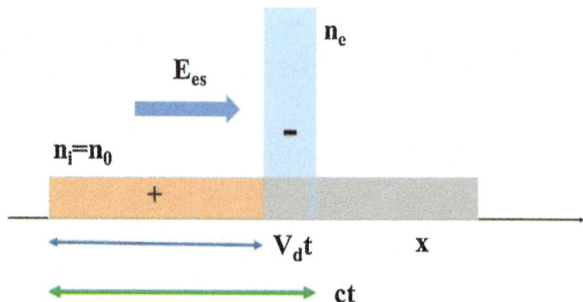

Fig. 6.1 Schematics of electron dynamics and generation of electrostatic field by charge separation, when directly applying the solution of free electron motion given in (5.3.23b). After a short time, the assumption of free electron brakes down and the effect of the electron static field should be taken into account as used for the electrons in plasmas

$$\frac{d}{dt} p_x = -eE_{es} \quad (6.1.1)$$

The electric field by the charge separation is enhanced with time as

$$E_{es} \sim \frac{en_0}{\varepsilon_0} ct \quad (6.1.2)$$

The electrons obtained the drift velocity by laser are decelerated and stop moving. This time can be calculated with (6.1.1) and (6.1.2) to be

$$\omega t = a_0 \frac{\omega}{\omega_{p0}} \quad (6.1.3)$$

Insert parameters, for example, laser intensity 10^{22} W/cm^2 (10^{20} W/cm^2) and $\omega/\omega_{pe} = 10$; the value of (6.1.3) is $\omega t = 10^3$ which corresponds to 500 fs (50 fs), which is comparable to standard pulse durations. However, when lasers are irradiated to high-density plasmas or to solid targets, the plasma density is high enough so that the plasma frequency is of the order of laser frequency, $\omega_{p0} = O(\omega)$. The establishment of the electrostatic field by charge separation is taken place very soon after $\omega t \sim 1$. Therefore, the drift motion of free electron derived in (5.3.27) is not realistic, and electron motion is reasonable to describe with no drift motion shown in (5.3.33) with condition (5.3.29).

Although, of course, the solution (5.3.24) is one obtained in the vacuum and not a solution in such an additional force in plasma, (6.1.3) gives us the qualitative image of Fig. 6.1. Roughly speaking, in the case where the laser pulse duration is about the time in (6.1.2), substantial amount of laser energy goes to the electrostatic field energy. Using (6.1.2) and (6.1.3), the energy density of the electrostatic field is obtained in the form:

6.1 Charge Separation in Low-Density Plasma

$$\varepsilon_{es} = \varepsilon_0 E_{es}^2 = a_0^2 mc^2 n_e \qquad (6.1.4)$$

On the other hand, the laser energy density is obtained as

$$\varepsilon_{em} = \varepsilon_0 \langle E_{em}^2 \rangle = \frac{1}{2} a_0^2 mc^2 \frac{\omega^2}{\omega_{p0}^2} n_e \qquad (6.1.5)$$

It is noted that the factor 1/2 in (6.1.5) disappears by taking account of the magnetic field energy. Comparing (6.1.4) and (6.1.5), the fractional energy going to the electrostatic field by charge separation is about

$$f_{ab} = \frac{\varepsilon_{es}}{\varepsilon_{em}} = \frac{\omega_{pe}^2}{\omega^2} \qquad (6.1.6)$$

It is found that with increase of the plasma density, the laser deposits its energy to plasma. Note, however, that the solutions of electron motion in the vacuum are used and more precise analysis is required. One is the generation of wake plasma wave discussed below.

6.1.2 Wake Field Generation and Energy Deposition

Such electrons accumulated at the laser front start overshooting to oscillate if the laser pulse is shorter than (6.1.3). The plasma waves produced by ultra-short pulse are called **plasma wake field by laser**. The mechanism of the wake field generation is easily seen by assuming an extremely short pulse, which can be modeled with impulse like delta function.

Since the wake field is produced in the laser propagation direction, assume one-dimension and include the electrostatic field as in (6.1.2):

$$\frac{d}{dt}\mathbf{p} = -e\mathbf{E}_{es} - e(\mathbf{E}_{em} + \mathbf{v} \times \mathbf{B}) \qquad (6.1.7)$$

The electrostatic field should satisfy Maxwell Eqs. (5.2.2) and (5.2.3):

$$\varepsilon_0 \frac{\partial E_{es}}{\partial t} = e n_e v_x, \qquad \varepsilon_0 \frac{\partial E_{es}}{\partial x} = e(n_0 - n_e) \qquad (6.1.8)$$

Combination of both in (6.1.8) gives the time derivative in the electron fluid frame:

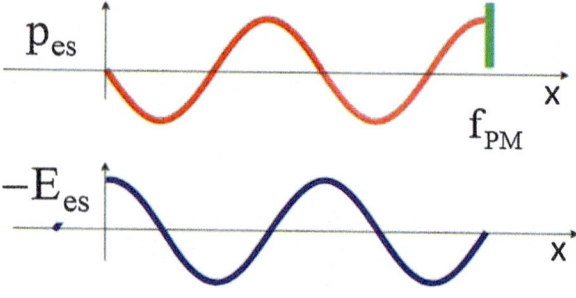

Fig. 6.2 Schematics for the wake field production by electron drift in long length and low-density plasmas. The laser pulse is modeled with a force to produce a displacement for a short time like a delta function in time

$$e\frac{dE_{es}}{dt} = \omega_{pe0}{}^2 v_x \qquad (6.1.9)$$

Assume that Lorentz force by laser field gives abrupt increase of the momentum in x-direction as shown in (5.3.27). Taking the time derivative of (6.1.7) and use (6.1.9) to eliminate E_{es}, the following model equation is obtained for the momentum due to the electrostatic wave in the x-direction $p_{es}(t, x_0)$:

$$\frac{d^2 p_{es}}{dt^2} = -\frac{\omega_{p0}{}^2}{\gamma} p_{es} + p_{em}\delta(t - x_0/c) \qquad (6.1.10)$$

In (6.1.10), x_0 is the initial coordinate of each electron layer in the slab plasma, so-called Lagrangian coordinate, and it is independent of the time t. In (6.1.10) the electron plasma frequency in relativistic regime is defined as

$$\omega_{p0}^2\Big|^R = \frac{e^2 n_0}{\varepsilon_0 m(\gamma)} = \frac{1}{\gamma}\omega_{p0}^2\Big|^{NR}, \qquad (6.1.11)$$

where "R" and "NR" mean relativistic and non-relativistic, respectively. This is the effect of **relativistic mass correction**. It is very important that the plasma frequency is reduced by Lorentz factor γ due to the relativistic effect in strong laser field. Although (6.1.11) is in general a nonlinear differential equation difficult to solve exactly, it is easily solved in the case of circularly polarized laser with a constant $\gamma = \gamma_0$. Then, (6.1.10) is easily solved to find the solution:

$$p_{es}(t, x) = \begin{cases} 0 & (x > v_g t) \\ p_{em} \cos\left\{\frac{\omega_{p0}}{\sqrt{\gamma_0}}(t - x/v_g)\right\} & (x < v_g t) \end{cases} \qquad (6.1.12)$$

In order to make the evaluation more realistic, we have replaced the speed of light c in (6.1.10) to the group velocity of laser pulse in plasmas v_g, which is less than the speed of light. In Fig. 6.2, schematics of the ultra-short pulse and wake field profile are shown. From (6.1.12) it is clear that the assumption of the delta function of laser

6.2 Laser Propagation in Plasmas

pulse is a good approximation if the wavelength of the wake plasma wave is longer than the pulse duration, namely, the density is low enough.

From (6.1.12), it is clear that the laser energy is used to generate the wake field. This is equivalent to laser absorption by plasmas. The energy densities of laser field and that of the wake field are easily calculated to be

$$E_{em} = a_0^2 mc^2 \frac{\omega^2}{\omega_{p0}^2} n_e c\tau_L$$

$$E_{es} = a_0^2 mc^2 n_e L,$$
(6.1.13)

where τ_L is the laser pulse duration and L is the length of the plasma slab. In obtaining (6.1.13), (5.3.23) is used and v_g is approximated to c. Therefore, the energy deposition fraction by generation of plasma wake field in the under-dense plasma is given:

$$f_{ab} = \frac{E_{es}}{E_{em}} = \frac{\omega_{pe}^2}{\omega^2} \frac{L}{c\tau_L}$$
(6.1.14)

In the case of gas jet plasma, for example, this fraction may be smaller than unity. As will be described in later chapter, the electron laser acceleration aiming at alternative accelerator requires the condition $\omega_{pe}/\omega < < 1$ and $L/c\tau_L > > 1$. This requirement is better for higher laser energy deposition to the wake field in the plasma.

This result is convenient to find the image. Since the laser triggers the wake field by depositing the same amount of momentum in the x-direction in (6.1.14) and the wake field oscillations start with this momentum, the energy deposited to one electron is constant for a given laser intensity a_0. As a result, the absorption fraction in (6.1.14) is simply proportional to the plasma density.

The present analysis is based on the solutions of electron motion in vacuum, and the laser propagation itself changes in the plasma; therefore, more precise and consistent analysis is required, especially when laser wake field electron acceleration is studied in details.

6.2 Laser Propagation in Plasmas

6.2.1 Relativistic Transparency

Consider the propagation of laser in under-dense plasmas. Electron motion makes the current in the same direction as **A**, transverse direction. The governing equation of laser propagation is (5.2.8), and we have to evaluate the electron current. In general, electrons have motion in the perpendicular and parallel directions to the wave propagation as seen in Sect. 6.3. In the circularly polarized case, the electron motion in plasma is only perpendicular plane as in (5.3.36) and no force by the magnetic field. In addition, Lorentz factor is constant; therefore, (5.3.23) is easily solved to obtain the current density:

$$\frac{\mathbf{j}}{\varepsilon_0} = -\frac{\omega_{p0}^2}{\gamma_0}\mathbf{A}, \qquad (6.2.1)$$

where γ_0 is defined in (5.3.16). Inserting this current to (5.2.8), we obtain the linear dispersion relation for electromagnetic waves in plasmas:

$$\omega^2 = c^2 k^2 + \frac{\omega_{p0}^2}{\gamma_0} \qquad (6.2.2)$$

Equation (6.2.2) shows that the effective plasma frequency becomes lower due to the relativistic mass correction by which the electron becomes heavier in the relativistic regime. This fact gives an important change of physics in laser-plasma interaction in relativistic case. The cutoff density n_{cr} determined only by the laser frequency becomes also a function of Lorentz factor.

It is clear that the cutoff density effectively increases from n_{cr} in non-relativistic case to the form:

$$n_{cr}^R = \gamma_0 n_{cr} \qquad (6.2.3)$$

When the intensity is high enough, the cutoff density possibly becomes higher than the plasma density of solid materials. Then, laser can penetrate though the solids, and they become transparent to the laser light. This phenomena is **relativistic transparency**. When the matter becomes transparent to strong laser field, the electrons start to drift as shown in Fig. 6.3. Then, for example, strong electrostatic field is produced on the rear surface of matter, and this electrostatic field is used for ion acceleration by relativistic lasers. In the ultra-relativistic limit, the contribution by the plasma or matter to the dispersion relation of (6.2.2) disappears, and the laser propagates as if the matters are transparent.

It is informative to estimate the critical laser intensity at which solid matters become transparent to the laser with 1 μm laser wavelength ($n_{cr} = 10^{21}$ cm^{-3}). From (6.2.3), we obtain the threshold value of a_0 so that solid material becomes transparent in the form:

$$a_0 \approx \sqrt{2}\frac{n_{solid}}{10^{21} \text{cm}^{-3}} \qquad (6.2.4)$$

The solid aluminum with the density 2.7 g/cm^3 has the electron number density 8×10^{23} cm^{-3}. The corresponding critical laser intensity for the relativistic transparency is 10^{24} W/cm^2 ($a_0 = 10^3$). The liquid hydrogen has the density 0.07 g/cm^3, and it becomes transparent at 2×10^{21} W/cm^2. Plastic target with the solid density of 1 g/cm^3 has the critical intensity of 10^{23} W/cm^2. It is surprising that even in solid

6.2 Laser Propagation in Plasmas

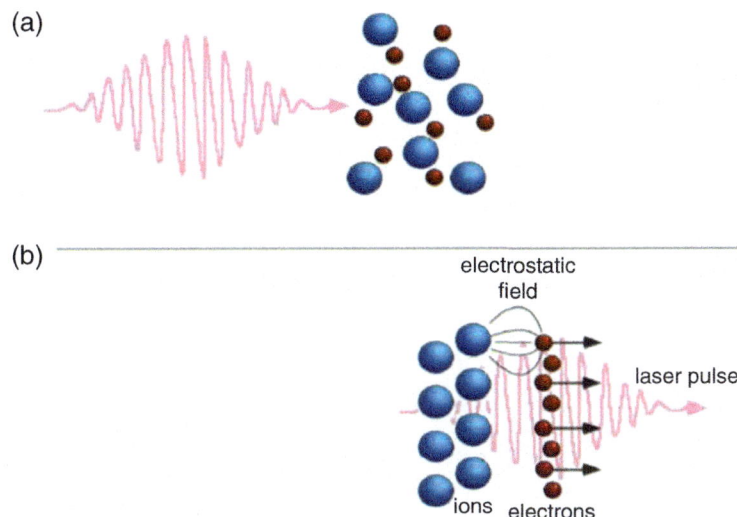

Fig. 6.3 The laser wake field mechanism. (**a**) A laser pulse interacts with a plasma, an ionized gas, composed of electrons and ions. (**b**) In the relativistic regime, the electromagnetic force acting on the electron pushes the electrons forward in the laser direction. The charge separation between the light electrons and the massive ions in a plasma produces a large longitudinal static electric field comparable to the transverse field of the laser. Note that the plasma acts as an efficient optical rectifier. [Figure 7 in Ref. 5]

targets, laser can propagate through thanks to the relativistic effect at higher laser intensities available experimentally these days.

In the case of linearly polarization, it is shown that the average Lorentz factor defined in (5.3.23) is applicable to the relativistic transparency.

6.2.2 Higher Harmonic Generation (HHG)

In linearly polarized case, not only the relativistic change of plasma frequency but also higher harmonic generation (HHG) is derived as follows. Since the electron motion is given in (5.3.33), the electron current in the perpendicular direction is easily obtained from (5.3.14):

$$\frac{\mathbf{j}}{\varepsilon_0} = -\frac{\omega_{p0}^2}{\gamma(t)} \mathbf{A} \qquad (6.2.5)$$

The time-varying Lorentz factor is given in (5.3.15a). The Lorentz factor is only a function of $\mathbf{a}^2$:

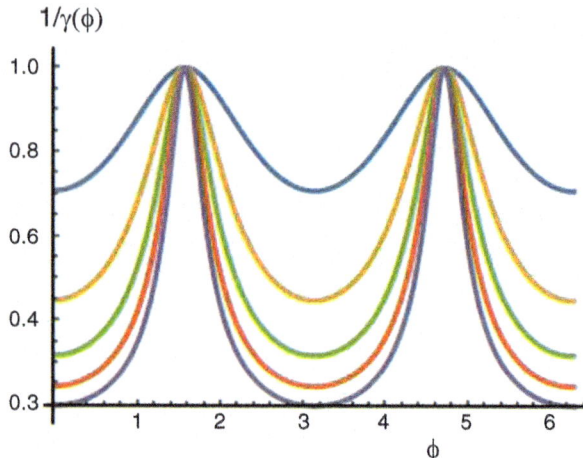

Fig. 6.4 The profile of $1/\gamma = 1/\sqrt{a^2+1}$ with $a^2 = 1/2 a_0^2 \cos^2\phi$ as a function of ϕ for the case of $a_0 = 1, 2, 3, 4,$ and 5. The left is linear plot, while the right is log plot. From the effective width of two spikes, we can evaluate the peak of Fourier components

$$\mathbf{a}^2 = \frac{a_0^2}{2}[1 + \cos\{2(\omega t - kx)\}] \tag{6.2.6}$$

The current term in (5.2.8) including the relativistic effect is proportional to

$$\mathbf{j}^R \propto \frac{\cos\phi}{1 + \eta \cos 2\phi} \propto \cos\phi \sum_{n=0} B_{2n} \cos(2n\phi), \tag{6.2.7}$$

where η is a function of a_0^2, $\phi = \omega t - kx$, and the B_{2n} is Fourier component calculated by

$$B_{2n} = \frac{1}{\pi}\int_0^\pi \frac{\cos 2\phi}{1 + \eta \cos 2\phi} d\phi \tag{6.2.8}$$

Note that $B_0 = <1/\gamma>$, and this term provides the relativistic transparency. The Fourier components $n = 1, 2, 3, \ldots$ are easily found that they provides nonlinear currents oscillating with the frequencies of $3\omega, 5\omega, 7\omega, \ldots$ Higher harmonics of the irradiated lasers appear because of the nonlinear motion of electrons due to relativistic effect. Note that n-dependence of B_{2n} is roughly imaged with the profile of $1/\gamma = 1/\sqrt{1+a^2}$ shown in Fig. 6.4 for $a_0 = 1, 2, 3, 4,$ and 5.

It is, however, pointed out in Ref. [1] that HHG due to only the relativistic effect overestimated the nonlinear current. The nonlinear density bunching due to the figure-of-eight motion by vxB force becomes important with the same magnitude of nonlinear currents with opposite sign as that due to the relativistic contribution shown above. As will be seen in the next section, the second harmonic motion in x-direction shown in (5.3.23) provides the electron density perturbation in 2ω. Therefore, as the nonlinear current in (5.2.12), additional term

6.2 Laser Propagation in Plasmas

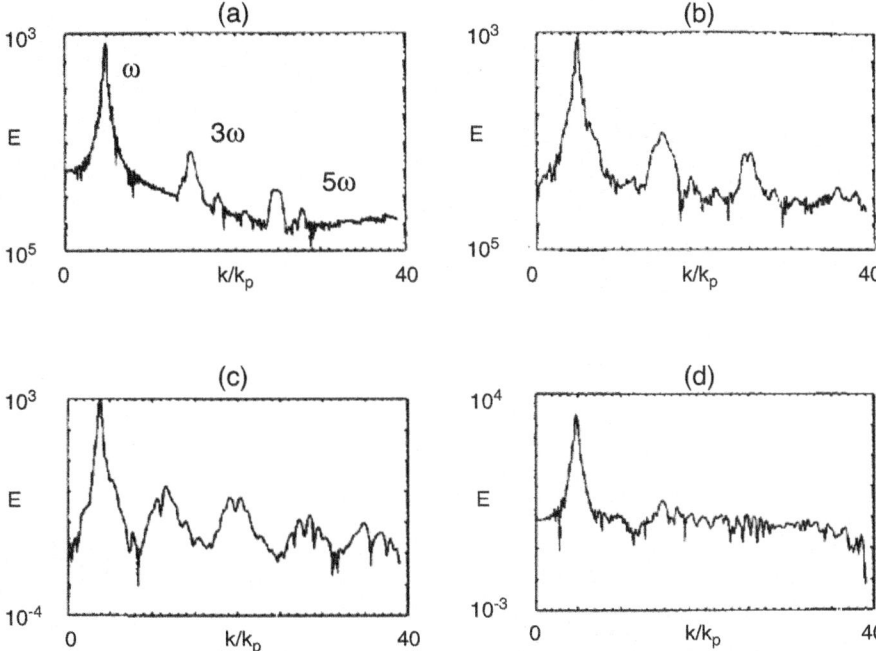

Fig. 6.5 The E field k spectra from simulations with $\omega_0/\omega_p = 5$. (a) $a_0 = 0.2$, (b) $a_0 = 0.3$, (c) $a_0 = 0.5$, and (d) $a_0 = 1$. [Figure 5 in Ref. 1]

$$\mathbf{j}_{db} = -\frac{e^2}{m\gamma_0}\delta n(2\phi)\mathbf{A}(\phi) \qquad (6.2.9)$$

Regarding the third harmonics, the nonlinear current (6.2.9) is almost the same value as $\mathbf{j}^R$ in (6.2.7) [1]. This is shown soon later by calculating both currents explicitly.

In Ref. [1], one-dimensional PIC simulation has also been done to observe the emission of HHG in under-dense uniform plasmas, where $n_0/n_{cr} = 1/25$ is set for laser strength $a_0 = 0.3$. In Fig. 6.5, the spectrum at the time $\omega_{p0}t = 50$ is plotted. As shown in the figure, strong peaks of ω, 3ω, and 5ω are observed. Simulation has been done up to $a_0 = 1$, and it is seen that with increase of a_0, the relative intensity of HHG increases, while the spectrum widths of HH line emission spread, and finally they became like white noise.

Consider the reason why the spectrum width of HHs becomes broader with increase of the laser intensity. Increase of laser intensity and source term of the nonlinear current inducing higher harmonics modify the amplitude of the laser in plasmas; consequently the source term also fluctuates since it is proportional to a_0^3. For example, assume that the amplitude of electric field of transverse mode is modified from a uniform in space and time to the fluctuating amplitude as shown in Fig. 6.6, Laplace or Fourier transformation of the electric field in the form:

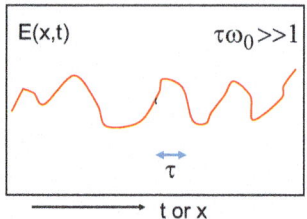

Fig. 6.6 Schematics showing broadening of the spectrum of HHG in Fig. 6.5. Because the pump laser energy is depleted to the HHG and other nonlinear interaction with the background plasmas, the amplitude of the laser and generated HHs are modified in time and space like the red curve. The typical time scale of such bumping is longer than the laser oscillation period but short enough. By Fourier or Laplace transformation of such amplitude results the broad spectrum with the width of about $(\Delta\omega)^2 = \langle(\omega - \omega_n)^2\rangle \sim 1/\tau^2$ at each HH spectrum and the fundamental spectrum as seen in Fig. 6.5

$$E = A(x,t) \exp[-in(\omega_0 t - k_0 x)], \qquad n = 1, 3, 5, \cdots$$

Are shown to be

$$E(\omega,k) = \iint dxdt A(x,t) \exp\{-i[\Omega t - Kx]\}$$
$$\Omega = (\omega - n\omega_0)$$
$$K = (k - nk_0)$$

If the amplitude A(x,t) fluctuates with a typical time interval of τ, it is clear that the spectrum has a typical width of $1/\tau$ in Ω space. This is the reason of the broadening of the spectrum, which is not taken into account in the above idealistic analysis where a_0 is assumed constant in time and space.

6.2.3 Electrostatic Field Excitation by vxB Force

It is already obtained that in the relativistic regime, 2ω oscillating motion of electrons appears in the direction of laser propagation as shown in (5.3.33). This is due to the **vxB** force in (5.2.23). The **vxB** force can be rewritten as

$$\mathbf{v} \times \mathbf{B} = \frac{e}{m\gamma} \mathbf{A} \times (\nabla \times \mathbf{A}) = -\frac{e}{2m\gamma} \nabla \mathbf{A}^2 \qquad (6.2.10)$$

The equation of motion to $\widehat{p}_x = p_x/mc$ is derived to be

6.2 Laser Propagation in Plasmas

$$\frac{d^2\widehat{p}_x}{dt^2} + \frac{\omega_{p0}^2}{\gamma}\widehat{p}_x = \frac{d}{dt}\left(\frac{2c}{\gamma}\frac{\partial}{\partial x}a^2\right) \tag{6.2.11}$$

The Lorentz factor should be contributed by both of laser and induced electrostatic motion:

$$\gamma = \sqrt{1 + a^2 + \widehat{p}_x^2} \tag{6.2.12}$$

In low-density plasmas, the a^2 in (6.2.12) is given in (6.2.6).

Inserting (6.2.6) to (6.2.11) and assuming $|\widehat{p}_x| \ll |a| \ll 1$, the following approximate equation is obtained:

$$\frac{d^2\widehat{p}_x}{dt^2} + \omega_{p0}^2\widehat{p}_x = \omega_0^2 a_0^2 \cos(2\phi) \tag{6.2.13}$$

Equation (6.2.13) is an equation of harmonic oscillator with an external driver term with different frequency from the natural frequency ω_{p0}. It is easy to obtain stationary oscillating solution of (6.2.13) in the form:

$$\widehat{p}_x = -\frac{1}{4 - \omega_{p0}^2/\omega^2} a_0^2 \cos(2\phi) \tag{6.2.14}$$

It is found that large amplitude electrostatic oscillation is induced. It is clear that this electrostatic oscillation is sustained by the electron density perturbation δn shown in (6.2.9). Let us evaluate the nonlinear current (6.2.9) within the perturbation theory.

6.2.4 Density Bunching Current

In order to obtain the density perturbation, the equation of continuity is used:

$$\frac{\partial}{\partial t}\delta n = -cn_0 \frac{\partial}{\partial x}\widehat{p}_x \tag{6.2.15}$$

Inserting (6.2.14) into (6.2.15)

$$\frac{j_{db}}{en_0 c} = \frac{a_0^3}{4 - \omega_{p0}^2/\omega^2} \cos(2\phi)\cos\phi \tag{6.2.16}$$

On the other hand, the perturbation of the relativistic nonlinear current in (6.2.5) with 3ω frequency is easily obtained as

$$\frac{j_R}{en_0 c} = -\frac{a_0^3}{4} \cos(2\phi) \cos\phi \qquad (6.2.17)$$

The total nonlinear current inducing 3rd HH wave is obtained:

$$\frac{j_R + j_{db}}{en_0 c} = -\frac{a_0^3}{16} \left(\frac{\omega_{p0}}{\omega}\right)^2 \cos(2\phi) \cos\phi \qquad (6.2.18)$$

The leading term of the order unity disappeared, and the total current is proportional to the small value of plasma frequency of low-density plasmas.

However, it is noted that for the case when a_0 approached unity, the perturbation analysis to the density bunch is not valid. From (6.2.16), the density perturbation δn is evaluated to be

$$\left|\frac{\delta n}{n_0}\right| \sim \frac{1}{4} a_0^3$$

The density perturbation is about 100% for $a_0 > 1$. Therefore, the cancelation of the leading term described above is not the case for the relativistic intensity region of $a_0 > 1$.

6.2.5 Simulation for $a_0 = 10$ in Low Density

With increase of the field strength of lasers, nonlinear effects cannot be regarded as perturbative nonlinearity, and the laser propagation is affected significantly by nonlinear coupling with the background plasmas. One-dimensional PIC simulation was done in Ref. [2]. Let us pick up the important result seen in the simulation.

Circularly polarized laser with $a_0 = 10$, $\omega_{p0}/\omega = 1/5$, and pulse width of 125 c/ω which for a 1 μm laser correspond to a 1.4×10^{20} W/cm^2, 60 fs pulse propagating through a n = 4×10^{19} cm^{-3}. Time evolution of propagating laser pulse shape shows that the pulse front erosion is evident. It is mentioned that after the propagation of only 0.08 mm, the front of the laser pulse has started to deplete. Analysis of the backscattered radiation indicates that this initial depletion is primarily due to Raman scattering as explained in Chap. 4. The back-scattered radiation grows very rapidly, and its Fourier spectrum is downshifted.

In Fig. 6.7, the snap shots after the propagation of ct = 0.32 mm are plotted for (a) the transverse electric field, (b) longitudinal electric field, (c) plasma density, and (d) transverse vector potential [1]. In Fig. 6.7a, almost the front half of the laser pulse is eroded away, and it shows a sharp rise profile. The sharp rise of the depleted laser pulse push the fresh electrons forward to form an electron density spike as seen in Fig. 6.7c. The high-density electron bunching at the front induces strong electrostatic field via charge separation as seen in Fig. 6.7b. It is also observed that the sharp front

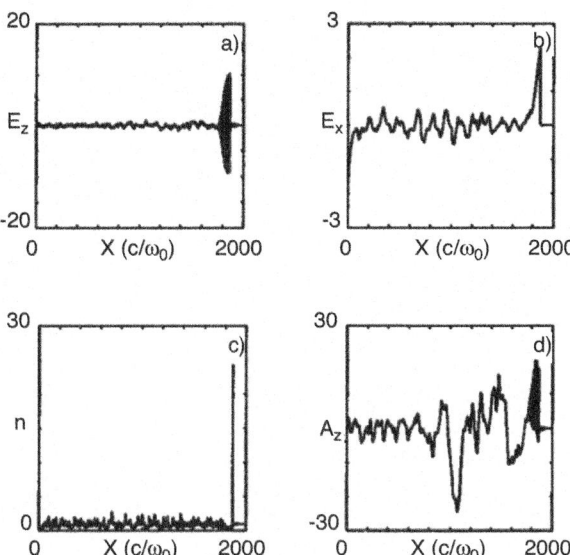

Fig. 6.7 (a) Transverse electric field $eE_z/m\omega c$, (b) longitudinal electric field $eE_x/m\omega c$, (c) electron density n/n_0, and (d) vector potential eA_z/mc versus x for ct = 50.32 mm from 1D simulation. [Figure 3 in Ref. 2]

of the laser excites a large amplitude plasma wake and large density spike as simply discussed and evaluated previously. The electric field by the plasma wake is seen in the region after the passage of the laser pulse. In addition, the low-frequency (wavenumber) transverse waves are seen after the passage of the laser. These photons are photon cascade and condensation [3] as seen below.

In Fig. 6.8, Fourier spectrum of the fields and particle momentum distributions are plotted at the same time as Fig. 6.7. In Fig. 6.8a, the laser spectrum at the initial time is plotted with dotted line, while it develops to the spectra of transverse electric field and vector potential as plotted with solid lines. Note that the electric field is lower than the vector potential in lower k region ($ck/\omega < 1$), while the vector potential is lower in the high-frequency region ($ck/\omega > 1$). The photon number [$n_p(k)$] at each frequency has the relation:

$$n_p(k) \propto \frac{E_k^2}{k} \propto kA_k^2 \qquad (6.2.19)$$

In Fig. 6.8a, the lower frequency radiation is due to photon deceleration, and as a result, the number of photons is roughly conserved in the lower frequency region. It is said that the number of photons in the system generated by stimulated cascade is conserved as roughly expected from Fig. 6.8a and (6.2.19). The physical reason is explained in Ref. [3] that nonlinear interplay between backward and forward stimulated Raman scattering and relativistic modulation instabilities produces strong spatial modulation of the laser pulse and the down cascade (inverse cascade) in the light frequency spectrum.

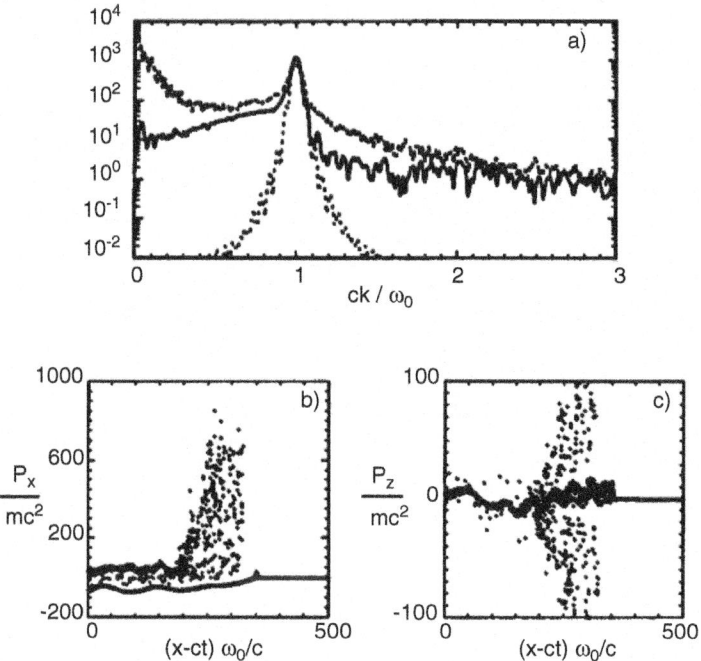

Fig. 6.8 (a) K-spectrum of transverse electric field $eE_z/m\omega c$ (solid line) and transverse vector potential eA_z/mc (dashed line) for $ct = 50.32$ mm from 1D simulation. For comparison we also plot the initial laser pulse spectrum (dotted line). (b) Longitudinal electron momentum p_x/mc and (c) transverse electron momentum p_z/mc versus x-ct. [Figure 4 in Ref. 2]

Of course, higher-frequency components are also generated, for example, as the anti-Stokes waves generated via Raman scattering and the other nonlinear coupling with the wake plasma waves. It is noted that HHG described previously is not seen in such system where plasmas are highly nonlinear and electrons are in chaotic motions. This was suggested from Fig. 6.8d.

In Fig. 6.8b, longitudinal electron momentum (p_x) is plotted for particles versus position x after the laser pulse propagated 0.32 mm. It is seen that the electrons are accelerated up to $p_x = 800$ mc (400 MeV). Since the peak field strength of the plasma wake is roughly $eE_x/(mc\omega) = -0.3$ and the acceleration distance is about 2000 c/ω from Fig. 6.7, a simple calculation is

$$\Delta E = eE_x L \approx 0.3 \times 2000 \ mc^2 \tag{6.2.20}$$

The obtained energy is about $600\ mc^2$, good agreement with the computational result in Fig. 6.8b. This agreement suggests that the maximum acceleration energy is not given by the dephasing between wake filed and particle orbits, so called **wave-breaking**.

In Fig. 6.8c, the transverse electron momentum p_z is plotted same as the longitudinal one. It is seen that the bulk electrons are oscillating in the z-direction even after the passage of laser pulse. This also indicates that the low-frequency electromagnetic waves are left behind the laser pulse with large amplitudes to keep the oscillation of p_z. In addition, some electrons accelerated by the wake field are rotating along the propagation axis by the fields of condensed low-frequency electromagnetic waves.

6.3 Ponderomotive Force in Relativistic Field

The ponderomotive force (PM force) is the force appearing after the time average of the force by oscillating field. This force is due to the inhomogeneity of the field strength and is very important to plasma dynamics interacting with short pulse and spatially lasers. The PM force is obtained as follows in the case of electromagnetic fields.

Begin with the equation of motion of an electron shown in (5.3.4):

$$\frac{d\mathbf{p}}{dt} = \frac{\partial \mathbf{p}}{\partial t} + \mathbf{v} \cdot \nabla \mathbf{p} = e\frac{\partial \mathbf{A}}{\partial t} - \frac{e}{m\gamma}\mathbf{p} \times (\nabla \times \mathbf{A}) \tag{6.3.1}$$

Since the generalized momentum **P** defined by

$$\mathbf{P} = \mathbf{p} - e\mathbf{A}$$

is conserved in the total system of an electron and fields, **A** in the second term in (6.3.1) can be replaced with **p**. Then, (6.3.1) reduces to

$$\frac{\partial \mathbf{p}}{\partial t} = e\frac{\partial \mathbf{A}}{\partial t} - \frac{1}{m\gamma}[\mathbf{p} \cdot \nabla \mathbf{p} - \mathbf{p} \times (\nabla \times \mathbf{p})] \tag{6.3.2}$$

Using a mathematical relation:

$$(\mathbf{p} \cdot \nabla)\mathbf{p} = \frac{1}{2}\nabla|\mathbf{p}|^2 + \mathbf{p} \times (\nabla \times \mathbf{p}), \tag{6.3.3}$$

(6.3.2) reduces:

$$\frac{\partial \mathbf{p}}{\partial t} = e\frac{\partial \mathbf{A}}{\partial t} - \frac{1}{2m}\gamma^{-1}\nabla|\mathbf{p}|^2. \tag{6.3.3a}$$

Taking the time average of (6.3.3a) over laser oscillation < > to remain only slowly varying terms, the following simple relation is obtained:

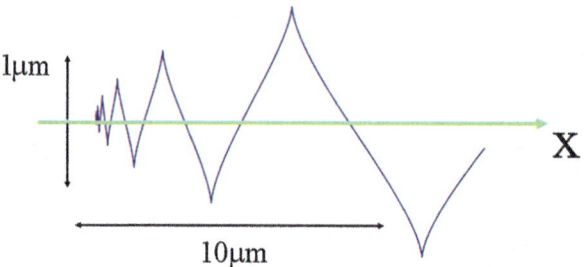

Fig. 6.9 An electron orbit is plotted when a linearly polarized laser with a Gaussian profile of 30 fs width and intensity 10^{22} W/cm^2 is passing through an electron initially at rest. The particle is accelerated by the ponderomotive force before the laser peak intensity arrives

$$\frac{\partial \langle \mathbf{p} \rangle}{\partial t} = -\frac{1}{2m} \left\langle \gamma^{-1} \nabla |\mathbf{p}|^2 \right\rangle$$
$$= -\left\langle \frac{mc^2}{2\gamma} \nabla (\gamma^2 - 1) \right\rangle = -mc^2 \nabla \langle \gamma \rangle \quad (6.3.4)$$

Since $\nabla \langle \gamma \rangle$ is defined at each fixed position and is now the physical quantity of a new field. Therefore, a local force acting to an electron much slower than laser oscillation time is given in the following form of potential already shown in (1.3.20):

$$\mathbf{f}_{PM} = -\nabla U_P, \quad U_P = mc^2(\langle \gamma \rangle - 1) \quad (6.3.5)$$

This force is called **ponderomotive force** and U_p is called **ponderomotive potential**. It is clear that the ponderomotive potential is the time averaged kinetic energy of an oscillating electron. If the amplitude of the vector potential change slowly compared to the oscillation period and give as $a_0 = a_0(x, t)$, the Lorentz factor can be replaced with a_0 by assuming the plasma case (5.3.30) as

$$\langle \gamma \rangle = \sqrt{1 + \frac{1}{2} a_0^2} \quad (6.3.6)$$

More simply the PM force is expressed in the form:

$$\mathbf{f}_{PM} = -mc^2 \nabla \langle \gamma \rangle \quad (6.3.7)$$

In Fig. 6.9, an electron orbit is plotted when a linearly polarized laser with a Gaussian profile of 30 fs and intensity 10^{22} W/cm^2 is passing through an electron initially at rest. It roughly shows a typical motion given in (5.3.24) and the amplitude increase according to the increase of the laser intensity. It is clear that the time averaged position shifts to the laser propagation direction and its velocity increases with the increase of laser intensity. This is regarded due to the ponderomotive force, and oscillation motion is due to the oscillating electric field.

6.3.1 PM Force by Electrostatic Wave

Let us consider the case of strong field produced by plasma waves. As mentioned previously, strong plasma waves are produced as wake field along with the propagation of intense laser. The equation of motion of an electron in such electrostatic field $\mathbf{E}(x,t)$ is given as

$$\frac{d\mathbf{p}}{dt} = \frac{\partial \mathbf{p}}{\partial t} + \mathbf{v} \cdot \nabla \mathbf{p} = -e\mathbf{E} \tag{6.3.8}$$

From (6.3.3), the advection term can be modified like

$$\mathbf{v} \cdot \nabla \mathbf{p} = \frac{1}{m\gamma}(\mathbf{p} \cdot \nabla)\mathbf{p} = \frac{mc^2}{2\gamma} \nabla (\gamma^2 - 1) = mc^2 \nabla \gamma \tag{6.3.9}$$

Taking the time average of (6.3.8), the following PM potential is obtained:

$$U_P = mc^2(\langle \gamma \rangle - 1) \tag{6.3.10}$$

In using the vector potential amplitude defining $\mathbf{E}$, the time averaged Lorentz factor has the same form as (6.3.6) like

$$\langle \gamma \rangle = \sqrt{1 + \frac{1}{2} a_0^2} \tag{6.3.11}$$

Since the PM forces of electromagnetic and electrostatic waves have the same form, the form (6.3.7) can be used any situation of the mixture of the both waves.

6.3.2 Validity of Electrons in Plasmas Assumption

In Sect. 5.3, it is assumed that no drift motion is better for analyzing electron motions in plasmas. This is because the electrostatic field by charge separation pulls electron to cancel the drift motion. Explain the drift velocity in (5.3.25) from the additional force loading to free electrons by ponderomotive force.

Including the ponderomotive force in (6.3.4), (6.3.11) may change the following form:

$$\frac{d}{dt}(p_x - \gamma mc) = -\frac{\partial}{\partial x} U_p \tag{6.3.12}$$

Time integration of (6.3.12) reduces as follows:

$$\int \frac{\partial}{\partial x} U_p dt = \int \frac{dt}{dx} \frac{\partial}{\partial x} U_p dx = \frac{1}{c} U_p \qquad (6.3.13)$$

Before the laser comes to an electron, evaluating the integral constant of (6.3.12), the following relation is obtained. Integrating to time (6.3.12), the following relation is obtained:

$$\widehat{p}_x - \gamma \big|_{-\infty}^{t} = -\widehat{U}_p \big|_{-\infty}^{t}$$

Considering the initial condition that laser is before interacting with an electron, we obtain

$$\widehat{p}_x - \gamma = -\langle \gamma \rangle$$

From (6.3.12), it is clear that including the ponderomotive force in the equation of motion we naturally conclude that

$$\alpha = \langle \gamma \rangle$$

6.4 Relativistic Raman Scattering

During an extremely short time for tens or hundreds of laser oscillation periods, sub-picosecond, only electrons can response to the laser fields, and Raman scattering becomes dominant in the parametric coupling in under-dense laser plasma interaction. As has been well described in Chap. 4, the Raman scatterings are induced through the coupling among three waves, incident laser, scattered electromagnetic waves, and the plasma waves. The plasma waves are induced by the nonlinear currents due to the ponderomotive force by the beat wave of incident and scattered waves. We can easily obtain the growth rate of the Raman scattering by changing the ponderomotive potential from non-relativistic form to the relativistic form, namely,

$$U_p^{NR} = \frac{1}{2} m \langle V_{os}^2 \rangle \quad \Rightarrow \quad U_p^R = mc^2 \langle \gamma_{os} \rangle, \qquad (6.4.1)$$

where V_{os} is the quivering velocity and γ_{os} is Lorentz factor of oscillating motion. Note that both oscillations include those by laser and scattered wave fields.

It is reasonable to assume that in analyzing the growth rate of the Raman instability, the amplitude of scattered wave is small enough. Evaluate the total ponderomotive force by laser, suffix "0", and scattered wave, suffix "1"

6.5 Relativistic Self-Focusing

$$\gamma_{os} = \left[\hat{p}_{x0}^2 + \left(\hat{p}_{y0} + \hat{p}_{y1} \right)^2 + 1 \right]^{1/2} \tag{6.4.2}$$

The ponderomotive force by the beat wave of laser and scattered wave is obtained by Taylor expansion as

$$\langle \gamma \rangle_{beat} = \frac{a_0 a_1}{\sqrt{a_0^2/2 + 1}} \tag{6.4.3}$$

Replacing the nonlinear coupling term obtained in non-relativistic case with (6.4.3), we obtain the growth rate of Raman scattering in the form:

$$\gamma_{SRS}^{R} = \frac{\omega_{p0} |\cos\theta|}{2\sqrt{\omega_{ek}(\omega_0 - \omega_{ek})}} kc \frac{a_0}{\sqrt{a_0^2/2 + 1}}, \tag{6.4.4}$$

where θ is the angle between the laser and scattered wave. From matching condition of three waves, the wavenumber k is smaller than the laser wavenumber k_0 for the forward scattering, while it is larger than the backward scattering. This is the reason why the backscattering is dominant compared to the forward scattering. Since the term kc in (6.4.4) is about the laser frequency ω_0 and the last term is almost unity for relativistic intensity, therefore, the growth rate is roughly evaluated as

$$\gamma_{SRS}^{R} \approx \left(\frac{\omega_{p0}}{\omega_0} \right)^{1/2} \omega_0 \tag{6.4.5}$$

This means that Raman scattering grows over several cycle of laser oscillation, the order of 10 fs.

If the plasma is long enough, the induced electromagnetic fields can be the source wave to induce dominantly the backward Raman scattering. Repeating the backward scattering many times, the photons are confined in plasmas. Since the Raman scattering conserves the number of photons, we can expect the photon cascade and condensation in lower frequency region as shown in Fig. 6.8.

6.5 Relativistic Self-Focusing

The refraction index of electron plasmas to laser field is given in the form:

$$N^2 = 1 - \frac{\omega_{pe}^2}{\omega_0^2}, \quad \omega_{pe}^2 \propto \frac{n_e}{m_e} \tag{6.5.1}$$

Fig. 6.10 Schematics of the radial distribution of the laser intensity (red) and resultant depression of the density and the inverse of the electron mass $1/m(\gamma)$, blue and black

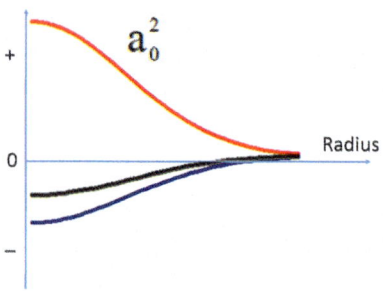

Relativistic mass correction increases the electron mass in the region of higher intensity of laser. The ponderomotive force is the force to repel the electron from the region where the laser intensity is higher. Assume that the electron mass change is Δm and the density change is Δn. It is clear that for a cylindrical beam propagating uniformly over long enough with the radial intensity distribution as shown in Fig. 6.10, the mass and density perturbations can be plotted as shown there.

The equation of laser propagation to a normalized laser field **a** is given in the form to a low-intensity laser:

$$\frac{\partial^2 \mathbf{a}}{\partial t^2} - c^2 \nabla^2 \mathbf{a} + \omega_{p0}^2 \mathbf{a} = 0 \tag{6.5.2}$$

This form is the same for linear or cylindrical polarizing lasers, and we solve (6.5.2) for scalar physical value a. The relativistic and ponderomotive effects modify the plasma frequency in (6.5.2) in the liner perturbation case:

$$\omega_{pe}^2 \propto \frac{n_e}{m_e} = \frac{n_0}{m_0}\left(\frac{\Delta n}{n_0} - \frac{\Delta m}{m_0}\right) \tag{6.5.3}$$

Since the density perturbation is negative, both effects work to reduce the plasma frequency in the region of higher laser intensity. As a result, the refractive index of the laser propagation channel becomes higher than that around the laser channel from (6.5.1). It is well-known that a light beam traveling a shallow angle near the surface with air is perfectly reflected at the surface. This is the phenomenon of total reflection. In the present case, the boundary is not surface, but refractive index changes continuously from the beam center to the outer radius. In addition, the profiles of the laser intensity, effective electron mass, and the electron density can change as a function of time. Let us find via rough evaluation the condition that the intensity increases in time due to the so-called **self-focusing**.

6.5.1 Self-Focusing Condition

Assuming that the laser beam is like Gaussian profile with the radius R, (6.5.2) can be reduced to

$$\frac{\partial^2 \mathbf{a}}{\partial t^2} - c^2 \frac{\partial^2 \mathbf{a}}{\partial x^2} + \frac{c^2}{R^2}\mathbf{a} + \omega_{p0}^2(1 - \delta_n - \delta_m)\mathbf{a} = 0,$$

where δ_n and δ_m are the normalized contributions by ponderomotive force and relativistic effect, respectively. The electron density perturbation in the radial direction induces the electrostatic field via charge separation to the ions at rest. We have to solve the electron fluid motion starting with the following equations to δn, flow velocity $\mathbf{u}$, and produced electrostatic field $\mathbf{E}$:

$$\frac{\partial \delta_n}{\partial t} + \nabla \cdot \mathbf{u} = 0 \tag{6.5.4}$$

$$m\frac{\partial \mathbf{u}}{\partial t} = -e\mathbf{E} - mc^2 \nabla \langle \gamma \rangle \tag{6.5.5}$$

$$\varepsilon_0 \nabla \mathbf{E} = -en_0 \delta_n \tag{6.5.6}$$

Taking time derivative of (6.5.4) and inserting (6.5.5) and (6.5.6), we obtain the following relation:

$$\frac{\partial^2 \delta_n}{\partial t^2} + \omega_{p0}^2 \delta_n = c^2 \nabla^2 \langle \gamma \rangle \tag{6.5.7}$$

Assume that the time evolution is much slower than the plasma frequency. This assumption means the ponderomotive force is always balanced by the electrostatic field by the charge separation; two terms in RHS in (6.5.5) cancel each other. Assuming the spatial derivative of $\langle \gamma \rangle$ is about the beam size R, the normalized density is given to be

$$\delta_n \approx \frac{\rho_s^2}{R^2}(\langle \gamma \rangle - 1), \tag{6.5.8}$$

where ρ_s is the **plasma skin depth** defined as

$$\rho_s = \frac{c}{\omega_{p0}} \tag{6.5.9}$$

The mass correction term is easily obtained as

$$\delta_m = 1 - \frac{1}{\langle\gamma\rangle} \quad (6.5.10)$$

For weak intensity limit, Taylor expanding (6.5.8) and (6.5.10) to the amplitude a_0, the divergence term and focusing terms by ponderomotive force and relativistic mass correction effect are given in the form:

$$\delta_r : \delta_n : \delta_m = \frac{\rho_s^2}{R^2} : \frac{\rho_s^2}{R^2}\frac{a_0^2}{4} : \frac{a_0^2}{4} \quad (6.5.11)$$

It is clear that in the low-intensity limit, the refraction is dominant, and the laser beam diverges, while the relativistic and ponderomotive term becomes dominant to focus the laser beam due to nonlinear change of the refractive index. This is called **self-focusing** phenomenon.

It is important to pay attention to the radius dependence of three contribution in (6.5.11). In the low density and $a_0 < 1$, the diffraction term dominates than the two focusing terms, and no self-focusing may not happen. However, as the increase of density and laser strength, the focusing terms become dominant to overwhelm the diffraction term. The physics of laser propagation appears to be complicated in propagating in the high-density with relativistic intensity.

It is important to point out that there is a critical laser power for self-focusing condition, P_{cr} in the limit of weak laser intensity ($a_0 \ll 1$). When the relativistic effect balances with the diffraction effect, the following condition should be satisfied:

$$a_0^2 R^2 = 4\rho_s^2, \quad P_{cr} \propto \omega^2 a_0^2 R^2 \propto \frac{n_{cr}}{n_0} \quad (6.5.12)$$

More precise calculation yields this critical power of cylindrical laser beams for self-focusing condition in the form:

$$P_{cr} = 17.5 \frac{n_{cr}}{n_0} \quad [\text{GW}] \quad (6.5.13)$$

This critical power is relatively low compared to the present-day power of ultra-short ultra-intense lasers, 100TW ~10PW. If such laser is irradiated on the plasmas, we have to note that the self-focusing always is taken place in the plasma.

6.5.2 Strong Laser Limit

The above evaluation is based on (6.5.4), where we have assumed that $a_0 \ll 1$. Let us consider the opposite case for $a_0 \gg 1$. In this limit, the ratio shown in (6.5.11) changes as

6.5 Relativistic Self-Focusing

$$\delta_r : \delta_n : \delta_m = \frac{\rho_s^2}{R^2} : \frac{\rho_s^2}{R^2} \frac{a_0}{\sqrt{2}} : 1 - \frac{\sqrt{2}}{a_0}, \quad (6.5.14)$$

where we note that $\delta_n < 1$ should be satisfied. The relativistic effect has a limit to overwhelm the diffraction effect at $R \sim \rho_s$, while the ponderomotive effect is dominant over the diffraction effect for $a_0 > 2$ even from (6.5.11), and the electron density is subject to empty in the laser channel for sufficiently strong laser field. This is called **hole boring**.

As we see above, the plasma skin depth defined in (6.5.9) is the critical length for discussing the physics of self-focusing of laser pulse.

6.5.3 Difference of 2D and 3D Focusing

It is important to note the fact that a planer 2D PIC simulation fails to predict the self-focusing phenomenon in some case. There are many studies with the assumption of two dimensions in geometry, say, assuming the physical system is uniform in the z-direction. Then, two-dimensional dynamics in the x-y plane is simulated for convenience of computation time. Some physics will be of no problem; however, qualitative difference is inherent in 2D compared to the real 3D dynamics. This self-focusing is a typical example because of the following reason.

In 2D plane geometry, most of the above analysis is the same. Replace R with $1/k_y$, where k_y is the typical wavenumber of filaments in the y-direction, provides the same equation as (6.5.4). However, the energy conservation in 2D is $P_0^{2D} \equiv a_0^2 L$ for $L = 1/k_y$, and the relation of relativistic and diffraction effects changes as

$$\delta_m - \delta_r \Rightarrow \left(1 - 4\frac{\rho_s^2}{P_0}\right) \Rightarrow \left(1 - 4\frac{\rho_s^2}{P_0^{2D}L}\right) \quad (6.5.15)$$

This represents that when the thickness of the beam approaches the critical value

$$L = \frac{1}{a_0}\rho_s, \quad (6.5.16)$$

the diffraction term becomes stronger than the nonlinear focusing, and the focusing terminates. This is **beam trapping**, instead of self-focusing. It is noted that this is technically achieved as glass fiber shown in Fig. 6.11. In the beam trapping, the diffraction and the refractive index structure is designed to balance to allow the long distant transport of beam in the glass fiber with constant intensity, brightness.

Fig. 6.11 A bunch of glass fibers. The glass fiber is manufactured so that the refractive index is high at the center in the glass fiber. Then, the incident light is totally reflected at the boundary of the central high refraction zone and the surrounding low refraction material

6.5.4 Frequency Shifts

It is noticed that the self-focusing and nonlinear frequency shift happen simultaneously, and the evaluation by paying attention only to the self-focusing shown above is difficult to apply to the analysis of the frequency shift of the laser in the traveling channel made by the nonlinearity of laser itself.

For the infinite plasma $R \to \infty$ in (6.5.4), the nonlinear terms seem to provide only the change of wavenumber k and phase velocity of the laser propagation.

$$\omega^2 - c^2 k^2 = \omega_{p0}^2 (1 - \delta_m) = \omega_{p0}^2 \frac{1}{\langle \gamma \rangle} \qquad (6.5.17)$$

Equation (6.5.17) is nothing without (6.2.2) showing the relativistic dispersion relation and transparency. In order to study the self-focusing and nonlinear frequency shift, we have to invoke more basic analysis for a nonlinear coupling of two weak electromagnetic waves propagating with slightly oblique angle along the laser beam. Then, we can also obtain the dynamic time scale of the self-focusing depending on the size of the intensity width of laser beam as see below.

6.5.5 Filamentation Instability

Let us evaluate the stability of the two waves propagating with the angles slightly oblique to the propagating laser wave due to the nonlinearity via the relativistic mass correction. Assume the original wave has $\mathbf{a} = \mathbf{a}_0$ and these two oblique waves are given as $\mathbf{a}_1$ and $\mathbf{a}_2$, namely,

$$\mathbf{a} = \mathbf{a}_0 + \mathbf{a}_1 + \mathbf{a}_2 \qquad (6.5.18)$$

The condition $|\mathbf{a}_0| > > |\mathbf{a}_1|, |\mathbf{a}_2|$ is also satisfied.

Inserting a linear polarized plane wave into the 0-th wave $\mathbf{a}_0$ with $k = k$ and $\omega = \omega_0$. Then, assuming weak nonlinearity such as

6.5 Relativistic Self-Focusing

$$\frac{1}{\gamma} \approx 1 - \frac{1}{2}\mathbf{a}^2, \qquad (6.5.19)$$

the following simple relation is obtained for the nonlinear term for $\mathbf{a}_1 = \mathbf{a}_2 = 0$.

$$|\mathbf{a}|^2 = a_0^2 \cos^2\varphi_0 = \frac{a_0^2}{2}\{1 + \cos(2\varphi_0)\}, \qquad (6.5.20)$$

where

$$\cos(2\varphi_0) = \frac{1}{2}\left(e^{2\varphi_0} + e^{-2\varphi_0}\right)$$
$$\varphi_0 = k_0 x - \omega_0 t \qquad (6.5.21)$$

The first term in the brackets on RHS in (6.5.20) is a nonlinear modification of the dispersion relation, and the second term is $2\omega_0$ oscillation term.

Obtain the relation for the two waves propagating with the wavenumber in the y-direction $+k_y$ and $-k_y$ and the frequency $\omega_0 + \delta$. So, their phases are

$$\varphi_1 = (k_0 x + k_y y) - (\omega_0 + \delta)t$$
$$\varphi_2 = (k_0 x - k_y y) - (\omega_0 - \delta)t \qquad (6.5.22)$$

Equation (6.5.2) can be reduced to the following coupled two wave equations for the oscillation with φ_1 and φ_2, respectively.

$$\left[(\omega_0 + \delta)^2 - c^2\left(k_0^2 + k_y^2\right) - \omega_{p0}^2\right]\mathbf{a}_1 = \frac{a_0^2}{4}\omega_{p0}^2 \mathbf{a}_1 + \frac{a_0^2}{8}\omega_{p0}^2 e^{2\varphi_0}\mathbf{a}_2$$
$$\left[(\omega_0 - \delta)^2 - c^2\left(k_0^2 + k_y^2\right) - \omega_{p0}^2\right]\mathbf{a}_2 = \frac{a_0^2}{4}\omega_{p0}^2 \mathbf{a}_2 + \frac{a_0^2}{8}\omega_{p0}^2 e^{2\varphi_0}\mathbf{a}_1 \qquad (6.5.23)$$

It is noted that LHSs of (6.5.23) are conventional linear dispersion relations to the waves propagating obliquely along the laser beam, while the first terms on RHS are the nonlinear modification to the linear dispersion relation due to the relativistic mass correction, and finally the second terms on RHS are the nonlinear coupling term between small perturbing waves propagating obliquely along the laser beam. These coupling terms appeared through the nonlinear coupling with the fundamental mode, the laser beam $\mathbf{a}_0$.

It is clear that (6.5.23) has non-trivial solution when the determinant for $(\mathbf{a}_1, \mathbf{a}_2)$ vanishes. It is found that the determinant has the frequency shift δ only in the form δ^2, and we set

$$Y = \delta^2 \qquad (6.5.24)$$

Then, we obtain the dispersion relation to the coupled waves:

$$Y^2 - \left(\omega_0^2 + 2c^2 k_y^2\right) Y - D = 0, \quad (6.5.25)$$

where

$$D = \left(\frac{a_0^2}{8} \omega_{p0}^2 - c^2 k_y^2\right) \left(\frac{a_0^2}{8} \omega_{p0}^2 + c^2 k_y^2\right) \quad (6.5.26)$$

It is clear that (6.5.25) has a negative solution $Y = \delta^2 < 0$ for $D > 0$. The negative solution means the frequency shift δ has pure imaginary and the system (6.5.22) becomes unstable.

The condition for the two waves unstable is the condition of the **filamentation instability**. Two waves consist of the standing wave in the y-direction as

$$\mathbf{a}_1 + \mathbf{a}_2 \Rightarrow 2 \exp(|\delta| t) \cos(k_y y) \cos(k_0 x - \omega_0 t) \quad (6.5.27)$$

The growth of the amplitude is due to the energy flow of the fundamental wave, namely, the plane homogeneous wave of the laser which is going to be modified sinusoidally in the perpendicular direction with exponential increase of its amplitude.

The condition of $D > 0$ provides a new condition for the beam size to be self-focusing:

$$k_y < \frac{a_0}{2\sqrt{2}} \frac{1}{\rho_s} \quad (6.5.28)$$

With increase of the laser intensity and/or density of plasmas, the critical beam size (6.5.28) gets smaller. It is informative to compare this critical beam size for filamentation instability with that obtained from (6.5.11). The condition that the relativistic mass correction is stronger than the diffraction effect gives the critical radius in the form from (6.5.11):

$$\frac{1}{\sqrt{2} R} < \frac{a_0}{2\sqrt{2}} \frac{1}{\rho_s} \quad (6.5.29)$$

It is a good coincidence that by replacing an effective wavenumber k_y with $1/\sqrt{2}R$, rough evaluation in (6.5.11) coincides with the precise analytic result.

For the case the wavenumber k_y is much smaller than the critical value (6.5.28), we can obtain the growth rate of such filamentation instability, $|\delta| \equiv \gamma_{FI}$ in an approximate form:

6.6 Relativistic Skin Depth

$$\gamma_{FI} = \frac{a_0^2}{8} \frac{\omega_{p0}^2}{\omega_0} \propto a_0^2 n_0 \qquad (6.5.30)$$

This growth rate indicates that with increase of laser intensity and the plasma density, the growth rate increases. In the present analysis, we have assumed that $a_0 \ll 1$, while this theory is valid even for $a_0 \sim 1$ or $a_0 > 1$, since the nonlinear coupling, viz., $2\omega_0$ oscillation, which can be kept even for a_0 gets large as shown in (6.2.7). For higher a_0, there would be another nonlinearity stemming from higher-order oscillation or ponderomotive effect which appears to be important and may couple to (5.3.23) in positive or negative contribution to the filamentation instability.

6.6 Relativistic Skin Depth

The dispersion relation of relativistic lasers propagating in plasmas is given in (6.2.2), and it has the plasma dielectric constant $\varepsilon(k,\omega)$ in the form:

$$\varepsilon(k, \omega) = 1 - \frac{\omega_{p0}^2}{\gamma_0 \omega^2} \qquad (6.6.1)$$

Here we assume that the dielectric media is immobile (electron oscillation amplitude is infinitesimally small), and its spatial distribution is fixed. Except the relativistic term γ_0, the skin depth of the penetration of the laser field into the solid plasma can be obtained by the same process as non-relativistic case. For the normal incidence, the amplitude of the reflecting electric field at the solid-vacuum boundary $x = 0$ is obtained.

$$\frac{a_s}{a_0} = \frac{\gamma_0 \omega^2}{\omega_{p0}^2}, \qquad (6.6.2)$$

where a normalized amplitude a_s is the value at $x = 0$. The amplitude of electric field summed by incident and reflected fields and its x-derivative should be continuous at the solid surface $x = 0$. The field penetrates into the solid ($x > 0$) in the form:

$$a(x) = a_s \exp(-Kx), \qquad (6.6.3)$$

where the skin depth λ_s is defined as

$$1/\lambda_s \equiv K = k_0 \sqrt{-\varepsilon} \qquad (6.6.4)$$

In the limit that the solid density is much higher than the critical density, the **relativistic skin depth** is given in the approximate form:

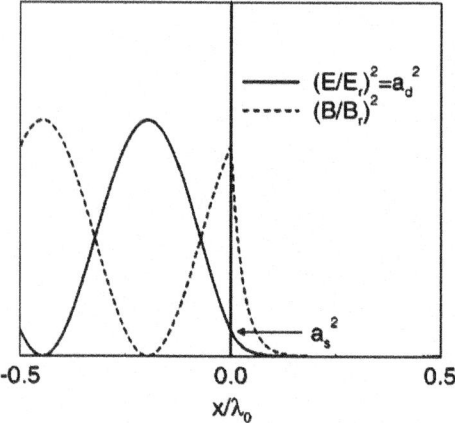

Fig. 6.12 Scaled electric and magnetic field of the standing wave at the plasma surface. A skin depth $\lambda_0/2\pi$ is assumed, corresponding to $n_0/n_c = 9$. [Figure 12 in Ref. 4]

$$\lambda_s \approx \sqrt{\gamma_0}\frac{c}{\omega_{p0}} = \sqrt{\gamma_0}\rho_s \tag{6.6.5}$$

It is noted that the skin depth is extended by the relativistic effect compared to the non-relativistic plasma skin depth $\rho_s = c/\omega_{p0}$.

The magnetic field in the solid region is easily obtained as

$$B_z = \frac{\partial A}{\partial x} = \frac{m}{e}\frac{\partial a}{\partial x} \tag{6.6.6}$$

Note that RHS in (6.6.6) is continuous at the sharp boundary $x = 0$ and the magnetic field also continuously changes across the boundary; however, the derivative is in general not continuous due to a surface current.

It is noted that to obtain the field structure from the vacuum to the solid region, it is necessary to solve wave equation including the phase of the waves. As an example, for the case of $n_0/n_c = 9$ and assumed skin depth $\lambda_0/2\pi$, the normalized electric field and magnetic field amplitudes are plotted as shown in Fig. 6.12 [4]. The skin depth is much shorter than the wavelength λ_0 in typical solid density plasmas. The following point should be noted. The relativistic skin depth is derived from the electron current in (6.2.2) or induced current **j** in (6.2.1), where we have assumed electron fluid model in evaluating the induced current. However, the electron motion at sharp density boundary is not in the direction of **A**. The JxB force pushes them to the forward direction to produce hot electrons. The appearance of oscillating current also induces electromagnetic field near the boundary, and real laser field structure may be complicated. So, the skin depth derived with local dielectric constant is not physically meaningful in discussing the relativistic laser dynamics.

6.7 JxB Force and Heating

When a linearly polarized laser with relativistic intensity is irradiated on a solid surface, it is well-known that the **v** × **B** force works to the direction of laser propagation. Since the velocity and magnetic field oscillate with the laser frequency ω, the ponderomotive force shown in (6.3.4) gives the momentum to the electrons near the surface. In addition, it is clear that the force oscillating with 2ω also works to the electrons. The latter force is called **JxB force** [5].

This force is calculated in general as follows:

$$\mathbf{f} = -e\mathbf{v} \times \mathbf{B}$$
$$\mathbf{v} = \frac{\mathbf{p}_\perp}{m\gamma} = -\frac{e}{m\gamma}\mathbf{A} \qquad (6.7.1)$$

Therefore, the force **f** reduces to

$$\mathbf{f} = -mc^2 \nabla\left(\sqrt{a^2 + 1}\right), \qquad \mathbf{f} = -mc^2 \nabla \gamma \qquad (6.7.2)$$

Compare (6.7.2) to (6.3.10). In weak relativistic case ($a^2 < <1$), the ponderomotive and the force **f** are clearly derived from (6.7.2) in the form:

$$\mathbf{f} = -\frac{mc^2}{4} \nabla |a_0(\mathbf{r})|^2 [1 + \cos(2\omega t)] \qquad (6.7.3)$$

The first term in RHS of (6.7.3) is the ponderomotive force defined in (6.3.5), while the second term is JxB force. Inserting sinusoidal oscillation of a in (6.7.2) for arbitrary amplitude, the relativistic nonlinear forces are given:

$$\mathbf{f} = -mc^2 \nabla \left[1 + \frac{1}{2}|a_0(\mathbf{r})|^2 (1 + \cos(2\omega t))\right]^{1/2} \qquad (6.7.4)$$

Note that in the relativistic case, not only 2ω but also 4ω and other higher harmonic forces appear. In the limit of highly relativistic intensity, (6.7.4) is approximated to

$$\mathbf{f} = -\frac{mc^2}{\sqrt{2}} |\cos(\omega t)| \nabla |a_0(\mathbf{r})| \qquad (6.7.5)$$

It is obvious that when the laser is irradiated on the solid surface, the **ponderomotive force** keeps pushing the electrons toward the inside of the solid. In the limit of ultra-short pulse, we can assume that the ions are immobile, and the charge separation creates the ambipolar electrostatic field as mentioned in Sect. 6.1. Including the pull-back force by the electrostatic field, the equation of the electron at the solid surface is approximately given from (6.1.10) in the form in the non-relativistic limit:

$$\frac{d^2 p_x}{dt^2} = -\omega_{p0}^2 p_x + \frac{d}{dt} f_x, \quad (6.7.6)$$

where x is the target normal direction and f_x is the x-component of (6.7.3). For simplicity, assume that the solid electron density is much higher than the cutoff density. Then, we can find the following oscillating solution of the electrons at the sharp density boundary with the position X(t) in the form:

$$X(t) \approx -\frac{\rho_s^2}{4} \frac{\partial |a_0(x)|^2}{\partial x} [1 + \cos(2\omega t)] \quad (6.7.7)$$

It is clear that the average position of the surface shifts inside the solid and it oscillates with frequency 2ω. Equation (6.7.7) indicates that the oscillation distance X(t) is larger with lower density of solid and/or higher laser intensity and depends on the realistic value of the penetration of the laser field, or we can regard it is the size of electron layer affected by the laser force. It will not be appropriate to assume the size is the skin depth in relativistic laser intensity in (6.6.5). Note that the inclusion of LHS in (6.7.6) alters the fraction of the solution (6.7.7) in the form:

$$X(t) = -\frac{\rho_s^2}{1 - 4\gamma\omega^2/\omega_{p0}^2} \frac{\partial \gamma}{\partial x} \quad (6.7.8)$$

In the above derivation of the JxB force, the realistic value is still open and depends on the effective laser intensity distribution in the solid, $a_0(x)$, namely, how the laser penetrates into the solid material and gives the force to bunch of electrons. If the laser intensity is weak enough so that the solid is not affected by any laser force, the penetration of the laser intensity via the skin effect as seen in Sect. 6.6 is reasonable. In contrast, however, the laser intensity is high, and the nonlinear force induces the electron motion into the solid; it is not clear how many electrons are affected by the laser force. For example, relativistic fluid equations are solved to the electron fluids in immobile ion system [6]. In this model, the electron density is assumed to be in an exponential form with the density slope of the skin depth to the solid region and a sharp boundary on the vacuum boundary. With this assumption, the trajectory of the boundary is easily calculated.

However, fluid description to electrons is not realistic at relativistic intensity regime. Let us see an example of 2D PIC simulation, where a cupper foil of 2.4 μm thickness is irradiated from the left by a linearly polarized short pulse laser with its half width of 40 fs and the maximum intensity of 10^{20} W/cm^2 ($a_0 = 6.8$) [7]. The electron density ratio of the solid density to the critical density is 1400, a realistic value. Laser is irradiated normally on the solid surface with focusing diameter of 4 μm. In Fig. 6.13, a snapshot of electron momentum at the laser center axis x is shown to clarify how JxB force produces high-energy (relativistic) electrons. Such electrons are also called, in general, **hot electrons**. It is clear that they are generated at the left and penetrate into the solid target. The electron distribution

6.7 JxB Force and Heating

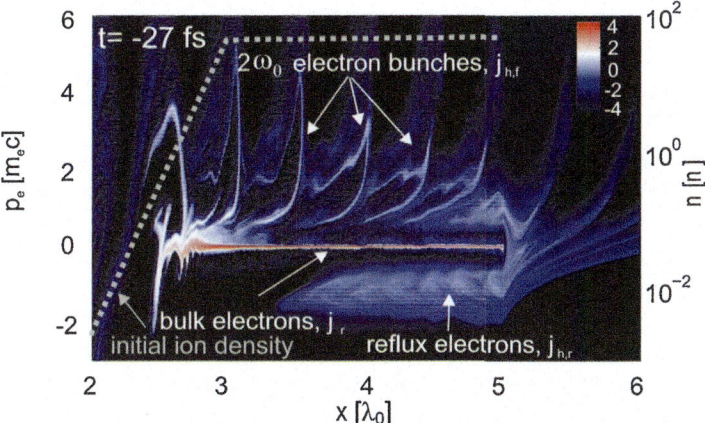

Fig. 6.13 A snap shot of electron momentum distribution in space, when a relativistic laser is irradiated on solid surface from the left. The JxB force produced the accelerated electron flux going inward at $2\omega_0$ frequency. Such intermittent relativistic electron flux is reflected by sheath potential at the rear side of the target so that the reflux electrons and the bulk electrons drift toward the left so as to keep charge neutrality of the system. [Figure 2 in Ref. 7]

function of the momentum p_x is shown as a function of target depth x along the laser beam center. It is clarified that the relativistic electrons are produced with 2ω cycle by the JxB force. Such electron bunches propagate into the solid target almost with the speed of light, and each of bunch is partially reflected backward at the rear (right boundary) of the foil by the ambipolar electric field generated by the hot electrons.

This computational result indicates that in high-intensity regime, the fluid description of the electrons is not acceptable, and laser energy is converted as the energy of the hot electrons in the vicinity of the solid surface, and these electrons run into the solid while passing the cold electrons in the solid. This is because the hot electrons are collision-free particles even in the solid density. Consider the physics of what happens during each cycle of 2ω oscillation, during a time of one cycle $\tau = 2\pi/2\omega$. A fraction of electrons near the surface are accelerated as hot electrons by the JxB force and escape into the inside of the solid over a half cycle ($0 < t < \tau/2$). Then, the direction of the JxB force changes to the opposite, outward direction from the solid surface. In addition, the sheath field generated by the escape of the hot electrons attract the bulk electrons to the surface region during $0 < t < \tau/2$. Then, during the time ($\tau/2 < t < \tau$), the electron-poor region near the surface is filled with the cold bulk electrons. These bulk electrons will be the electrons to be accelerated in the next half cycle ($\tau < t < 3/2\tau$). The physical process described above is repeated in each cycle to generate the cycle electron bunch of the hot electrons shown in Fig. 6.13.

Let us check if such physical process is correct or not. In this process, electrons near the surface are accelerated by the JxB force. It is reasonable to assume that the maximum energy that the hot electrons obtain by the force in (6.7.4) is

$$P_{JxB} \sim mc(\gamma)_{2\omega} \sim \frac{mc}{\sqrt{2}} a_0, \tag{6.7.9}$$

where we took the potential difference over 2ω oscillation of the force, that is, γ. Since $a_0 = 6.8$, the accelerated momentum may have $p_x \sim 4.8$ mc, roughly in agreement with the value in Fig. 6.13. Such scaling law for the hot electron temperature

$$T_h \approx \Delta E = mc^2(\gamma - 1), \quad \gamma = \left(\frac{1}{2}a_0^2 + 1\right)^{1/2} \tag{6.7.10}$$

is called **Ponderomotive scaling** of hot electron temperature. It is noted that such simple scaling provides a rough estimate of the temperature of hot electrons. As will be discussed below, however, this simple scaling is applicable only to limited cases. Another electron acceleration physics becomes important in the other cases. Especially in the case of longer pulse, large spot size, and targets with pre-formed plasmas, the further interaction of re-circulating hot electrons with laser fields and longitudinal plasma fields heats again and again the hotel electrons to increase the hot electron temperature even for the same laser intensity.

6.8 Moving Mirror Model and Higher Harmonic Generation from Solid Surface

In the case where the solid surface reflecting an incident laser is oscillating by the laser force such as JxB force, the frequency of reflected laser is affected by the Doppler shift as schematically shown in Fig. 6.14. Then, it is well-known that the reflected light by relativistic oscillating surface has a group of **higher harmonics (HH)**. The physical model starting with a relativistic oscillating mirror by the laser

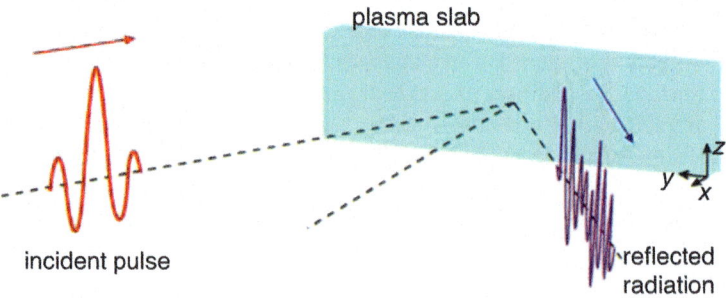

Fig. 6.14 Schematics of the relativistic laser reflection from a sold surface. Due to the motion of the surface by the laser pressure, the reflected laser frequency is highly modulated including higher harmonic components. [Figure 1 in Ref. 10]

6.8 Moving Mirror Model and Higher Harmonic Generation from Solid Surface

force is called **relativistic oscillating mirror** (ROM) model first proposed by Bulanov et al. [8]. Consider the Doppler shift when a laser coming from the left is reflected by a perfect mirror moving with a velocity V_m (> 0 for the case of moving to the right). Then, the frequency of the light in the frame moving with the mirror is Doppler shifted as in (5.2.44):

$$\omega' = \sqrt{\frac{1-\beta_0}{1+\beta_0}}\omega, \qquad \beta_0 = V_m/c \tag{6.8.1}$$

This is a red shift for the case $V_m > 0$. The reflected laser should have the same frequency as (6.8.1) in the moving frame. When the reflected light is observed in the laboratory frame, the reflected light is again red-shifted, and the frequency is given in the laboratory frame as

$$\omega'' = \sqrt{\frac{1-\beta_0}{1+\beta_0}}\omega' = \frac{1-\beta_0}{1+\beta_0}\omega \tag{6.8.2}$$

In the surface being pushed by the JxB force, the mirror is oscillating, and the maximum frequency in (6.8.2) is obtained at the negative peak of $V_m < 0$. The maximum frequency ω_{max} of the reflected light is given to be

$$\omega_{max} = \frac{1+\beta_0}{1-\beta_0}\omega \approx 4\gamma_0^2\omega, \tag{6.8.3}$$

where Lorentz factor γ_0 is defined by the amplitude of oscillation velocity V_m.

The above simple evaluation seems to suggest that the reflected light is continuum with the maximum frequency in (6.8.3). It is, however, shown mathematically that the reflected light is discretized like a combination of the fundamental and its higher harmonic components. For the perfect reflection mirror, there is no filed penetration into the solid region. That is, the following boundary condition should be satisfied for the vector potential A of the incident A_i and reflected A_r components, where both are complex values:

$$A_i \; \exp\left[i(kx - \omega t)\right] + A_r(x,t) = 0 \quad \text{at } x = X_m(t), \tag{6.8.4}$$

where X_m is the position of the moving mirror surface. As shown above, it is reasonable to assume that the mirror surface is oscillating with a given frequency Ω:

$$X_m(t) = X_0 \sin(\Omega t) \tag{6.8.5}$$

Inserting (6.8.5) into (6.8.4), the reflected component A_r is proportional in the form:

$$A_r \sim \exp\left[ikX_0 \sin(\Omega t)\right] \exp(-i\omega t) \sim \sum_{n=-\infty}^{\infty} J_n(kX_0) \exp\left[i(n\Omega - \omega)t\right], \quad (6.8.6)$$

where the formulae of Bessel function in (6.4.3) is used. For the case of the JxB force, $\Omega = 2\omega$, and it is found that only the odd modes are generated. It is noted that the amplitude of higher harmonics is proportional to the value of Bessel function with the argument kX_0. It is clear from Fig. 5.7 that with increase of the oscillation amplitude, the generation of higher harmonics is enhanced.

It is useful to compare the present physical mechanism of the **higher harmonic generation (HHG)** to the case discussed in Sect. 5.4, where we considered only an orbit of single particle to derive HHG. In the perfect mirror, surface current is induced to satisfy the boundary condition (6.8.4). Then, the electron motion at the mirror surface consists of two components; one is the motion by the JxB force with 2ω oscillation in the x-direction, while the other is the perpendicular direction by laser E field oscillating with ω. Then, it is clear that the electron orbit on the surface of the mirror is the same as (5.4.10) in the nonlinear Thomson scattering. This is the reason why only odd HHG is obtained in the JxB moving mirror case.

Solving Maxwell equation with tangential current on the surface of an oscillating perfect mirror, detailed analysis has been carried out to show the intensity dependence of generated higher harmonics on the harmonic number [9]. It is derived that for the harmonic number whose number is smaller than the cutoff number n_c defined with (6.8.3)

$$n_c = w_{max}/w, \quad (6.8.7)$$

the intensity of each harmonics decreases following the power law:

$$I_n \propto n^{-5/2} \quad \text{for} \quad n < n_c \quad (6.8.8)$$

It is also pointed out by the same group that more harmonics are generated for higher n than the n_c. The power law and the critical harmonic number in this extended region is derived from [10]

$$I_n \propto n^{-8/3} \quad \text{for} \quad n < \sqrt{8\alpha\gamma_0^3}, \quad (6.8.9)$$

where α is a numerical factor. Note that this does not change the result dramatically, while it is important relating to the atto-second (10^{-18} s) pulse generation.

These theoretical predictions are compared with experiment with Vulcan laser, RAL in UK, and the first evidence of X-ray harmonic radiation extending to 3.8 keV (order n > 3200) is demonstrated by Petawatt class laser-solid interaction experiment [11]. The experimental data for the harmonics more than n = 1000 is plotted in Fig. 6.15. The values of Lorentz factor of the used lasers are 10 (red line) and 13 (blue line), corresponding to the "rolls over" harmonic number, which is equal to

6.8 Moving Mirror Model and Higher Harmonic Generation from Solid Surface 237

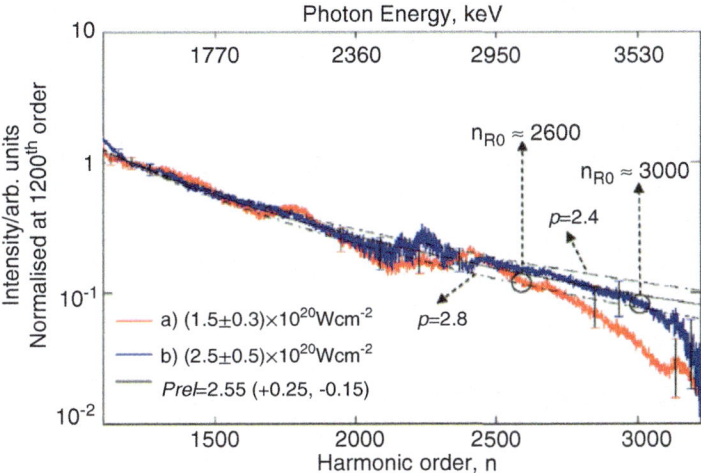

Fig. 6.15 High-frequency spectrum of reflected laser observed experimentally in the range of photon energy of keV. Intensity shows power law dependence, and more than 1000 harmonics is widely observed. [Figure 1 in Ref. 11]

n_c in (6.8.7), $n_{RO} \sim 500$. The experimental data clearly show the extended harmonics generation over n_c ($=n_{RO}$) and the power law:

$$I \propto n^{-p}, \quad p = 2.4 - 2.8 \qquad (6.8.10)$$

It is shown experimentally that the laser intensity dependence of the rolls over number is well identified [11]. In the comparison with the experiment, $\alpha = 1$ is assumed in (6.8.9).

So far, we have discussed by assuming the normal incident of laser on the target surface. Of course, the discussed physics changes for the oblique incident case. In the case of the s-polarization, it is straightforward to extend to the case of the normal incidence. The force of the JxB into the normal direction is $\cos\theta$ time weaker, and just taking account of the angle effect, $\cos\theta$, guides us rough understanding of the JxB effect. It is, however, very different in the case of the p-polarization. In what follows, we discuss the **relativistic vacuum heating** and HHG for the p-polarized oblique incident laser and solid interaction, where the solid has a sharp density jump.

In the case of the p-polarization, the vacuum heating shown in Chap. 4 occurs when a relativistic laser interacts with a sharp density solid surface. Since the electric field has the x-component in the p-polarization, it directly oscillates the electrons near the solid surface. It is obvious in this case that the oscillation frequency of the electron surface is the same as the electric field frequency ω. Then, Ω in (6.8.5) is equal to ω, and all higher harmonics are possibly generated. In addition, the hot electrons are generated, and they are generated near the surface with the frequency of ω. It is useful to evaluate the maximum energy of the hot electrons due to such heating. Since each acceleration is done in a short time during one cycle by the laser electric field E, the effective work done by laser field to accelerate a hot electron is evaluated to be over only one wavelength. Then, the energy one electron can get is

$$T_h \sim -\frac{eE_0}{k} = mc^2 a_0 \qquad (6.8.11)$$

Note that (6.8.11) is the same as the ponderomotive scaling (6.7.10) in the relativistic limit, where the velocity of electrons is assumed $v \sim c = \omega/k$.

References

1. W.B. Mori, C.D. Decker, W.P. Leemans, IEEE Trans. Plasma Sci **21**, 110 (1993)
2. C.D. Decker et al., Phys. Plasmas **3**, 2047 (1996)
3. K. Mima et al., Phys. Plasmas **8**, 2349 (2001)
4. R. Lichters, J. Meyer-ter-Vehn, A. Pukov, Phys. Plasmas **3**, 3425 (1996)
5. W.L. Kruer, K. Estabrook, Phys. Fluids **28**, 430 (1985)
6. Z.Y. Ge et al., Phys. Rev. E **89**, 033106 (2014)
7. L.G. Huang, T. Kluge, T.E. Cowan, Phys. Plasmas **23**, 063112 (2016)
8. S.V. Bulanov, N.M. Naumova, F. Pergoraro, Phys. Plasmas **1**, 745 (1994)
9. S. Gordienko et al., Phys. Rev. Lett. **93**, 115002 (2004)
10. T. Baeva, S. Gordienko, A. Pukov, Phys. Rev. E **74**, 046404 (2006)
11. B. Dromey et al., Phys. Rev. Lett. **99**, 085001 (2007)

Chapter 7
Relativistic Laser and Solid Target Interactions

7.1 Pre-formed Plasma in Laser-Solid Interaction

7.1.1 Pedestal of Laser Pulse

In the early time of the experimental research of relativistic laser-solid interaction, it was very difficult to obtain reproducible data from experiments, because it was very hard technically to obtain the same intensity profile after the pulse compression by the chirped pulse amplification (CPA). The main technical challenge is so-called **pedestal** before the main pulse, ultra-short pulse of relativistic intensity. The pedestal is orders of magnitude longer pulse with orders of magnitude lower intensity pre-formed light before the main pulse. In many experiments, the pre-plasmas produced by the pedestal are formed as seen below. As a result, the idealistic theory such as the vacuum heating and higher harmonic generations are not well applicable to a lot of data observed experimentally. A typical pedestal and main pulse intensity is plotted as a function of time in Fig. 7.1 [1]. The measured contrast ratio of the initial pulse from a commercial Ti:Sapphire CPA laser is about 2×10^{-8} at tens of picoseconds before the main pulse as shown with the black curve. Such a contrast ratio can be improved to 10^{-10} like the red curve. It is noted that theoretically even more improvement to 10^{-13} to 10^{-14} is possible.

The effect of the pedestal pulse on the generation of pre-formed plasma before the arrival of the main pulse is studied by use of hydrodynamic code with reasonable physics modeling [2]. The parameters of the pedestal and target are taken from a real experiment, where the aluminum foil is irradiated by a main laser which is 10^{19} W/cm^2, the contrast ratio of pedestal is 2×10^{-9}, and length of pedestal is 1 ns. It is studied by a hydrodynamic simulation code like the same as explained in Chap. 3. The intensity of the pedestal beam is 2×10^{10} W/cm^2. It should be pointed out that the electron-ion collision frequency at low temperature is very high, and the laser photons are absorbed every 10^{-15-16} s in aluminum solid as shown in Fig. 3.16. This

Fig. 7.1 Measured temporal contrast of the commercial Ti:Sapphire CPA laser (black curve) and the front end (red curve). The inset shows the measured temporal contrast on the 15 ps scale. [Figure 4 in Ref. 1]

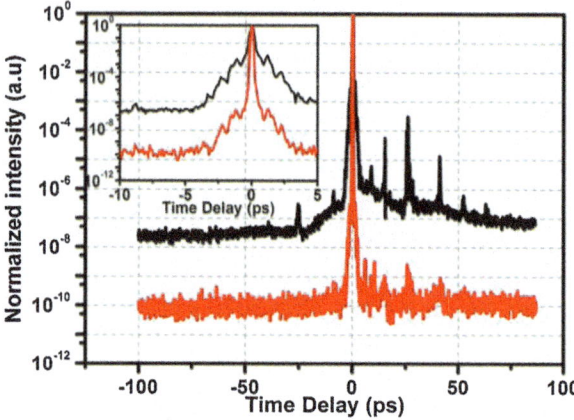

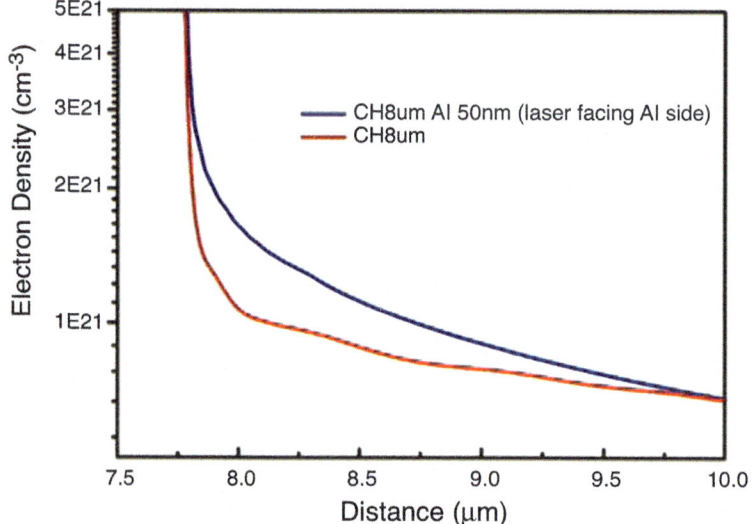

Fig. 7.2 Electron density profile after the ASE pre-pulse interaction obtained from the HELIOS hydrodynamic code. [Figure 4 in Ref. 2]

classical absorption is effective to generate pre-formed plasma before the main pulse. The hydrodynamic simulation is carried out for two cases; (a) CH of 8 μm and (b) same CH coated with aluminum of 50 nm on laser side. It is expected that the target (b) is easily heated in the Al layer, because Al is metal and has a lot of free electrons collisional heating by laser field. The density profiles produced by the 1 ns pedestal irradiation are shown in Fig. 7.2. It is clear that the pre-formed plasma with scale length of about 1 μm is produced and its density is around the critical density. This is an example of pre-formed plasma, and the relativistic laser is usually irradiated on such density profile. It is concluded that such targets show the pre-formed plasmas with scale length of L ~ 0.1–1 λ. It is found that even if the pedestal is the same, the pre-formed plasma condition depends on the target materials. We have to know more about the physics of laser cutting as shown in Fig. 2.11.

7.1.2 Model Experiments with Controlled Pre-formed Plasmas

Recently, with advancement of technology, the pedestal is well eliminated. Well controlled experiment is done and compared to PIC simulation [3]. In the experiment, optically polished silica foils with the electron density 400 times critical density for 800 nm light are irradiated obliquely (angle 55 degree from the target normal) with p-polarization by laser of 20–25 fs pulse duration with its intensity around 2×10^{19} W/cm^2 (a_0 ~ 3.5). The intensity contrast is achieved around 10^{-13} for > 100 ps before the main pulse. In order to control the scale length of the pre-formed plasma, the other laser is irradiated while altering the time interval before the main laser irradiation. The density scale length of the pre-plasma L is measured with interferometry technique. Typical experimental data are shown in Fig. 7.3, where many data obtained by changing the normalized scale length L/λ of the pre-plasmas (λ: laser wavelength) are plotted for the hot electron energy distribution (a) and the intensity of higher harmonic signals (b). In Fig. 7.3a, the x-axis is the angle of the electron ejection from the targets measured from the specular reflection

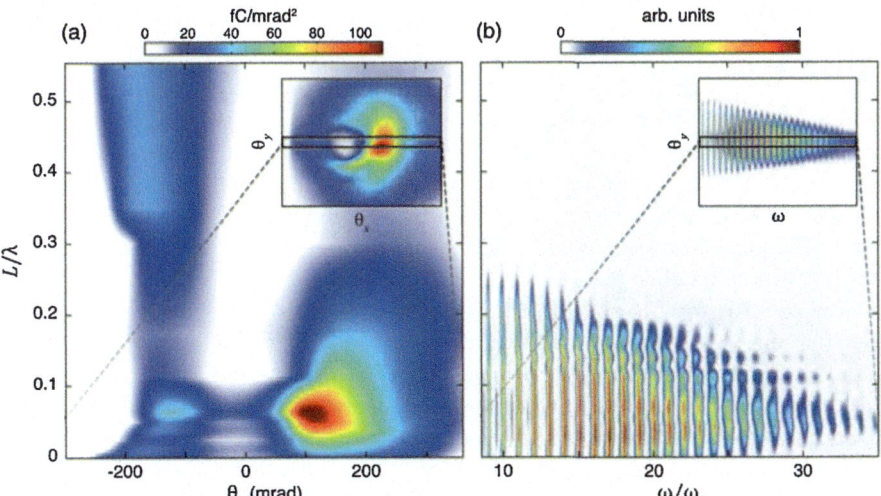

Fig. 7.3 Evolution of the experimental observables with the density gradient scale length. The angular profile of the relativistic electron beam in the incidence plane (**a**) and the emitted harmonic spectrum (**b**) are plotted as a function of L, for a p-polarized laser field. [Figure 3 in Ref. 3]

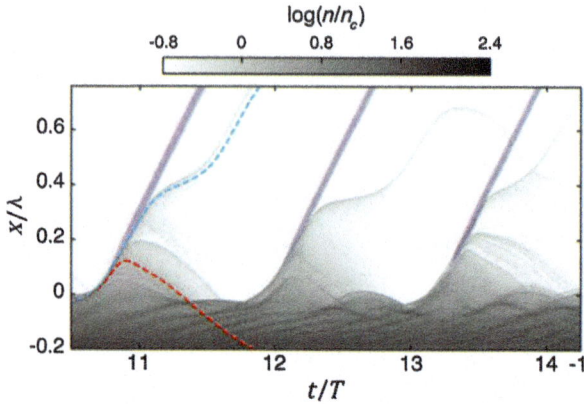

Fig. 7.4 Temporal dynamics of a dense plasma exposed to an ultra-intense laser field in the Brunel regime. This graph displays results from a particle-in-cell simulation performed for $a_0 = 2$ ($I = 8.5 \times 10^{18}$ W/cm^2, $\lambda = 800$ nm), $\theta i = 55°$, and a density gradient scale length $L/\lambda = 10$ (with λ the laser wavelength). The gray scale color map shows the temporal evolution, during three laser optical periods, of the plasma electron density around the target surface. [Figure 1 in Ref. 3]

angle, and the plus angle is that to the target normal direction, while the minus is opposite.

In Fig. 7.3a, a large amount of hot electrons is observed for sharp density profile case, while this signal disappears with increase of L/λ. It is clear that the **relativistic vacuum heating** effectively works for $L/\lambda < 0.2 \sim 0.3$. It is seen, however, that for $L/\lambda > 0.3$, the hot electrons are ejected to the different direction (angle $\sim$ -200 mrad). The physical mechanism of the electron heating will be discussed later, and here we just pointed out that in Ref. [3] it is concluded that stochastic heating produces hot electrons in such regime.

Associated with the hot electrons generation via the vacuum heating, HHG are observed for $L/\lambda < 0.2$ as seen in Fig. 7.3b, where the intensity distribution of the spectra of the scattered light reflected to the specular direction is plotted. The higher harmonics of more than 30 are observed. Note that the maximum frequency given in (3.12.3) is about 50. It is clear that with increase of the density scale length, the maximum of the harmonic number decreases, and they disappear for the normalized density scale more than 0.2.

For comparison, 3D and 2D PIC simulations are also carried out [3]. The hot electron ejected outward and inward directions of the solid targets are well reproduced and shown in Fig. 7.4 for a case of $a_0 = 2$ and the density scale length $L/\lambda = 0.1$, for which the strongest harmonic generation is seen in Fig. 7.3b. This is the idealistic **Brunel regime** and vacuum heating is clearly seen in the electron dynamics during the three cycles in Fig. 7.4. A small fraction of surface electrons is accelerated to the outward by p-polarized laser electric field as shown in the blue dotted line, while another small fraction is pushed back toward the solid region after the change of the sign of the electric field as shown in the red dotted line. In addition, the return current flow toward the surface is seen in the bulk plasma of the solid. The

7.1 Pre-formed Plasma in Laser-Solid Interaction 243

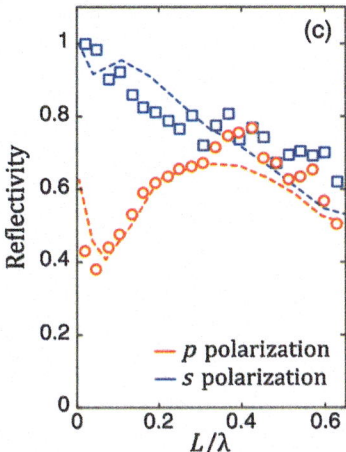

Fig. 7.5 Reflected fundamental beam and evolution of the plasma reflectivity. Using a scattering screen, the spatial intensity profile of the laser beam reflected by the target can be measured. From the spatial integration of these images, the reflectivity of the plasma for the fundamental laser frequency can be determined and is plotted as a function of L for both s and p-polarizations (squares and circles). The lines show the corresponding results of 2D particle-in-cell simulations. [Figure 5c in Ref. 3]

return current electrons flow from the inside to the surface. For example, the return current flow starting from the time of 12 to 12.5 is seen the black region in Fig. 7.4. Suck cycle schematically shown in Fig. 3.33 is confirmed in the simulation. Then, the same process repeats for the new fraction of electrons supplied as return current form the solid region. The hot electrons observed in Fig. 7.3a are the component accelerated to the direction of the p-polarized electric field.

The electron dynamics in Fig. 7.4 looks very coherent over three period. The pulse duration is 20–25 fs, and it is very short so that almost no ion motion couple to the electron dynamics. As pointed out later, for the pulse duration more than 100 fs, roughly speaking, the ion motion during the main pulse makes the pre-formed plasma, and HHG and absorption of laser are significantly affected. As we see later, the laser intensity, pulse length and pre-formed plasmas are very important to know the physics of laser-matter interaction.

In Fig. 7.5, the laser reflectivity (absorption rate) as a function of the density scale length of pre-formed plasma is plotted, where the red and blue are p-polarized and s-polarized, respectively, and the marks are from experiment and dotted lines are from 2D PIC simulation [3]. In Fig. 7.5, it is seen that about 50% of laser energy is absorbed via Brunel mechanism for $L/\lambda < 0.2 \sim 0.3$, where strong emission of higher harmonics is also observed in Fig. 7.3. For $L/\lambda > 0.3 \sim 0.4$, there is no difference of the absorption fraction between p- and s-polarizations. If the resonance absorption in Sect. 3.6 becomes important in the p-polarization, there should be a difference in the absorption fraction. This fact suggests the density modification is dominant in such relativistic intensity. It is also imagined that the density modification by ponderomotive force cause turbulent state in the pre-formed plasma, and there is another type of dominant interaction physics between laser and the plasmas, being rather independent of the laser polarization. In [3], it is concluded that the laser energy is transferred to the hot electron energy via **stochastic heating** process [4].

For a given laser intensity, higher plasma density of solid implies a higher restoring force by charge separation due to electron oscillations. This may deduce the oscillation amplitude of the surface and degrade the HHG according to the ROM model. In contrast, if the plasma density is too low, large amplitude surface oscillation becomes unstable, resulting the non-synchronized oscillation to cause broadening of generated higher harmonic spectrum. It is also noted that if the density is lower than the relativistic critical density defined in (6.2.3), the plasma becomes relativistic transparent and no reflection is expected. The optimized density is also a function of laser intensity.

The generation of HH also depends on the scale length of pre-formed plasma as shown in Fig. 7.3b. It is noted that with 1D PIC code, HHG is more efficient for the case with pre-formed plasma with an appropriate length, being equal to $L/\lambda = 0.2$ (figure c) in Fig. 7.6 [5]. The target is 5 μm thickness with solid density of $49n_c$ (n_c: critical density), and the laser is irradiated with p-polarized oblique incidence at the intensity 10^{18} W/cm^2 ($a_0 = 0.3$). For three different L/λ cases, the HHG is seen also at $L/\lambda = 0.02$ (case b), while the Fourier components of the transverse currents generating HH are seen in the higher density positions whose plasma frequencies resonate with the HH frequencies in the case (b). The transverse currents are shown in the middle of figure b with the black lines along with the density profile (red). Such HHG is also driven by the nonlinearity of plasma oscillations as seen in Sect. 6.2. The HHG in case b is due to such resonance in the over-dense region and as a result HHG is limited by the plasma frequency (ω_p) of the solid density. For $\lambda/L = 0.2$ (case c), the higher harmonic currents are seen to be localized at the same position of the critical density x_{cr}, and the amplitude in the bottom well explains the PIC result when the amplitude of ROM is assumed to be $X_0 = 0.06\lambda$ in (6.8.5) as plotted with red circles.

As clear from comparison between the results in Figs. 7.3 and 7.6, the optimum density scale length is different. It is obvious that the optimum condition of HHG is depends on the laser intensity, density scale length, 1D PIC or 2D PIC simulation, ion-electron mass ratio, the number density of solid, etc. It is not valuable to compare the difference but to keep in mind the physics of HHG is important to analyze computational and experimental data, since it is not always the case that we are able to measure the energy fraction carried out as higher harmonics lights.

7.2 Laser Absorption at Solid Targets

Laser absorption of ultra-intense laser is very sensitive to the laser and target conditions. Especially, the laser-matter interaction physics strongly depends on the property of the pre-formed plasma. Roughly saying, the physics of laser absorption

7.2 Laser Absorption at Solid Targets

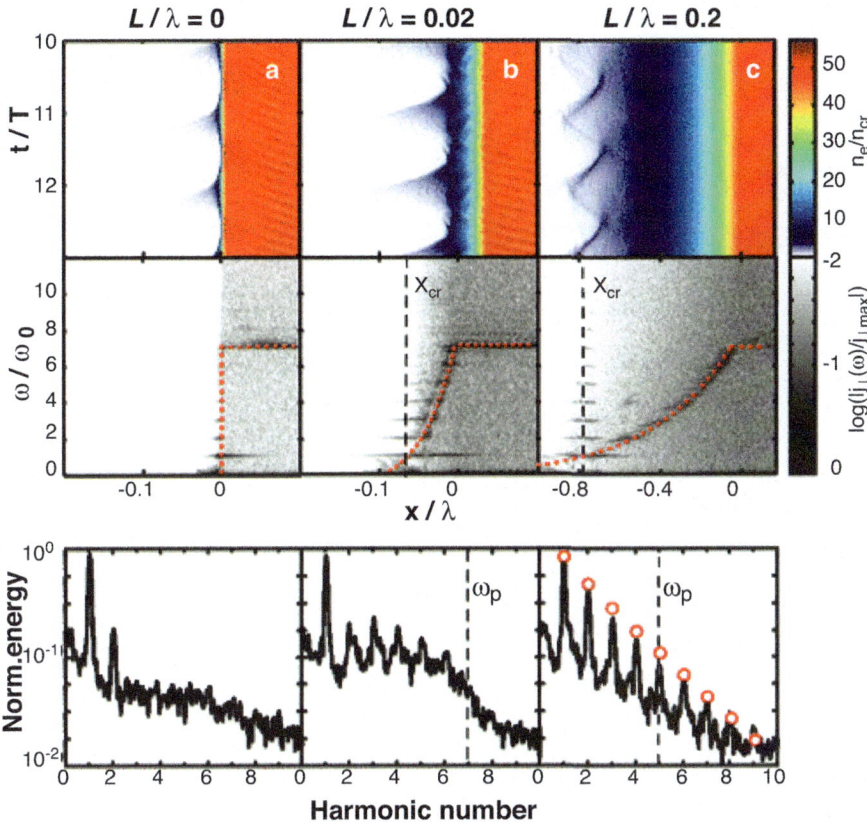

Fig. 7.6 Electron density (top), spectra of the transverse current (middle), and harmonic spectra (bottom) for p-polarized excitation with $a_0 = 0.3$. The circles (bottom left) represent the spectrum calculated for a harmonically oscillating mirror with $X_0/\lambda = 0.06$. The dashed lines indicate the critical surface (middle) and ω_p (bottom). Red dotted lines represent $\omega_p(x)$. [Figure 3 in Ref. 5]

depends on the scale-length of pre-formed plasma as suggested in Fig. 7.5 for the case of laser intensity 2×10^{19} W/cm^2 and very short pulse 20–25 fs. A systematic comparison among three different lasers has been done to compare the intensity dependence of laser absorption, etc. [6]. Attention is focused on the effect of intensity contrast and consequent energy fraction of the pedestal by ASE to the main laser energy. Experimental data for the best contrast case with Astra Gemini laser are obtained under the following condition [7].

The laser energy is 12 J and the pulse duration of 50 fs. Its wavelength is 800 nm. The laser is irradiated on a flat thin aluminum foil of 0.1 μm with p-polarization at angle of 35 degree, and intensity is varied from 10^{17} W/cm^2 to 10^{21} W/cm^2. The focusing diameter is 2.5 μm. The intensity contrast ratio of pedestal to the main pulse is 10^{-10}, while the energy ratio is measured to be 2×10^{-5}, which value was almost the best in 2009. In high contrast and very short pulse, some fraction of laser energy

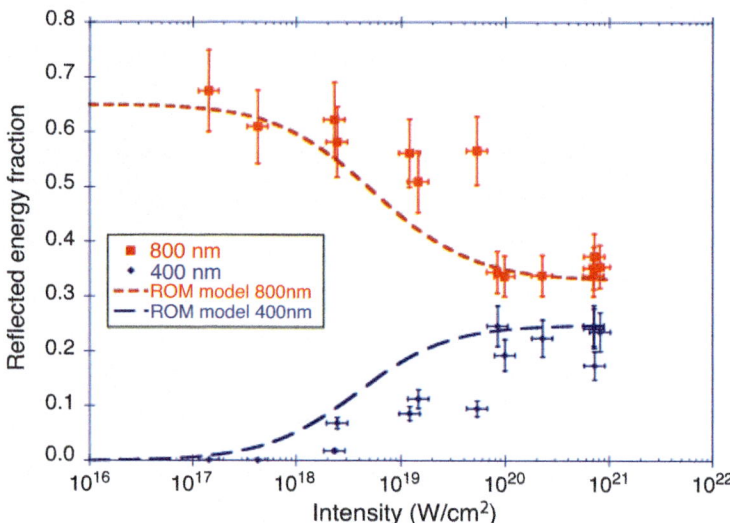

Fig. 7.7 Energy incident onto the scattering screen as observed at 800 nm (squares) and 400 nm (diamonds). Vertical error bars are combined calibration and measurement errors. In the region near the focus, the intensity was calculated as the mean within the FWHM radius; away from the focus, it is the mean intensity within the beam radius. The dashed lines represent the results of the ROM model. [Figure 3 in Ref. 7]

is converted to the second, and higher harmonic signal and both of the fundamental (800 nm) and second harmonics (400 nm) light reflected from the target haven been measured. In Fig. 7.7, the experimental data are plotted with marks with error bars for 800 nm (red) and 400 nm (blue) as function of laser focused intensity. It is clear that once the laser intensity exceeds the relativistic intensity, the fraction going to the second harmonics increases, and then the fraction of the reflecting laser (red) decreases, roughly the total being constant. If the second harmonics is about 20%, 3rd harmonics may be about 7% according to (6.8.8).

In Fig. 7.7, the dashed lines shows theoretical values calculated with ROM. In the ROM model, perfect mirror is assumed and no absorption is included. So, the resultant values are reduced to a factor 65% assuming the 35% absorption [7]. The both colors correspond to the experimental ones and a good agreement is seen. It is noted that in ROM model, the solid surface motion is assumed to be the same as the eight-figure motion obtained in (5.3.34).

In Fig. 7.8, the intensity-dependent reflection fraction measured in the above experiment is plotted with solid circle with red error bars. It is concluded that the absorption fraction is almost constant about 30–40% in the rage of 10^{17}–10^{21} W/cm^2. In Fig. 7.8, the intensity dependence of absorption fraction from three experiments with three different lasers is plotted for different energy ratio of the ASE pedestal to the main pulse [6, 8]. It is clear that a dramatic change happens for the case with the energy ratio increases from 2×10^{-5} to 2×10^{-4} and also to 3×10^{-2}. These data seem to suggest that enhanced pre-formed plasma enhances the

7.2 Laser Absorption at Solid Targets

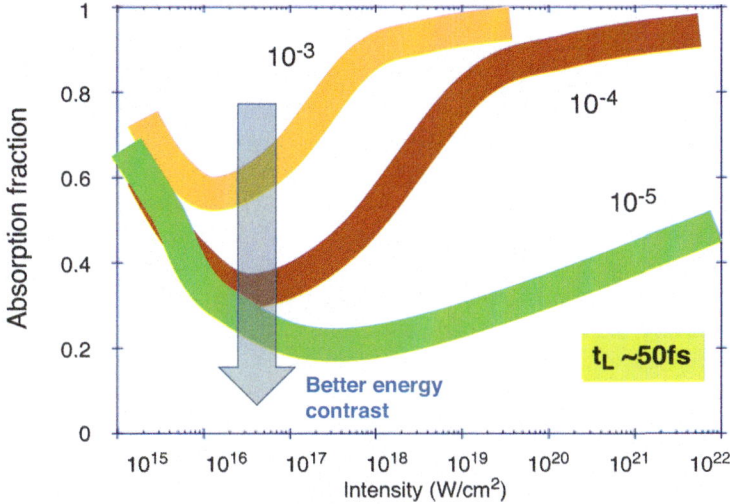

Fig. 7.8 Absorption fraction as a function of intensity from the experiments with three different lasers shown in [6, 7]. Note the contrast quoted here is the ratio of energy in the ASE to that in the main pulse (opposed to intensity contrast). [Based on Fig. 6 in Ref. 7]

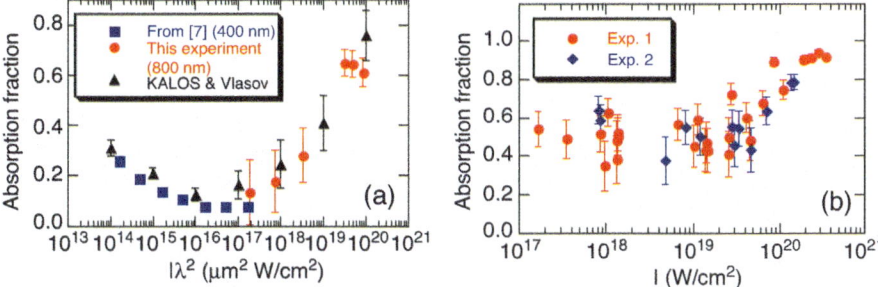

Fig. 7.9 Absorption fraction of the laser energy as a function of laser intensity at an incident angle of 6 (**a**) and 45 (**b**). Each point represents the result of a single laser shot. [Figure 1 in Ref. 8]

absorption rate to almost 100%. This is very important fact that the control of the laser energy contrast is essential to carry out reliable experiment repeatedly.

Systematic experimental study of absorption for relativistic laser has been carried out with *Callisto* laser in the Jupiter facility in LLNL [8]. The laser is 150 fs at 800 nm with energy up to 20 J focused on 5 μm width. The target is aluminum foil with thickness 1.5–100 μm. In this experiment, the energy ratio of the ASE pedestal to the main pulse is 2×10^{-4}, one order of magnitude larger than the case of Fig. 7.7. The experimental results are compared to 2D PIC simulation. In Fig. 7.9, almost normal incident experimental data with angle 6 degree (Fig. 7.9a) and p-polarized oblique incident case data (Fig. 7.9b) are plotted for the intensity-dependent absorption fraction.

It is noted that the absorption for the intensity less than 10^{17} W/cm^2 is mainly due to the classical absorption already mentioned in Chap. 3. The rapid increase of the absorption fraction over 10^{19} W/cm^2 is consistent with the absorption fraction in Fig. 7.8 for the energy contrast 2×10^{-4}. For explaining the experimental data, the pre-formed plasma and ion motion effects are studied with 2D PIC code. It is informative to show the result of absorption fraction η_{ab} for 3×10^{20} W/cm^2 resulted by the PIC code. Case A, $\eta_{ab} = 37\%$ for ion fixed, no pre-plasma. Case B, $\eta_{ab} = 68\%$ for mobile ion, no pre-plasma. Case C, $\eta_{ab} = 63\%$ for ion fixed, pre-plasma with $L > \lambda$. Comparing these three simulation results, it is said that the pre-plasma enhances the absorption fraction and the ion motion help to form the plasma cloud near the surface of solid target. Of course, the pre-formed plasma and mobile ions help the increase of the absorption rate.

In Fig. 7.8, the comparison has been done for three different lasers with the short pulse duration of about 50 fs, while the pulse duration of the *Callisto* laser is 150 fs, three times longer. Let us roughly evaluate how long it takes that the ions move over about laser wavelength λ due to the ambipolar field by electrons under the ponderomotive force. The ion motion is governed by the ambipolar field force, and it should balance with the ponderomotive force as follows:

$$m_i \frac{du_i}{dt} = +eE$$
$$0 = -eE - \frac{\partial}{\partial x} U_{PM} \quad (7.1.1)$$

Inserting the ponderomotive potential in (3.7.10) into (7.1.1), the dimensional analysis provides the following characteristic velocity of the ion fluid:

$$V_{PM} \tilde{c} \sqrt{\frac{m_e}{m_i} \langle \gamma \rangle} \quad (7.1.2)$$

The time for ions to move over $\lambda = 800$ nm is typically about 160 fs for the root of the mass ratio 60 and $\langle\gamma\rangle \sim 1$ from (7.1.2). This means the pulse duration of 150 fs is very critical value for the ion motion becomes important and the expanding plasma is created during the laser duration.

7.3 Absorption Enhancement by Hot Electron Re-circulation

It became clear that the absorption rate of ultra-short intense laser is sensitive to pre-formed plasma scale-length and pulse duration. In addition, it is experimentally and computationally demonstrated that the absorption rate is also sensitive to laser focal spot size, target thickness, and pulse duration [9]. In order to clarify the

7.3 Absorption Enhancement by Hot Electron Re-circulation

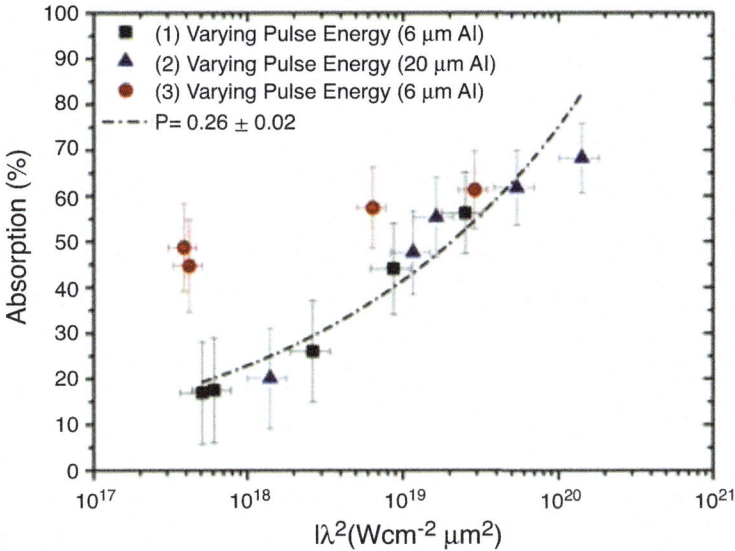

Fig. 7.10 Measured total laser energy absorption as a function of intensity. The black square and blue triangular data points are measurements made when the intensity is changed by varying E_L for target thickness 6 μm and 20 μm, respectively. The dot-dash line is a fit to this data using the empirical model presented in [9]. The red circular data points are measurements made by varying laser intensity for constant E_L

sensitivity, *FHELIX* laser at GSI with longer pulse of 700+−100 fs and larger laser energy of 4–180 J at $\lambda = 1.053$ μm was used for absorption experiment. The targets are aluminum foil, and the laser is irradiated normally on the target in the same condition as the experiment shown in Fig. 7.9a. In order to study the sensitivity to the focusing diameter ϕ, it was varied over the range, $\phi = 4$–270 μm. The effect of the target thickness was studied by varying the thickness from 6 to 50 μm. The laser energy contrast of the ASE pedestal was kept very small about 1.4×10^{-5} same as in the case of Fig. 7.7. Although the energy of the pedestal is reduced, the expanding plasma by ion motion is produced during the relatively long pulse duration.

In Fig. 7.10, the intensity dependence of laser absorption is plotted for the intensity range from 10^{17} W/cm^2 to 10^{20} W/cm^2. The black square and blue triangular data are measurements taken when the intensity is changed by varying the laser input energy for the target thickness 6 μm and 20 μm, and keeping the focused diameter ϕ is constant, $\phi = 4$ μm. Compared to Fig. 7.9a, the intensity dependence is well reproduced for this small focusing spot case. However, the red circle data taken for the case with same laser energy by varying the focus spot size shows higher absorption at lower intensity. The spot size is much larger than 4 μm, and it is 270 μm at intensity 5×10^{17} W/cm^2. It is speculated that larger spot size enhances the absorption rate at most three times the smaller case at this intensity.

The most possible physical reason is considered as follows. It is plausible that the hot electrons are confined by sheath (ambipolar) potential covering the both surfaces

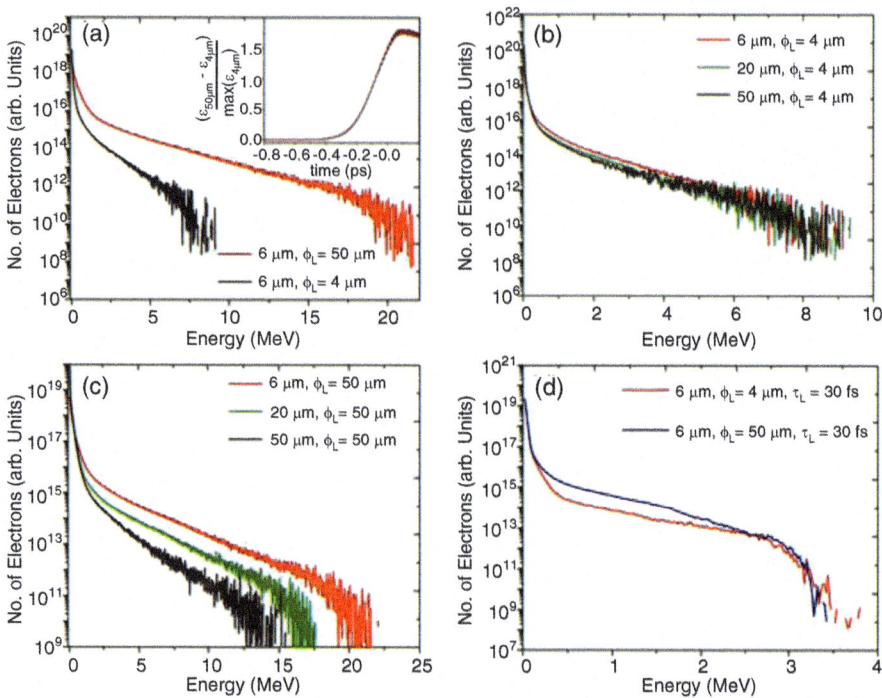

Fig. 7.11 Electron energy spectra for (**a**) focal spot sizes 4 μm and 50 μm, for fixed target thickness 6 μm and $\tau_L = 550$ fs; (**b**) variation in target thickness from 6 to 50 μm, for fixed spot size 4 μm and $\tau_L = 550$ fs; (**c**) variation in target thickness from 6 to 50 μm, for fixed spot size 50 μm and $\tau_L = 550$ fs; (**d**) focal spot sizes 4 μm and 50 μm, for fixed thickness 6 μm and $\tau_L = 30$ fs. The inset in (**a**) shows the total energy difference as a function of time (for electrons above the initial electron thermal energy) between the spot size 4 μm and 50 μm cases, normalized to the energy for spot size 4 μm (for fixed thickness 6 μm and $\tau_L = 550$ fs). [Figure 3 in Ref. 9]

of the front and rear of the aluminum foil. In the case of larger focal spot, such sheath barrier confines the hot electrons in the wide area of the target surface. 2D PIC simulation has been done to confirm this at the intensity of 10^{19} W/cm². The resultant electron energy spectra after the laser shot are shown in Fig. 7.11 with red line and black line for ϕ = 50 μm and ϕ = 4 μm, respectively. It is clear in Fig. 7.11a, c that large spot size helps the further heating of hot electrons. For the small spot size, however, the accelerated electrons easily escape from the laser interaction region, and the maximum energy is much smaller than the case of large spot size. This indicates that the high-energy electrons escape from the laser interaction region laterally, while they also escape or scattered in the solid region when the target thickness increases as shown in Fig. 7.11c for larger focused diameter case.

In Fig. 7.11b, the target thickness is changed while keeping the focusing diameter 4 μm. In this case, even the thinner target cannot confine the hot electrons, and no difference is merely seen. In order to see the proof of the hot electron confinement

7.3 Absorption Enhancement by Hot Electron Re-circulation

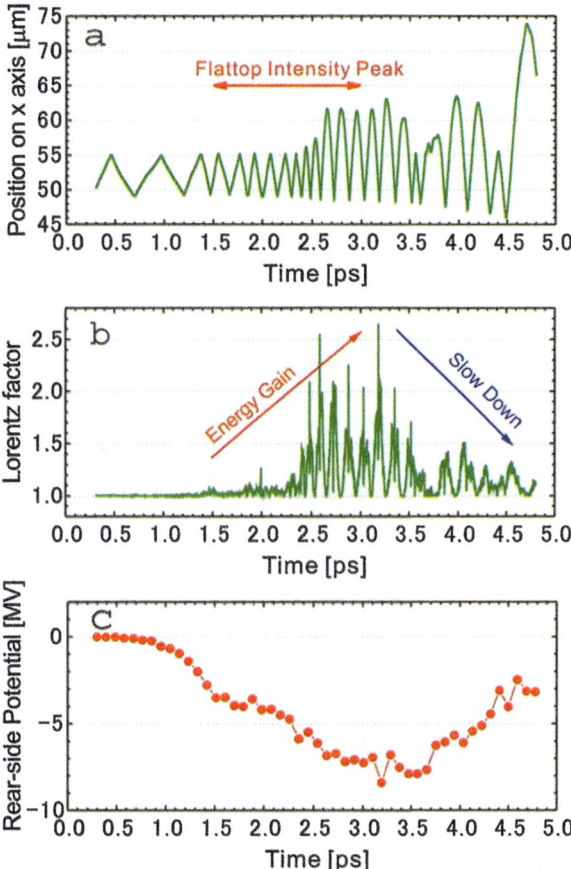

Fig. 7.12 (a) Trace of typical electron trajectory in the 1D PIC simulation. The target foil is initially at the position $x = 50$–55 μm, and the laser (2-pulse train) is incident on the surface at $x = 50$ μm. (b) Time evolution of the Lorentz factor of the electron shown in (a). (c) The temporal evolution of the potential generated on the rear side. [Figure 3 in Ref. 10]

and re-circulating acceleration in the interaction region, difference of the spectra for a short pulse heating has been simulated. In Fig. 7.11b the resultant spectra for the same condition as Fig. 7.11a except for the laser pulse length being 30 fs are shown. The same spectra are obtained, since there is not enough time for re-circulation of high-energy electrons in the target even for the larger laser-focusing diameter.

The **re-circulation** of high-energy electrons are simulated with 1D PIC code to compare with the experimental data of relatively long pulse (3 ps) at intensity of 2.3×10^{18} W/cm^2 [10]. The experiment has been done with *LFEX* laser consisting of four beams of 200 J/beam and 1.5 ps pulse duration. The purpose of the experiment is to demonstrate higher energy and higher efficiency of multi-tens MeV proton beam production. The energy increase of proton beam is caused by the energy increase of hot electrons. In order to enhance the hot electron temperature via the re-circulation heating, two beams are combined with a delay to form the flattop peak of about 2 ps. In Fig. 7.12, the increase of electron energy by long pulse is shown by tracking a typical hot electron from PIC code. In Fig. 7.12a, the electron is confined in the target region (50–55 μm) initially and accelerated by the first beam whose peak intensity is at 1.5 ps; then the second beam arrives to accelerate the electron mainly in the expanding plasma region (x > 55 μm). During this time the electron gains large amount of energy as seen in the time evolution of its Lorentz factor in Fig. 7.12b. This re-circulating acceleration is maintained by very high sheath potential, and the time evolution of the rear side of the potential is shown in Fig. 7.12c. It is noted that the laser-focusing diameter was about 200 μm and the electron bounce distance is about 30 μm; therefore, 1D PIC simulation is applicable to this case.

Summarize the present understanding about the laser absorption for the ultra-short relativistic intensity pulse irradiated on solid targets. In the case of ultra-short pulse with the pulse duration less than 100 ~ 200 fs, the absorption rate is very sensitive to the pre-pulse contrast ratio as shown in Fig. 7.8. With enough pre-formed plasmas, almost 100% of absorption is measured. Once the pulse duration becomes longer, the absorption rate is high as seen in Fig. 7.10, even if the pedestal is suppressed. For long pulse the ion motion is effective to form the plasma during the main pulse, and the absorption is enhanced if the re-circulation of high-energy electrons is expected for the large focal spot size, where the electrons are confined by sheath potentials at the front and rear side of targets.

7.4 Hole Boring by Ponderomotive Force

We have discussed the experimental and computational study on the laser absorption near the solid surface with and without the pre-formed plasma, ion motions, re-circulation, and so on. Here consider how the laser ponderomotive force interacts with the plasma relatively lower density than the normal solid density. Strong hole-punching phenomena, now-called **hole boring**, has been demonstrated with 2D PIC simulation [11]. Plasma slab with electron density of $4n_c$ (critical density) is irradiated at 1.2×10^{19} W/cm^2 ($a_0 = 3$). The laser is p-polarized in x-y of the 2D plane and irradiated normally on the slab plasma. No resonance absorption is observed, while the laser ponderomotive force of JxB shown in (6.3.7) affects the plasmas non-adiabatically, resulting the laser absorption via hot electron generation. In the simulation, the focusing width is 14 c/ω (~2.5 μm) in the simulation size in y-direction of 40 c/ω. Due to the non-oscillatory component of the ponderomotive force in (6.3.7), the plasma surface in the laser focal spot is found to move to the laser propagation direction. This is the boring by the light beam pressure. Evaluate the pressure of the irradiated laser. The photon momentum p_{ph} and the laser energy flux I_L are defined as

$$p_{ph} = \hbar k = \hbar\omega/c, \quad I_L = \hbar\omega N_{ph} c \qquad (7.4.1)$$

Then, the photon pressure P_L is defined and given as

7.4 Hole Boring by Ponderomotive Force

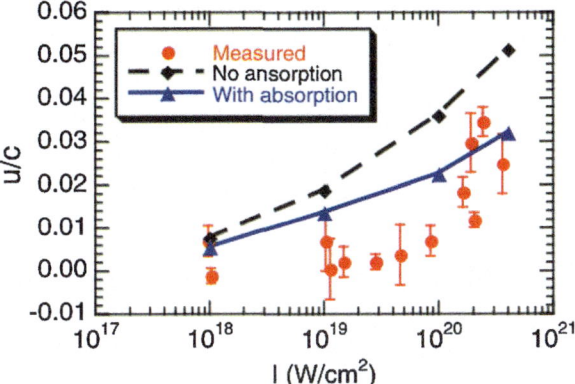

Fig. 7.13 Comparison of the surface velocity calculated from the Doppler red shift of specular-reflected 2ω to a simple model with and without absorption. [Figure 3 in Ref. 8]

$$P_L = p_{ph} N_{ph} c = I_L/c \tag{7.4.2}$$

The pressure of perfectly reflecting laser is give to be $P_L = 2I_L/c \sim 6 \times 10^3$ Mbar for, say, $I = 10^{19}$ W/cm^2. Since the plasma density is low, the ions are easily moved via ambipolar field toward the slab plasma region in the present density profile. Assuming that the ion fluid moves with the speed of u, then the increase of the ion momentum should be supported by the pressure (PdV) work at the contact surface of laser and plasma. This simple momentum flux conservation law provides the ion hole boring velocity u as

$$(\rho u^2) u = P_L u \quad \Rightarrow \quad \frac{u}{c} = \left[\frac{n_c}{n_e} \frac{Zm_e}{m_i}\right]^{1/2} \frac{a_0}{2}, \tag{7.4.3}$$

where n_e is the electron density of the plasma in front of the laser piston and Z is the charge state. In the present simulation, $n_e = 4n_c$ and $u = 0.025c$. In the simulation, the hole depth is about $d = 1.5\lambda$ ($=1.5$ μm) at $t = 300$ fs, which is reasonable distance compared to the above evaluation, $d = ut = 2.2$ μm.

In Ref. [8], the hole boring is experimentally measured by measuring the red shift of the second harmonics (2ω) emission generated near the critical density in pre-formed plasmas. From the measured red shifts, the inward-moving velocity is evaluated in Fig. 7.13 for the laser intensity from 10^{18} W/cm^2 to 10^{21} W/cm^2. In this experiment, short pulse with 150 fs pulse duration is irradiated on the spot size of 5 μm on Al foils. The velocity of the hole boring is measured to be $u/c \sim 0.01$–0.03 in this intensity range. The result is compared to (7.4.3) by assuming that the electron density is equal to the relativistic critical density, $n_e = \gamma_0 n_c$ as shown in (6.2.3), since the red shifted second harmonics are generated at the resonance density. It is assumed that the inward motion does not come to stop at the solid surface. By taking into account the oblique angle effect and absorption effect to the laser pressure, the black and blue lines derived with (7.4.3) is found to well explain the experimental result (red marks) at higher intensity case as seen in Fig. 7.13.

The hot electrons are measured to have an effective temperature of about 1.6 MeV. In Ref. [12], the scaling of the hot electron temperature based on the ponderomotive potential given in (6.3.6) is compared to several simulation results. It is reported that a good agreement with the following relation is obtained.

$$T_{hot} \approx mc^2(\langle\gamma\rangle - 1) = mc^2\left(\sqrt{\frac{1}{2}a_0^2 + 1} - 1\right) \quad (7.4.4)$$

In the present simulation, the spot size is relatively small so that the heated electrons escape from the laser interaction region after the first acceleration (kicking) by the JxB ponderomotive force and no further heating by re-circulation is expected. This is the reason why the ponderomotive scaling well explains the computational result.

7.4.1 Density Dependence of Hole Boring

The density dependence of the hole-boring phenomena is studied by irradiating *Vulcan Petawatt* laser, RAL, on a variety of foam target with different initial density [13]. The pulse duration is 550 fs, and the laser is focused on the surface with focusing diameter of 5 μm. Intensity on target is 8×10^{20} W/cm^2. The laser strength is $a_0 \sim 36$ and the relativistic cut-off density is $n_{rc} \sim 25 n_c$. The foam density used in the experiment is $n_e/n_c = 0.9, 3, 4.5, 6, 13.5$, and 30. The thickness is 250 μm for all targets. The ASE pedestal is expected to pre-ionize the foam surface and to form pre-formed plasmas. In the experiment, it is found that produced high-energy ions ejecting from the rear side of the targets are well collimated as the form density decreases. This means that the hole boring is expected in the target and the magnetic and electric field surrounding the hole helps to collimate the hot electrons, resulting the better energy conversion to the collimated ions.

2D PIC simulation is done to study the experimental data as shown in Fig. 7.14. The initial focusing diameter is 8 μm and pulse duration is 500 fs. The all snapshots are the ion density at the time of 2 ps, enough after the laser irradiation. It is seen that at lower density, the hole is bored through the targets (a and b), while at medium density, the hole is bored to stop inside the target (c and d). For higher density, the ponderomotive force is not enough to make holes inside the target and shallow region near the focal spot area are ablated out. For given laser parameter, a simple model in (7.4.3) for the hole-boring velocity is proportional as

$$u \propto 1/\sqrt{n_e} \quad (7.4.5)$$

If this velocity is kept during the laser pulse, then the length of the hole is proportional to (density)$^{-1/2}$, which qualitatively agrees with the simulation result. It is noted, however, that this experiment results an enhanced length of the bored hole for the region of the foam density higher than $10 n_c$. This enhancement is

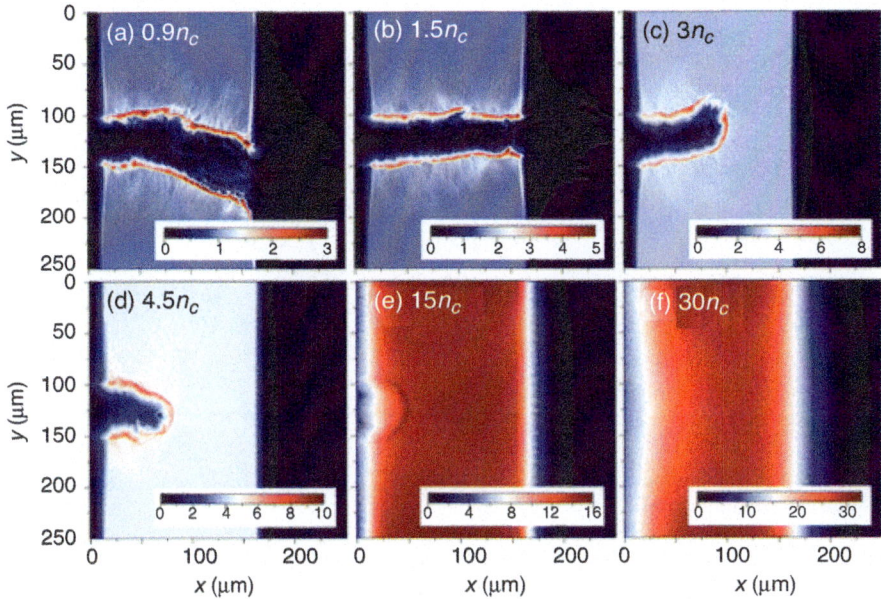

Fig. 7.14 Ion densities at 2.0 ps into the simulation. [Figure 3 in Ref. 13]

explained by the fact that static electric and magnetic fields in the laser channel act to collimate the electrons and therefore the hot electrons emerge from a narrow region and the electron energy is higher. This reason is speculated because each hot electron gains higher energy in the low-density case and enhances the speed of hole boring due to the enhanced charge separation force.

7.5 Laser Interaction in Long Pre-formed Plasmas

So far we have discussed the case where the pre-formed plasma has a scale length of about the laser wavelength. What different physics is seen when relativistic ultra-short pulse is irradiated, on the other hand, on long-ramped plasmas on purpose or accidentally? In Fig. 7.15, the density profile (solid line) and electron momentum in x-direction (particles) are plotted at about 400 fs from 2D PIC simulation, where the plasma is initially assumed with ramped density from 0 to $4n_c$, over a distance of 4λ, with an additional plasmas of density $4n_c$ behind the ramp. The laser intensity is 5×10^{18} W/cm^2 at $\lambda = 1$ μm. The laser is irradiated with s-polarization normally on the plasma surface, meaning computation space is x and y, and laser is linearly polarized in z-direction.

The ponderomotive force is due to JxB and pushes the low-density plasma to soon deform the plasma density as seen in Fig. 7.15. It is observed that plasma waves with large amplitude are generated due to the charge separation by the

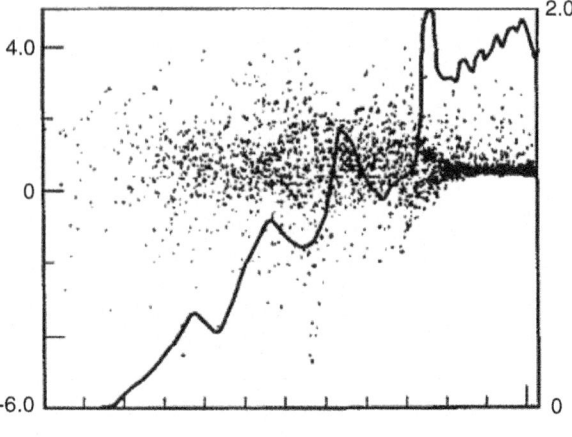

Fig. 7.15 Longitudinal momentum, p (particles), and density (solid line) versus longitudinal position. [Figure 6 in Ref. 12]

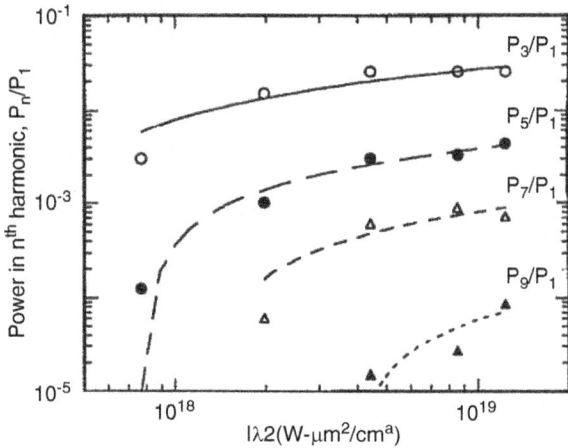

Fig. 7.16 Power in various odd harmonics (normalized to the backscattered signal at the incident frequency) as a function of intensity, showing saturation of the lower harmonics as the intensity is increased. The lines are a logarithm fit used to guide the eye. [Figure 5 in Ref. 12]

ponderomotive force, propagating to the lower-density region. The plasma waves accelerate electrons to the vacuum direction, while the electrons return to the right because of the sheath potential near the vacuum boundary. Such electron has the energy of about 1–2 MeV, while the ponderomotive scaling gives about 500 keV. Both accelerated electrons penetrate into the over-dense region, while they do not return to the interaction region because the laser spot diameter is 2.5 μm, very small.

In addition, the density steepening by the static term of the ponderomotive force is also seen in Fig. 7.15. A sharp density jump across the critical density is quivered by the 2ω oscillation force of JxB to generate higher harmonics even in such plasma with long-distance ramped density profile. In Fig. 7.16, relative power of the odd harmonics (normalized to the backscattered ω signal intensity P_1) is plotted as a function of the laser intensity. Different from the case of sharp boundary discussed in Sect. 6.8, each harmonics has a different threshold intensity to appear.

7.5 Laser Interaction in Long Pre-formed Plasmas

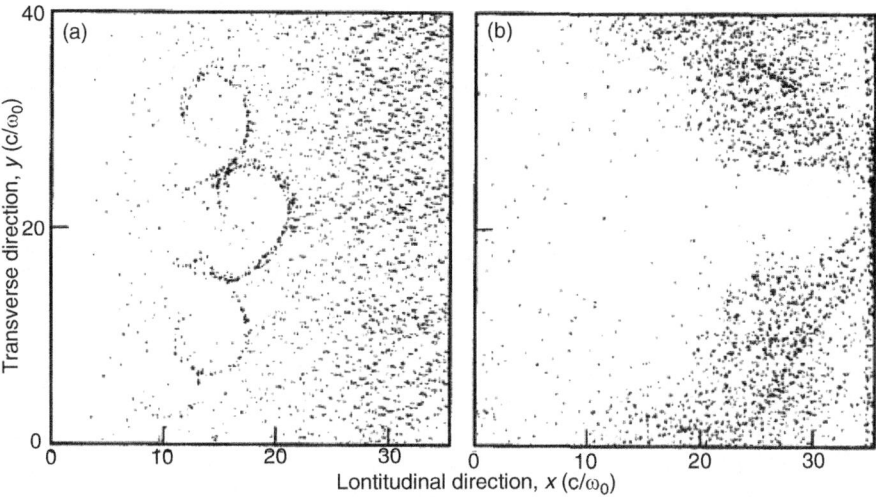

Fig. 7.17 (a) Ion real space at t = 400 fs showing Rayleigh-Taylor-like rippling of the critical surface due to s-polarization laser. (b) At a time t = 600 fs, the central bubble has bored through the plasma. [Figure 5 in Ref. 12]

It is also shown that the critical surface with sharp density jump seen in Fig. 7.15 is not stable and bubble-like structure appears as shown in Fig. 7.17a. Note that the laser field penetrates to the inner surface of the bubbles. This is **Rayleigh-Taylor instability** due to the photons (the light fluid) effectively accelerating the over-dense plasma (the heavy fluid). It is interesting to know that the diameter of the bubble at the center is about 1.3λ and the focused laser profile is broken up to the size of its wavelength due to the instability. Figure 7.17a is the plot of ion particles at time of 400 fs. In the later time at 600 fs (Fig. 7.17b), the bubble structure is merged to form a hole at the center, and the hole-boring type laser penetration is observed.

Hole boring dynamics is studied experimentally with the *Titan* laser at LLNL, and the experimental data are compared to 2D PIC simulation [14]. The laser is the pulse width of 1.4 ps with 150 J, corresponding to the intensity of 5×10^{19} W/cm^2. The ASE pedestal has 17 mJ over 3 ns, which corresponds to the energy ratio of 1.2×10^{-4} almost same as the experiment of Fig. 7.9. This energy ratio means substantial pre-formed plasma. However, the pulse length is about order of magnitude longer than the case in Fig. 7.9. The laser is irradiated on a solid aluminum plate with 1 mm thick slab. In order to model the pre-formed plasma, 2D radiation hydrodynamic code HYDRA is used to appropriate density and charge state distribution as the initial plasma profile for 2D PIC simulation. The time evolution of the Doppler shift of the second harmonics 2ω mainly generated at the relativistic critical density surface is measured at the laser incident direction, so the motion toward the forward to the over-dense region provides red shift, $\Delta\lambda/\lambda$.

The measured data are plotted in Fig. 7.18, where the black solid line with error bars is the measured time evolution of the Doppler shift. The time history of the

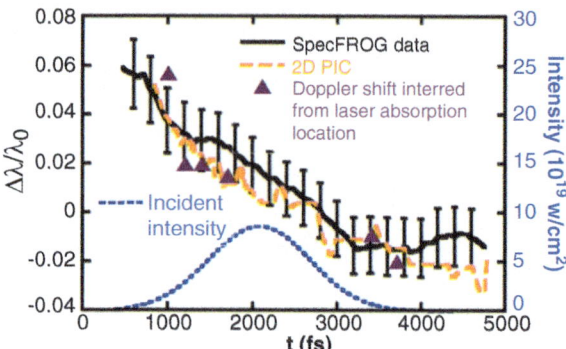

Fig. 7.18 Comparison of measured and simulated time history of spectral shift. Black solid line: experimental data. Orange dashed line: central wavelength of specular 2ω spectrum in the 2D PIC run. Purple triangles: expected Doppler shift inferred from motion of laser absorption point in the 2D PIC run. Blue dashed line: incident laser intensity profile. [Figure 2 in Ref. 14]

Fig. 7.19 (a) Snapshot of laser energy flux (P_{laser}), electron density (contour lines), and electron energy flux (F_e) at the peak power in the 2D PIC simulation. The red dashed contours of the electron density are at t = 0, and the blue contour lines are at t = 2.2 ps. The inset is a typical on-shot Titan vacuum focus laser intensity profile. [Figure 3 in Ref. 14]

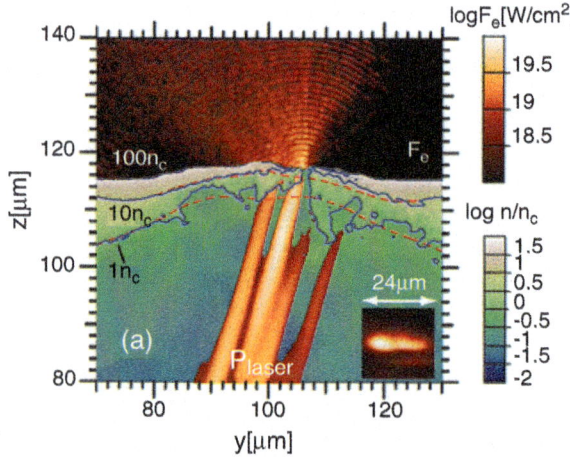

irradiated laser pulse is also plotted with blue dotted line. It is clear that the critical surface initially moves forward boring a hole in the pre-formed plasma with the velocity of about 6% of the speed of light as evaluated from (7.4.3). Then, the Doppler shift passes the null point after the peak of laser intensity, and the critical surface moves backward. This means the solid density is too heavy for the laser keeps the hole boring during the ps range. The Doppler shift obtained in 2D PIC simulation is also plotted with yellow line, suggesting that the laser breaks up to several filaments in the pre-formed plasma as shown below and the strongest filament mainly contribute the hole boring.

2D PIC simulation is carried out to compare and analyze the experimental result. Since the laser intensity distribution at the focused point is elongated in 2D space with 4:1 ratio with the long side of 24 mm as seen in the inset in Fig. 7.19, this condition is modeled in 2D code. In addition, for well computing the nonlinear laser propagation in pre-formed plasma, laser field is calculated with the vacuum boundary, 100 μm distant place from the critical surface. A snapshot of laser energy flux (P_{laser}; pointing flux), electron density (contour lines), and electron energy flux (F_e) is shown in Fig. 7.19 for the time of laser power peak. It is seen that the laser beam are subject to the filamentation instability before arriving at the critical surface. The brightest filament generates hot electrons into the solid density with wide angle. The red dotted line is the initial density contours, and the critical surface is found to have moved about 5 μm, consistent with the distance evaluated from the average value of the Doppler shift, velocity, in Fig. 7.18. At the time of the laser peak, a sharp density jump is seen at the critical surface of the brightest filament.

It is pointed out in Ref. [14] that the following simple theoretical model for energy and momentum balance between laser and plasma well explains the 2D PIC result. Forget about 2ω component of JxB, and require that the boundary pressure at the critical surface should satisfy a quasi-static condition:

$$I_+/c + I_-/c = P_e + P_i \qquad (7.5.1)$$

In addition, absorbed energy should go to the energies of electrons and ions:

$$I_+ - I_- = F_e + F_i, \qquad (7.5.2)$$

where $I_{+/-}$ denotes the incident and reflected laser intensities, $P_{e/i}$ electron and ion momentums, and $F_{e/i}$ the energy flux densities of electrons and ions, respectively. In the limit of relativistic electrons, the relation $\gamma \gg 1$ and $F_e = P_e c$ can be assumed. Assuming the ion energy flux is small in (7.5.2), the following relation is obtained:

$$P_i = 2I_-/c \qquad (7.5.3)$$

In the case of high absorption fraction, the ion pressure in (7.5.3) is much smaller than that in (7.4.3). Note that if this P_i is used in (7.4.3) instead of P_L, the hole-boring velocity given in (7.4.3) becomes much smaller. This reason can suggest the over estimation of the hole boring velocity in Fig. 7.13.

In Fig. 7.20a, the trajectory of the absorption point from 2D PIC simulation result is compared to the simple model given by (7.5.1) – (7.5.3). In Fig. 7.20b the time development of LHS and RHS in (7.5.1) is compared with the red line and blue plus marks. The relation (7.5.3) is compared with yellow line and green circles, where all data are taken from PIC simulation. The fine size of the both marks represents uncertainties due to 2D structure of plasmas. It is surprising to know that although the model is based on the simple balance relation, it well explains 2D PIC simulation result. In addition, the trajectory of the critical surface (laser piston) plotted in Fig. 7.18a is also well explained with (7.5.3) in (7.4.3). It is possible to evaluate the time evolution of the absorption fraction η_{ab} from Fig. 7.18b via the relation

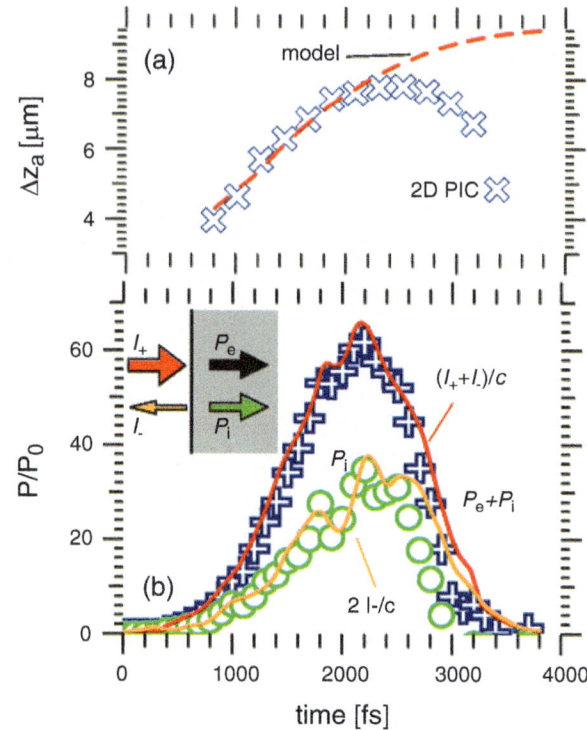

Fig. 7.20 (a) Motion of the laser absorption point in 2D PIC simulation vs simple model. The symbols are the locations where the laser field vanishes (uncertainties due to 2D structure are represented by the finite size). The line represents the motion predicted by the model equations. All positions are relative to the point of the critical density before the main pulse irradiation. (b) Momentum flux balance between laser light and plasma in units of $P_0 \equiv n_c m_e c^2 \approx 1$ Gbar for 1 µm wavelength light. The lines represent the laser momentum, and the symbols represent the electron or ion momentum. [Figure 4 in Ref. 14]

$$\eta_{ab} \approx \frac{I_+ - I_-}{I_+} = \frac{(I_+ + I_-) - 2I_-}{(I_+ + I_-) - I_-} \quad (7.5.4)$$

Then, the absorption rate near the laser peak is calculated about 60%. In high absorption case, the electrons obtain not only most of the laser energy but also take away the substantial amount of laser momentum as indicated in (7.5.1) and (7.5.3).

7.6 Absorption Efficiency Based on Conservation Laws

In the relativistic intensity regime, the absorption efficiency strongly depends on the interacting plasma density profile and a variety of nonlinear laser-plasma interactions. It is very hard to predict the absorption rate as a function of laser intensity, although in the non-relativistic regime discussed in the Chap. 2, the basic physics of the absorption process is relatively clear in both of the collisional and collisionless regimes. So, more robust theory based on one-dimensional conservation law is proposed [15] and compared with many experimental results [16]. The scattered experimental and simulation data are predicted within the

7.6 Absorption Efficiency Based on Conservation Laws

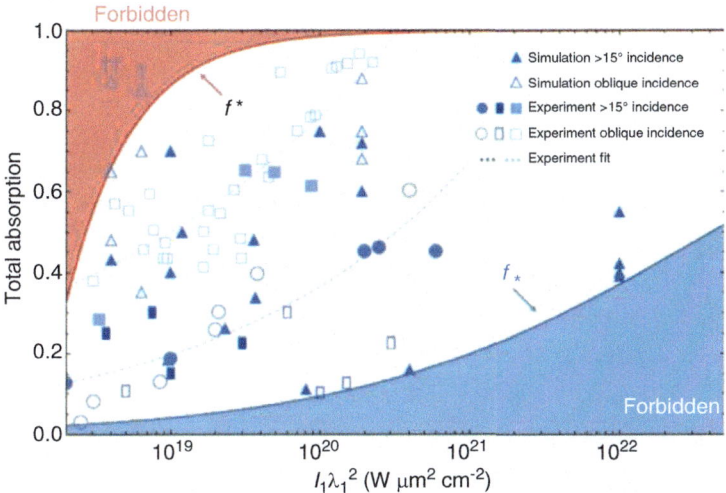

Fig. 7.21 Comparison between absorption bounds and published data. The complete data set compiled in [40] is reproduced here, spanning experimental and simulation data published over the past two decades, across a variety of laser and plasma conditions. Dashed lines corresponding to fits of selected experimental data are shown to guide the eye. Additional high-intensity simulation data are reproduced from [17]. The upper limit on absorption f* is depicted in red and the lower limit f∗ in blue, with forbidden regions indicated using shading. [Figure 3 in Ref. 16]

maximum and minimum absorption rate in the range of 10^{18} W/cm^2 ∼ 10^{22} W/cm^2 with this conservation law. The absorption rate is not given by a line; however, it is given by the upper and lower bands as seen in Fig. 7.21 [16], and the micro-physics of absorption is still open question.

Let us briefly survey the theory giving Fig. 7.21. In the conservation law, a stationary propagating hole-boring system is assumed to be seen in simple evaluation of hole-boring velocity such as in [14]. The conservation law in [15] is derived by stating with the kinetic description of hole-boring ions, bulk electrons, and hot electrons. The hole pushing boundary is assumed the interface of solid and vacuum as schematically shown in Fig. 7.22. This model is an extended **Rankine-Hugoniot relation** for a stationary propagating piston front and rear. The incident laser deposits its energy partially to the hot electrons and hole-boring electrons, and the reflecting light also deposits momentum to the electrons. The hole-boring ions are pulled by the stationary electrostatic field produced by both of the hole-boring electrons and hot electrons. The leakage of the charge of the hot electrons is assumed to be supplied by slowly drifting return current electrons.

After a precise mathematics, the following energy and momentum conservation relations are derived [15]:

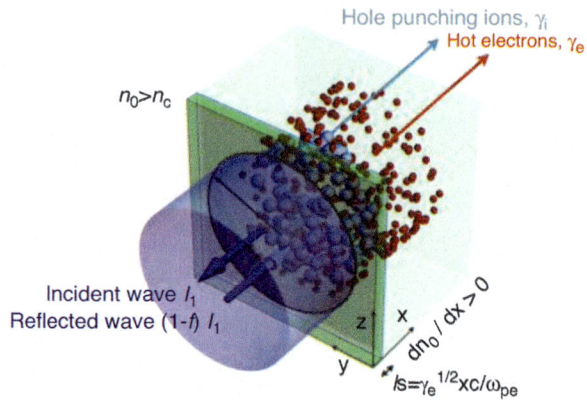

Fig. 7.22 Schematic showing key features of the relativistic laser and solid interaction. A high-power laser with strength parameter $a_0 > 1$ is shown striking an over-dense target, interacting over the Lorentz-transformed collisionless skin depth (dark green region), and exciting a highly relativistic electron flux (red spheres) and moderately relativistic ion flux (blue spheres). Laser and excited particle properties are connected across by applying relativistic Rankine-Hugoniot-like relations at the laser-matter interface. The laser is impinged and reflected (blue region). Depiction uses a frame of reference co-moving with the interface. [Figure 1 in Ref. 16]

$$(1 - R)\left(1 - \beta_p\right) I_L = (\gamma_h - 1) m_e n_h c^3 + (\gamma_i - 1) M n_i \beta_p c^3 \quad (7.6.1)$$

$$(1 + R)\left(1 - \beta_p\right) \frac{I_L}{c} = \gamma_h m_e n_h c^2 + \gamma_i M n_i u_i c, \quad (7.6.2)$$

where R, β_p, I_L, n_h, n_i, and u_i are reflection fraction, hole-boring piston velocity divided by c, incident laser intensity, hot electron density, ion density, and ion velocity, respectively, and $M = m_i + Z m_e$. In both relations, γ_i and γ_h are Lorentz factors of hot electrons and ion motion, respectively. Equation (7.6.1) is the conservation law of energy flux, where LHS is the absorbed laser energy flux, the 1st term in RHS is the hot electron energy flux, and the 2nd term is the ion energy flux. Equation (7.6.2) represents the momentum flux conservation by these three components.

It should be noted that the coupled relations (7.6.1) and (7.6.2) can be reduced to the forms of energy conversion rates to the hole-boring ions (f_i) and hot electrons (f_h), while they are given as functions of not only the incident laser intensity but also the laser reflection fraction and the plasma densities. The absorption fraction $f_{abs} = f_i + f_h$ has been obtained for various target condition and is plotted in Fig. 7.21, where the red and blue zones are forbidden region in the theory. In addition, it is clarified in [16] that the energy conversion to the hole-boring ions is enhanced at lower-density target. In addition, the energy conversion to the hot electron is high as long as the laser intensity is high enough. Typical experimental data and PIC simulation results summarized in [17] are also plotted in Fig. 7.21. In this chapter, it is clarified that the absorption efficiency is very sensitive to the plasma condition. It is not simple to show how to design the laser and solid target to obtain absorption efficiency as high as possible.

Regarding the effect of pre-formed plasmas, it is concluded that the pre-formed plasmas help increasing the absorption rate. Comparing the absorption rate from Fig. 7.8 with pre-formed plasma by pedestal beam to that shown in Fig. 7.21, it is possible to conclude that at a given intensity, the absorption rate increases with the decrease of the pre-formed plasma density, where the laser interacts with the plasma. This is roughly speculated from (7.6.1) to (7.6.2) as follows. From both relation, we can obtain an approximate relation:

$$R \approx \frac{n_h}{2I_L}\left(1 + \frac{Mn_i u_i}{m_e n_h c}\right) m_e c^2 \qquad (7.6.3)$$

Roughly speaking, the decrease of the density n_h and n_i reduces the reflection fraction for a given laser intensity I_L. The main reason for the scattering of the experimental and computational data of laser absorption in Fig. 7.21 is speculated to be due to the density of laser-plasma interaction region.

It is, therefore, suggested that multilayer target with a low-density foam layer in front of the main solid target may increase the energy conversion from relativistic laser to the plasmas. Recently, structured targets are used to control not only laser absorption but also other characteristics such as hot electron beam collimation, enhanced X-ray, gamma-ray emission, and so on. Let us see briefly the physics of the interaction of relativistic ultra-short laser with such targets.

7.7 Enhanced Coupling with Foam Layered Targets

We now know that clean ultra-short relativistic lasers irradiated on solid targets are convenient to the generation of higher harmonics via solid surface oscillation. However, it is not convenient to obtain higher laser energy absorption. In general, higher laser absorption is reported, however, for the case with pre-formed plasma by the pedestal (see Fig. 7.8). With an advancement of the laser technology, the fraction of the pedestal to the main laser, the contrast ratio, decreases dramatically so that the pre-formed plasma effect due to the pedestal is not significant in the recent experiments. As a result, it becomes clear that the laser absorption without pre-formed plasma is not efficient in irradiating on the conventional solid targets.

Now, a lot of research is focused on a variety of target design so that higher absorption efficiency is expected. As known well, laser energy is absorbed via a variety of collisionless nonlinear plasma process. Therefore, the increase of laser absorption means how highly laser energy is converted to the generation of hot electrons, the number of particles, and the temperature of the hot electrons. One simple way is not to reduce the energy of pedestal and provides pre-formed plasma by pedestal or another laser beam on purpose. Even in this case, controlling the

Fig. 7.23 Target morphology. Scanning electron micrographs of carbon foams with densities of about (**a**) 7 and (**b**) 25 mg cm^{-3}. The scale bar is common for both frames. [Figure 3 in Ref. 19]

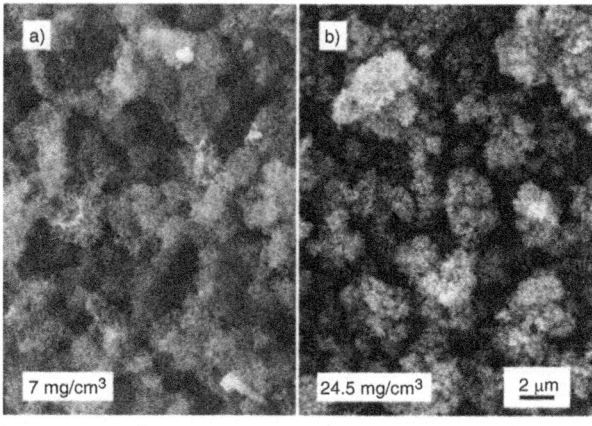

plasma density structure is not easy and the use of multilayer targets, for example, a foam layer attached with the solid target, is proposed [18], where enhanced energy coupling from laser to plasmas with use of high contrast laser pulses is demonstrated with PIC code.

Here we show real pictures of the form used for experiments. In Fig. 7.23, two different weight form materials are shown [19]. The ingredient is carbon and porous morphology is seen. The densities of (a) and (b) in Fig. 7.23 are 7 and 25 mg/cm^3, and their averaged electron densities are 1.2 and 4.3 times the critical density n_c, respectively. The foams are structured by the solid carbon with the electron density about $50n_c$, and the porous structure is characterized by an occupation factor of the solid carbon of about 2% for (a) so that the smoothed out density is equal to n_c. Typical size of the high-density part of the porous is about 0.1 μm, shorter than the laser wavelength. The average size of the vacant space of the porous is about 0.5–1 μm. It is very difficult problem to solve relativistic laser propagation in such porous dielectric matter. Many simulations have been done at first by assuming the foam as uniform low-density plasma with the electron density equal to the averaged one of the real foam.

A comprehensive physics of laser-plasma interaction for the case with the foam layer whose density is designed near the laser critical density n_c is reported in [20]. The PIC code *ALADIN* is used for 3D and 2D simulations. The targets have three layers, a low-density foam (C^{6+}; 1–12 μm), a thin metal foil (Al^{9+}; 0.5 μm), and

7.7 Enhanced Coupling with Foam Layered Targets

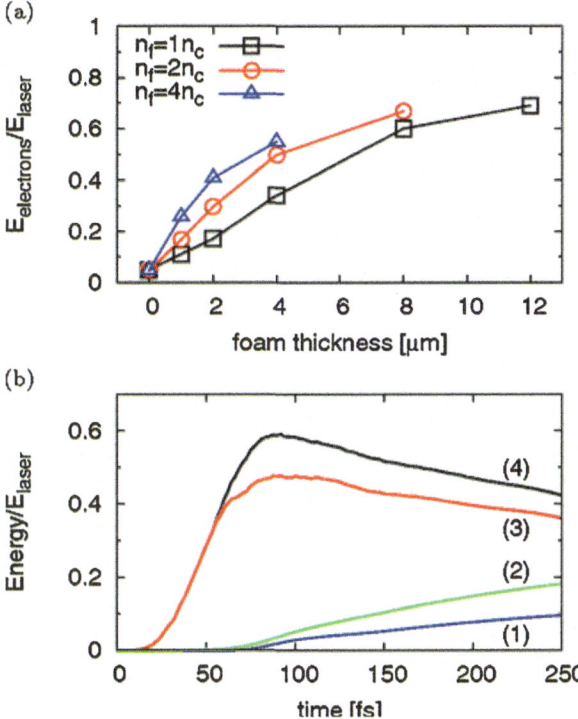

Fig. 7.24 (a) Maximum value of the electron energy absorbed by the target for different cases. (b) For the case with form thickness 8 μm and density of the critical one, the time evolution of the energy normalized to the initial laser energy of all electrons (black, 4), foam electrons only (red, 3), all ions (green, 2), and contaminant protons only (blue, 1). [Figure 3 in Ref. 20]

a thin proton contamination layer (H$^+$; 0.05 μm). The laser wavelength is 0.8 μm, pulse duration is 25 fs, and the focusing diameter is 3 μm. Focused laser intensity is varied as $a_0 = 3$–20, corresponding to $P_L = 2.8$–128 TW and laser energy 0.075–3.2 J. In the simulations, even for the case where the electron density of the form is higher than the critical density, $n_e > n_c$, the laser can propagate in the foam layer due to relativistic transparency effect as shown in (3.6.5) and ponderomotive channeling effect as shown in (3.9.1).

In Fig. 7.24a, the energy conversion fraction to the electrons from the laser is plotted as a function of foam thickness for three different densities of the foam, $n_e = n_c$, $2n_c$, and $4n_c$. It is found that without the foam layer, almost no absorption is obtained, while the laser energy coupling to the plasma electrons monotonically increases with the thickness of the foam layer, being the maximum about 70%. In Fig. 7.24b, the time evolution of the laser energy partition is plotted for the case with an 8 μm foam to the all electrons (black, 4), foam electrons (red, 3), all ions (green, 2), and contaminant proton only (blue, 1). Note that the laser peak intensity is around $t = 50$ fs and the pulse was over before $t = 100$ fs. It is clear that almost all of the laser energy is used to heat or accelerate the electrons inside the foam and the ions obtain their energy slowly via coupling through the electric field by charge separation. It is surprising that almost 10% of laser energy is transferred to the contaminated protons. Such scheme for accelerating ions by the hot electrons is

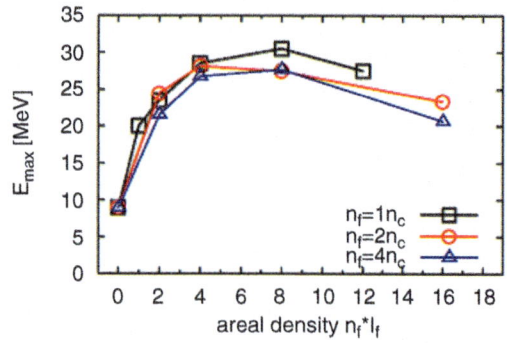

Fig. 7.25 (a) Proton maximum energy for $a_0 = 10$, different foam thicknesses, and foam densities (black squares, $n_f = n_c$; red circles, $n_f = 2n_c$; blue triangles, $n_f = 4n_c$) as a function of the areal density ($l_f\, n_f/n_c$). [Figure 5 in Ref. 20]

called **target normal sheath acceleration (TNSA)**, the physics of which will be discussed in Volume 3. It has, however, pointed out the critical issues of the ion acceleration, for example, an increase of energy of total ions and how to increase the average energy of the ions.

In Fig. 7.25, the maximum energy of the hot electrons is plotted as a function of the areal density (density)x(thickness) of the form layer. The laser intensity is $a_0 = 10$. It is interesting to know that the maximum energy does not depend on the thickness and density and can be scaled by the areal density. The **ponderomotive scaling** of the hot electron temperature shown in (6.7.10) is calculated as 3.1 MeV in the present parameters, while the E_{max} increases more than 3 MeV with the increase of the foam plasma areal density. This indicates that the ponderomotive is not enough and there are another acceleration process including the wake field, laser direct accelerations, and so on. If we assume that the hot electrons are accelerated by the wake field generated in the foam plasmas, the gained energy roughly equal to (force) × (length) should be constant. This means the electric field amplitude of the wake field is inferred to be proportional to the density.

Figure 3.26 shows snap shots of plasma at $t = 100$ fs after the end of laser pulse: (a) electron density and (b) longitudinal electric field distribution. The simulation is done for $a_0 = 10$, the form density equal to the critical density, and the thickness of the form 8 μm. The boundary of foil and foam is at $x = 0$. The laser is irradiated from the −x-direction with normal incidence. The laser field is in the p-polarization with laser electric field in 2D plane. Since the laser is focused with diameter of 3 μm, it is seen that the density perturbation waves are generated like laser-focusing cone structure accompanying with strong electrostatic electric field. It looks waves propagating to the boundary of the solid foil. The density perturbation of the waves is of the order of unity, and it may be regarded as the large amplitude plasma wave we studied in Sect. 3.9, although the present one is highly relativistic plasma waves.

Let us try to roughly evaluate the relation between the density perturbations to the electric field in the foam region in Fig. 7.26. Using the simple relation of Poisson Eq. (1.3.3):

7.7 Enhanced Coupling with Foam Layered Targets

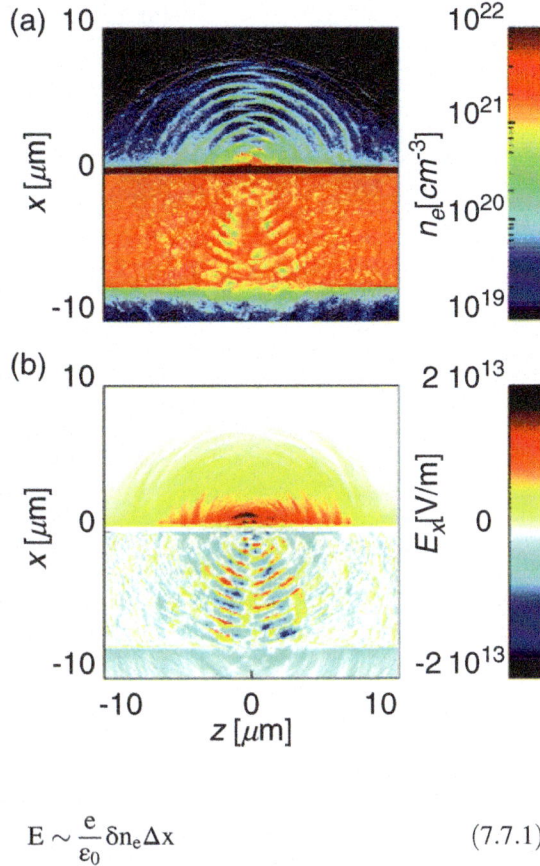

Fig. 7.26 P-polarized laser pulse ($a_0 = 10$) incident on a target with foam layer of critical density and thickness 8 μm at normal incidence: (**a**) electron density and (**b**) longitudinal electric field. [Figure 8 in Ref. 20]

$$E \sim \frac{e}{\varepsilon_0} \delta n_e \Delta x \qquad (7.7.1)$$

From Fig. 7.26a, we can roughly put $\delta n_e \sim n_c$ and $\Delta x \sim 1$ μm. Then we obtain a typical amplitude of the electrostatic field as

$$E \approx 1.8 \times 10^{13} \quad [\text{V/m}] \qquad (7.7.2)$$

This rough value can explain the maximum value in the foam plasma shown in Fig. 7.26b. If an electron is accelerated by the field strength in (7.7.2) over the form layer 8 μm, a simple evaluation of DC acceleration gives

$$E_{DC} = eEL_{\text{foam}} \approx 140 \quad [\text{MeV}] \qquad (7.7.3)$$

This value is of course over-estimated, and the simulation result is about 30 MeV as seen in Fig. 7.25. However, 30 MeV is larger than the potential energy of one wavelength being about 1 μm. Inserting $L = 0.5$ μm in (7.7.3), we obtain $E \sim 10$ MeV. This fact also suggests that there should be some stochastic process in the mechanism of accelerating the hot electrons in the foam region. The physics of

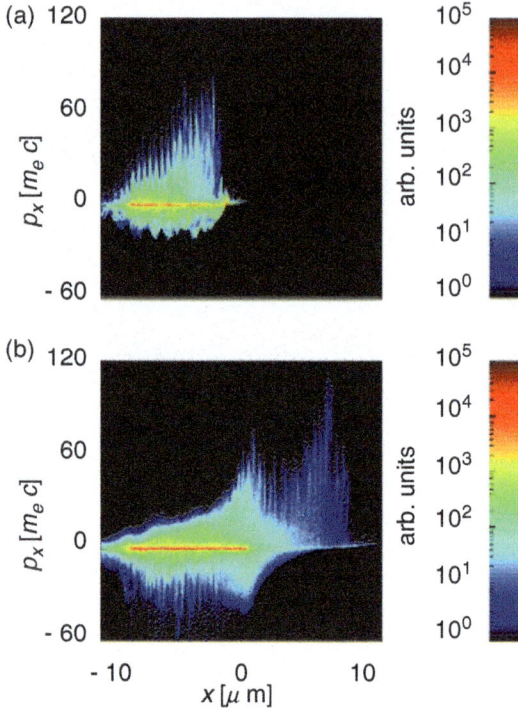

Fig. 7.27 Longitudinal phase space of the electrons of the foam in the same simulation as Fig. 7.26 at two different times, (a) 50 fs and (b) 83 fs; px in normalized mcc units. [Figure 9 in Ref. 20]

hot electron acceleration and/or generation in such stochastic system will be discussed later.

In Fig. 7.26a, expanding electrons are seen in the rear side of the target. The electrons are expanding hemispherically, accompanying relatively smooth structure of strong longitudinal electric field. The electric field is clear to be produced by the charge separation, and sheath potential confines the expanding electrons. Theory of the plasma sheath shows that the hot electrons are bouncing many times in the hemisphere and transfers small amount of energy to the ions, roughly a fraction of mass ratio in each bounce.

In Fig. 7.27, the momentum distribution of the x-component is shown as a function of space for two timings, (a) t = 50 fs (laser intensity peak) and (b) t = 83 fs (laser termination). It is clear in Fig. 7.27a that the electrons are accelerated in the foam plasma and the acceleration starts from the vacuum boundary to increase the maximum electron energy to the inside of the foam plasma. If the acceleration is due to the large amplitude plasma wave as assumed above, accelerating electrons soon run with almost the speed of light, while the plasma wave may propagate with the phase velocity faster than the speed of light. As a result, dephasing prevents the DC acceleration in (7.7.3), and electrons continue to be accelerated repeating many dephasing. As time average, with increase of the time average energy, most of accelerating electrons are drifting toward the inside of the foam. Such hand-wave argument may explain the special acceleration dynamics seen in Fig. 7.27.

7.7 Enhanced Coupling with Foam Layered Targets

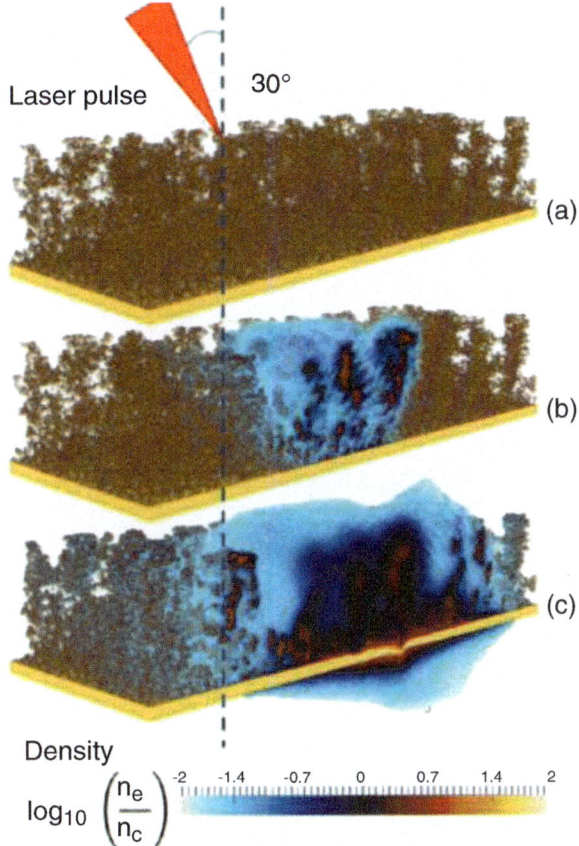

Fig. 7.28 3D PIC simulations of the electron density dynamics in the interaction between a laser pulse and a nanostructured foam at different times: (**a**) 0 fs, (**b**) 67 fs, and (**c**) 134 fs. The red cone represents the incident laser, and the yellow layer indicates Al substrate. [Figure 4 in Ref. 21]

In Fig. 7.27b, the number of hot electrons increases, and many of them are reflected by the sheath potential as we can see many electron with negative momentum. They are also confined in the target by the sheath potential generated at the vacuum boundary of the foam plasma. By repeating the bounce due to the reflection by both sheath fields, the ions are gradually accelerated. Since the mass of the proton is relatively light compared to the carbon ions in the foam plasma, the energy transfer is more efficient in the proton contamination plasma, showing about 10% energy gain in Fig. 7.24b.

Here, it is useful to see the validity of the assumption that the foam plasma can be modeled with low-density uniform plasma in PIC simulation. A 3D simulation is carried out to see the propagation of laser in entire nanostructured foam layer [19, 21]. Realistic nanostructure foam is modeled as a material composed by a random collection of 50 nm radius over-dense (n = $50n_c$) plasma nanoparticles. The initial configuration is shown in Fig. 7.28a. The porous structure is characterized by an occupation factor of about 2%, consistent, and an average electron density is approximately equal to the critical density n_c. The laser with pulse duration of 30 fs

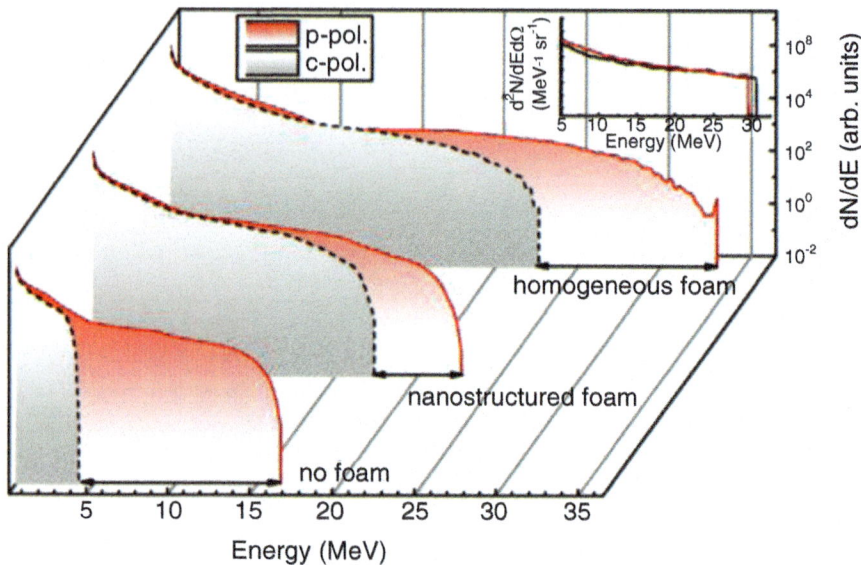

Fig. 7.29 Calculated proton energy spectra from homogeneous foam, nanostructured foam, and Al foil using p- and c-polarized laser pulse. The inset shows the corresponding experimental spectra. [Figure 5 in Ref. 21]

and a focal spot of 4 μm is irradiated on the foam at the intensity 3.5×10^{20} W/cm^2 ($a_0 = 18$). The target is of 8 μm thick porous foam on 0.8 μm thick aluminum foils ($n = 40n_c$) with 0.02 μm low-density ($10n_c$) CH layer on the rear surface.

The laser is irradiated with p-polarization as shown. It is noted that the space dimension is about 50×20 μm [19]. In Fig. 7.28b, the electron density distribution is shown at time 67 fs during the laser heating. It is seen that a relatively wide range about 20 μm diameter is ionized, although the laser-focusing spot is 4 μm. The irradiated laser is scattered and bended by the porous structure and seems to diffuse to the surrounding region. Figure 7.28b is at $t = 134$ fs, after the laser irradiation, and the density structure of the foam layer becomes almost uniform. It is concluded in [21] that comparing the 3D PIC result to the uniform density foam simulation, the energy conversion to the protons is reduced as shown in Fig. 7.29, although both are of course better than the case without foam layer on the aluminum foil. It is also necessary to note that the laser polarization dependence among p- and c (circular)-polarizations is very large for homogeneous foam and without foam in Fig. 7.29, while their difference is reduced for the case of nanostructured foam simulation. This is because of the fine structure of the surface of the nanostructures foam. The inset shows the corresponding experimental ion spectra showing almost no polarization dependence for the foam targets.

7.7.1 Experimental Result

Experiments mainly aiming at the effect of the foam layer to the accelerated proton beam energy from the contamination layer have been carried out with 1 PW laser system at IBS, Korea [19, 21]. At first, it is reported that the maximum energy of accelerated protons shown in Fig. 7.25 with PIC code is well reproduced in the experiment [19]. The value of the maximum energy of 30 MeV is well confirmed experimentally for 8 mm thick foam target [Fig. 5 in Ref. 19]. The maximum proton energy is compared for different polarization and intensity cases with 8 μm thick foam layer and aluminum foil targets experimentally [21]. The experimental result is shown in Fig. 7.30. It is surprising that the solid lines for the case with three different polarization irradiations on the foamed targets result almost the same maximum proton energy scaling to the intensity. This suggests the laser-matter interaction becomes independent of the laser polarization because the porous structure of foams. The dotted line data of aluminum foil targets without foam layer, however, show strong dependence of the polarization configuration on the solid aluminum surface.

We have discussed about the maximum energy of accelerated electrons in the case with foam layered targets. Specifically, Fig. 7.26 was used to evaluate the accelerated energy of electrons by the longitudinal electric field. The hot electron energy spectra are observed in the experiment with the *Gemini* laser at RAL, UK [22]. The laser condition is that the pulse duration 50 fs, f/2 plasma mirror, focal spot

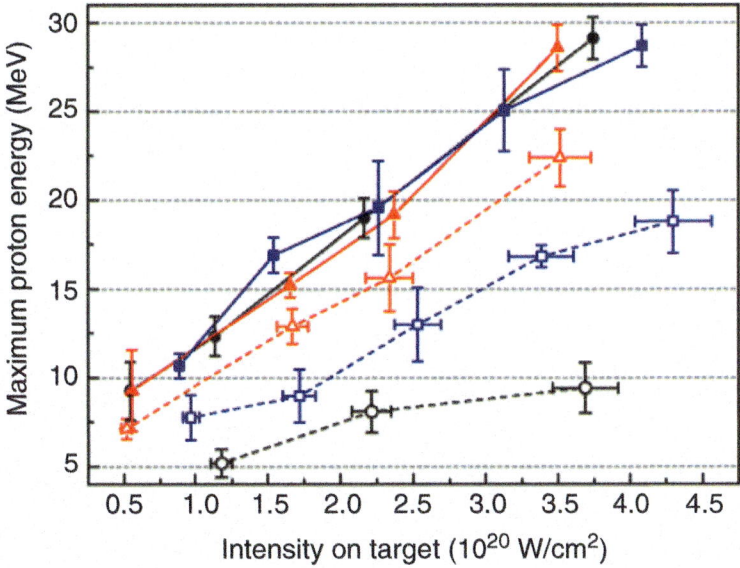

Fig. 7.30 Role of laser polarization and intensity. The maximum energy of measured protons as a function of laser intensity with s- (blue squares), p- (red triangles), and c-polarization (black circles) for solid target (empty symbol) and the double-layer target with 8 μm thick foam (full symbol). [Figure 3 in Ref. 21]

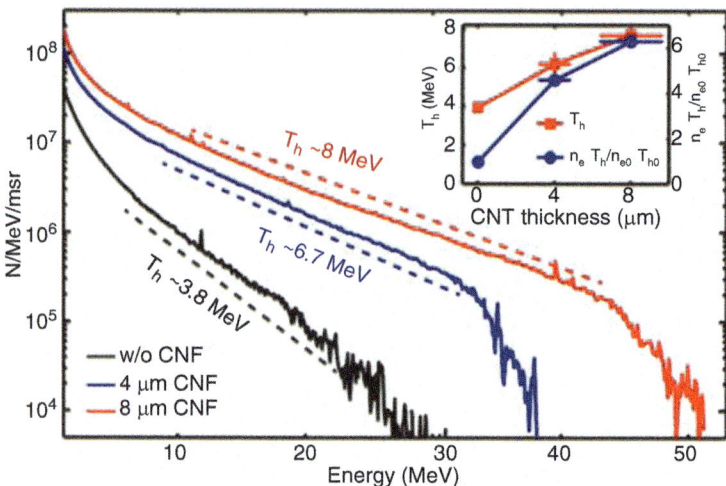

Fig. 7.31 Measured electron energy spectra along target normal (laser propagating direction) with linear-polarized pulses interacting with foam targets with different thickness attached to a 20 nm solid foil. The inset shows the comparison of corresponding electron temperature T_h and total electron energy as presented by the production $n_e T_h$ after normalization to the value $n_{e0} T_{h0}$ from 20 nm solid foil solely. Both quantities n_e and T_h are extracted from the fitting curve. [Figure 3 in Ref. 22]

3.5 μm in diameter, peak intensity 2×10^{20} W/cm² ($a_0 = 10$). The linearly polarized light is irradiated normally on the targets. The targets are a double-layer target with nanometer thin diamond-like carbon coated with carbon nanotube foam with near critical density. It is concluded that with use of near critical foam layer on a thin layer solid target, the maximum energy of proton beam is significantly enhanced from 12 MeV to 30 MeV in the experiment. This values well coincide with the results shown in Fig. 7.25, for example. Note that the laser and target conditions of the simulation and experiment are very similar.

In Fig. 7.31, hot electron energy distributions are plotted for three cases: the black line for no foam layer, blue line for 4 μm foam layer, and red line for 8 μm foam layer [22]. It is noted that the ponderomotive scaling in (7.4.4) predicts $T_h = 3$ MeV at $a_0 = 10$, while the hot electron temperatures at the cases with foam layer are **super-ponderomotive**. It is clear that the hot electrons are accelerated in the foam plasma with another physical process than the JxB force. It is also pointed out experimentally that not only the temperature but also total amount of the hot electron energy are significantly enhanced by the addition of the foam layer as shown in the inset of Fig. 7.31. Compared to without foam, the temperature increases two times, and the total energy of hot electrons increases about six times. This indicates the better nonlinear coupling of relativistic laser and the near critical foam plasma.

In Ref. [22], 2D PIC simulation is carried out to analyze the experimental data on the hot electron acceleration physics. It is concluded via PIC simulation that the efficient laser direct acceleration is the mechanism to accelerate the electrons to higher energy in the foam plasma. The physics is as follows. Focus laser on the foam surface makes well-defined plasma channel confining most of the laser energy because of the plasma self-focusing in Sect. 6.5. Then, the repelled electrons from the channel by the ponderomotive force induce radial ambipolar field. Then, some electrons oscillate in the potential of the ambipolar field. This is called **betatron oscillation**. Since the electrons have also the velocity in the channel direction, some of them can resonate with the laser electric field to keep accelerated by the laser electric field while increasing the amplitude of the betatron oscillation [23]. The authors in [21] concluded that the channel formation and betatron acceleration enhance the hot electron generation in the foam plasma region.

7.8 Efficient Absorption in Structured Targets

How to increase absorption efficiency of relativistic ultra-short lasers is a critical issue for any kind of application via plasma formation. Since the laser-plasma interaction in this regime is due to collisionless interactions and most of absorbed energies are converted to the energy of relativistic high-energy electrons. A simple strategy to enhance the collisionless interactions is to increase the surface and volume of the interaction region. However, pre-formed plasmas are not controllable as seen already. Since the laser cannot penetrate to the solid density, the use of low-density material like type of foam is beneficial for higher absorption, higher number of hot electrons, and their temperature. However, the interaction of relativistic laser and near critical density plasma is subject to many nonlinear instabilities, such as parametric instabilities, self-focusing, density modification, etc. For better coupling of laser and targets, several structured targets are proposed to keep the higher energy conversion to plasma by controlling plasma density profile interacting with lasers.

7.8.1 Micro-pillar Array Targets

Micro-pillar or nano-pillar arrays are proposed for laser-matter interaction region for efficient absorption and generation of higher energy hot electrons [24]. In Fig. 7.32, a schematic is shown for target structure used for micron size pillar target experiment. The target has a pillar of 1.5 μm in diameter spaced with 2.5 μm to the next pillar as shown in a stretched view. The laser is *PHELIX*, GSI, 1 μm wavelength, 500 fs pulse duration. Intensity is 1×10^{17} W/cm^2 (spot size; 300 μm) and 2×10^{18} W/cm^2 (spot size; 100 μm); thus, the number of pillars in the spot is 7000 and 800, respectively. ASE pedestal contrast is 10^{-10} to keep ultra-high contrast in the experiment. It is demonstrated in this experiment that the total amount of hot electrons are substantially enhanced by use of the micro-pillar targets compared to the flat solid targets.

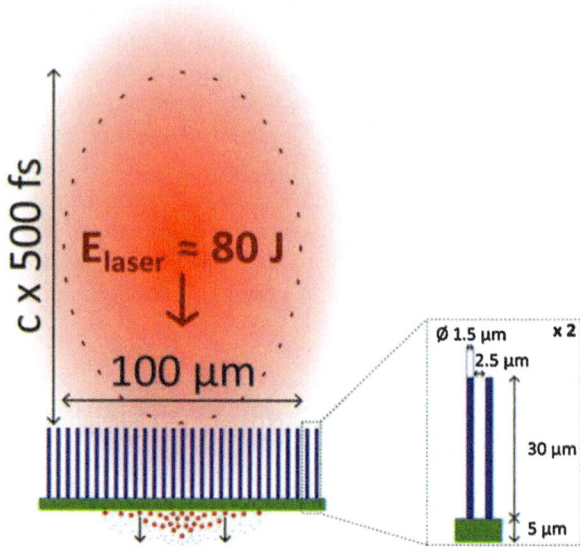

Fig. 7.32 A pulse from the PHELIX laser (red) irradiates a micro-pillar array (blue), inducing ion acceleration from the rear side of the film layer (green). All dimensions are to scale. Scanning electron microscopy (SEM) is used to characterize the targets before laser irradiation. [Figure 1 in Ref. 24]

Let us see the physics demonstrated by 2D PIC simulation. In Fig. 7.33, laser electric field distributions are plotted for (a) 1×10^{17} W/cm^2 ($a_0 = 0.24$) and (b) 2×10^{18} W/cm^2 ($a_0 = 1.2$) cases at the same time when the irradiated laser intensity is the half of the peak intensity in the front half of the pulse. The vacuum space between pillars is 2.5 µm, and the thickness of the pillar is 1.5 µm. The figure is stretched to the direction of y and the length of the pillars is 30 µm. The corresponding electron density profiles are plotted in (c) and (d), respectively. It is clear that in the space we don't see much of electrons in the low-intensity case (c) so that the laser fields almost freely propagate in the space. On the other hand, at higher intensity in (d), the electrons with the density less than the critical density are filled in the space; consequently the laser field propagates under the affection of phase modification and absorption by the electrons in space.

The advantage of the micro-pillar target is that the accelerated electrons are confined in the inside of the pillar structure and do not escape, because high-density ion charge confined the electrons produced in the space region. This has been demonstrated in Fig. 7.34, where three time snap shots of hot electrons (E > 100 keV) are plotted. The laser intensity is 2×10^{18} W/cm^2 ($a_0 = 1.2$) and its pulse duration is 500 fs. The three times in Fig. 7.34 are (a) −450 fs, (b) −250 fs, and (c) 0 fs (intensity peak). The sheath electric field surrounding each pillar surface attracts the electrons in the space to the inside of the pillar and keeps them inside the pillars by the same sheath field. In Fig. 7.34, it is seen that the hot electron density is higher in the pillar structure compared to the space as clear in (c).

7.8 Efficient Absorption in Structured Targets

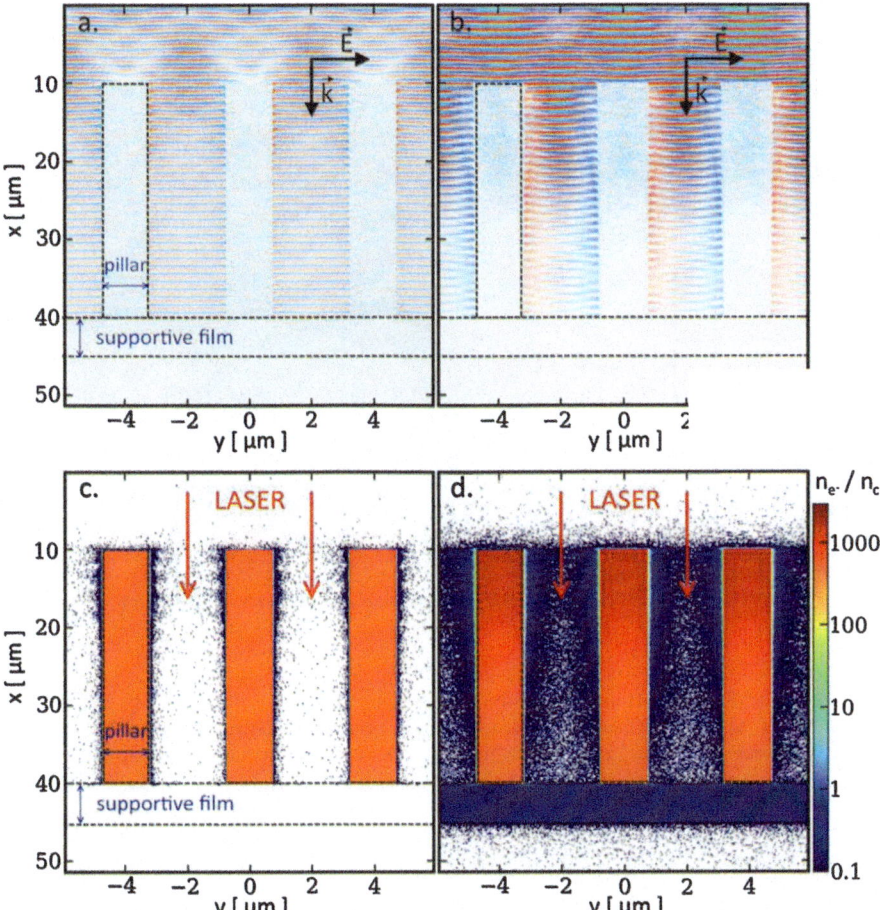

Fig. 7.33 (**a**) and (**b**) the electric field of the laser propagates between the micro-pillars in the PIC simulation (field strength in arbitrary color scale). Electrons heated up by the laser expand into the interstices. (**c**) and (**d**) the numerical calculation provides a spatial distribution of the electron densities normalized to the critical density value below which laser propagation is possible (dark blue regions). All plots correspond to the time when the laser intensity reaches half its maximum for two different laser strength parameters a_0 of 0.24 (left-hand panels) and 1.2 (right-hand panels). Attention to the difference of x and y scales. Real scale is as plotted in the enlarged view in Fig. 7.32. [Figure 4 in Ref. 24]

7.8.2 Nano-pillar Array Target

Nano-pillar arrays target is also used for experiment to see the enhancement of the hard X-ray line emissions [25]. The 2D PIC simulation of the time evolution of (a) laser electric field and (c) electron density is plotted in Fig. 7.35. The laser pulse is plotted with blue in (b), and the reflected light pulse is plotted with blue. Then, the

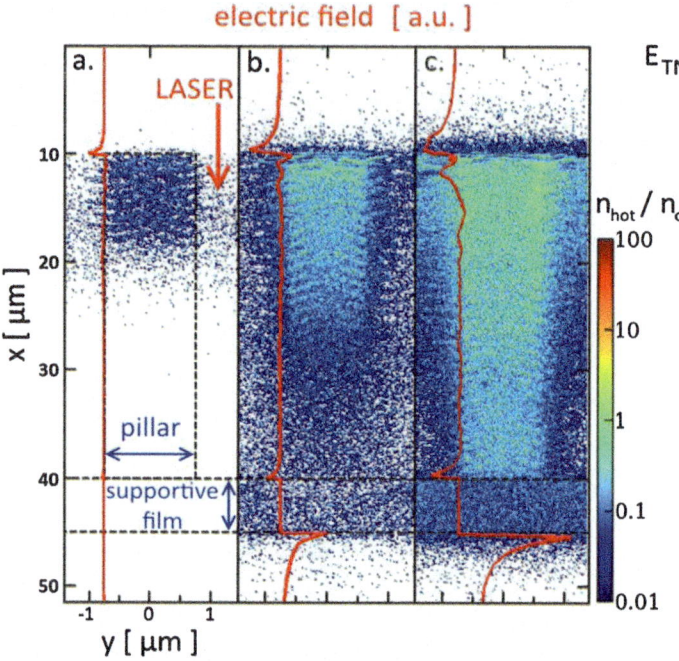

Fig. 7.34 The PIC simulation follows the evolution of the spatial distribution of hot electrons ($E > 100$ keV) at (**a**) -450 fs, (**b**) -250 fs and (**c**) 0 fs with respect to the time when the pulse peak ($a_0 = 1.2$) reaches the film layer. As the laser pulse moves forward into the pillar array, further energetic electrons are generated, and an acceleration sheath field (red curve) builds up in the pillar direction. [Figure 5 in Ref. 24]

fundamental question laser can laser propagate in the space much shorter than the laser wavelength? The nickel target with pillars made of 55 nm and spacing 100 nm and height 15 μm is irradiated by laser of 5×10^{18} W/cm^2, 400 nm wavelength, and 60 fs pulse duration. It is demonstrated experimentally that He-like Ni line emission is observed about 100 times stronger than the case of flat nickel foil experiment. It is inferred that the enhanced number of hot electrons and enhanced laser absorption are demonstrated in this experiment.

It is shown with 2D PIC simulation that the laser absorption efficiency is almost 100% and the laser electric field penetrates in the group of pillars almost uniformly. Since the diameter of the pillar is 55 nm much shorter than the laser wavelength 400 nm, the electrons inside the pillars cannot oscillate over the quivering distance by laser field and the resultant current cannot cancel the laser field in the pillars. As a result laser can penetrate into the solid pillar region. Then, the laser-electron interaction is very efficient to result enhanced absorption and higher conversion efficiency to hot electrons.

The target optimization for energy conversion to the hot electrons from irradiated laser energy, of course, depends on the laser condition such as laser intensity, focal spot size, pulse duration, etc. Compared to the large spot size and relatively

7.8 Efficient Absorption in Structured Targets

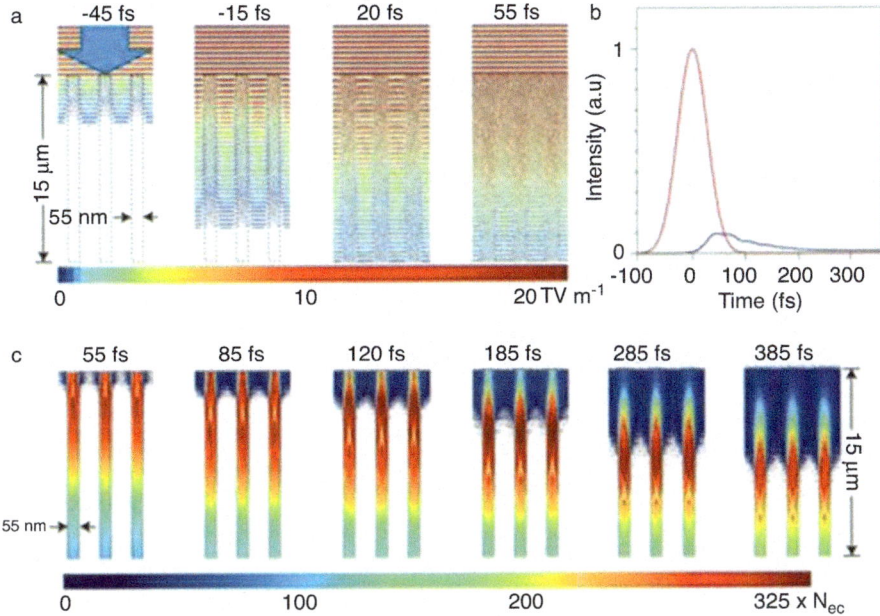

Fig. 7.35 (a) PIC simulations of the penetration of the laser beam electric field in an array of 15 μm long Ni wires with an average atomic density of 12% solid density irradiated at an intensity of 5×10^{18} W cm^{-2} by a $\lambda = 400$ nm, 60 fs (FWHM) duration laser pulse. Times are measured with respect to the peak of the laser pulse. The laser field scale is in units of TV m^{-1}. (b) Computed impinging (red) and reflected (blue) laser intensity. (c) Computed electron density evolution. The electron density scale is in units of critical density ($n_{ec} = 7 \times 10^{21}$ cm^{-3}). [Figure 1 in Ref. 25]

low-intensity experiment in [24], a small spot size and high-intensity experimental result is reported [26]. In [26], the spot size is 1.7 μm in diameter and the laser intensity 1.5×10^{21} W/cm^2 ($a_0 = 28$). In the experiment, targets with different thickness of pillar of diameter 55, 400, 600, and 1000 nm are irradiated, and it is concluded that the case with 1 μm diameter pillar case is the best performance. With 1.7 μm focal spot, only one pillar of 1 μm is in the spot. It is also reported that 55 nm nano-pillar array does not have a good performance. It may be considered that the difference of the result in [26] is mainly due to the laser intensity which makes the density of space between pillars higher than the critical density in early time.

7.8.3 Micro-tube Plasma Lenses

It is demonstrated with PIC simulation that laser irradiation in a hole of a micro-tube made of carbon manufactured by the 3D laser printing micro-machining technology can enhance the laser intensity in the hole, coupling of laser and matter, average

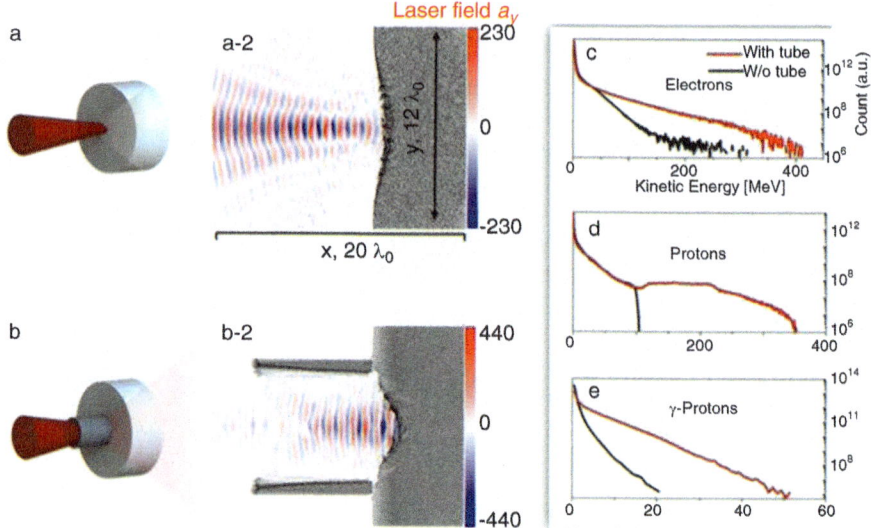

Fig. 7.36 (**a, b**) are the target setup and the corresponding distributions of the laser field and electron density. (**c–e**) shows the spectrum comparison for electrons, protons, and gamma-photons at 160 fs, respectively. [Figure 3 in Ref. 26]

energy of hot electrons, hot ions, gamma-ray, etc. [27]. In Fig. 7.36, the conventional flat target and micro-tube target are shown, where (a, b) are the target setup and the corresponding distribution of laser fields and the electron density, while (c-d) are resultant electron, ion, and γ-ray energy distributions for flat target (black) and micro-tube target (red). In the plane target (a-2), the spatial dimension is 12 μm and 20 μm, and the laser of 5×10^{22} W/cm^2 ($a_0 = 150$) is irradiated on the flat surface of plastic (CH) target.

On the other hand, the laser intensity at the solid surface after passing through the tube of 8 μm long in (b-2) is enhanced about ten times to 4.3×10^{23} W/cm^2 ($a_0 = 450$). Let us consider the physics happening in the tube. When the leading part of the laser pulse propagates in the vacuum tube, the laser and wall material interaction generates plasmas expanding to the center of the tube to fill the inside of the tube. When this density increases around the critical density and the peak of the laser pulse propagates this cylindrical plasma channel, self-focusing helps to increase its intensity. The electrons in the tube continue to interact with the laser electric field and obtain more energy after many interaction with the stochastic laser field confined in the cavity. As a result, the generated electrons, protons, and γ-photons are enhanced both in energy and amount as shown in Fig. 7.36c–e.

7.8.4 Common Rule of Better Coupling

It is challenging to predict the absorption fraction and hot electron temperature based on some theory even for simple targets. It is more challenging when the relativistic lasers are irradiated on a variety of structured targets. The physics of such relativistic laser and plasma interaction will be reviewed later. Within the experimental and computational results, nevertheless, it is seen that laser-matter coupling is enhanced and the hot electron generation is also enhanced in energy conversion rate and average energy (temperature) with use of complicated targets. Such resultant knowledge helps the engineering applications of ultra-intense lasers. Let us summarize the common role in the structured targets. We can enumerate the following advantage compared to the flat target case:

1. The structured plasma can confine the electrons for a long time in the laser interaction region.
2. The laser is also controlled so that the effective interaction surface with plasmas is enlarged by a complicated structure.
3. The laser is reflected in the structure, and a complicated laser intensity distribution is formed, enhancing the stochastic heating and direct laser.
4. The sheath field and self-generated DC magnetic field, to be explained below, are generated and are used to confine the hot electrons inside the structured region so that they gain more energy via interaction with laser electric field.

7.9 Magnetic Field Generation

Strong non-oscillating magnetic field generation is observed when intense or ultra-intense lasers are irradiated on any kind of materials. In the case of relativistic intense lasers, the strength of non-oscillating magnetic field is roughly evaluated by a dimensional analysis as follows:

$$\frac{d}{dt}\mathbf{p} = -e\mathbf{v} \times \mathbf{B} \Rightarrow B \sim \frac{m\omega}{eN} \sim \frac{107}{N} \text{ [MG]} \quad (\text{for } \lambda = 1 \text{ μm}), \quad (7.9.1)$$

where we assumed $p \sim mc$ and $v \sim c$. The time derivative is replaced with the pulse duration ($=N/\omega$), where N is the number of oscillation. Anyway, roughly speaking magnetic field of the order of tens of mega-gauss is possibly generated as DC field over the laser pulse duration.

It is better compared to the strength of laser magnetic field. It easily shown

$$B_0 = E_0/c = \frac{m\omega}{e}a_0 = 107 a_0 \text{ [MG]} \quad (7.9.2)$$

107 MG is the strength for $a_0 = 1$.

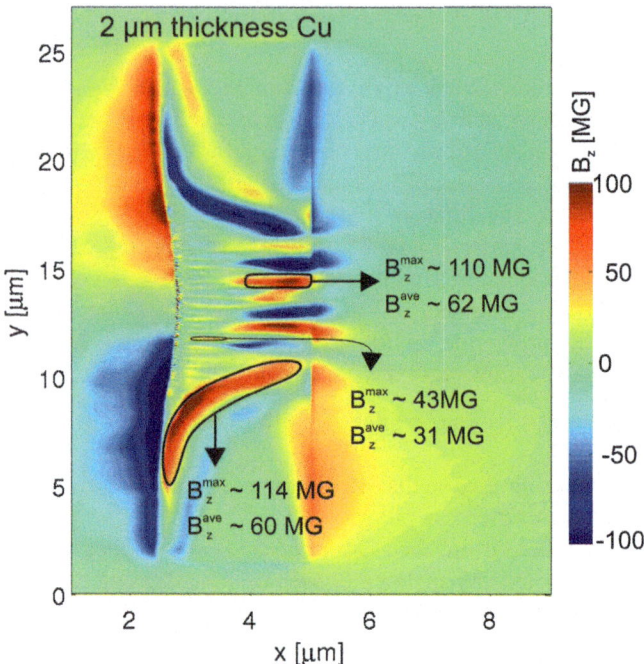

Fig. 7.37 The spatial distributions of free electron density and self-generated magnetic field in a laser irradiated Cu target with 20 μm and 2 μm thicknesses at t = 24 fs. B^{max}_z and B^{ave}_z are the maximum and spatially averaged magnetic fields in the indicated regions. For this specific simulation, we assume the laser peak intensity 10^{20} W/cm^2 and exponential pre-plasma with scale length of 0.1 μm in front of the target. This simulation uses the Thomas-Fermi ionization model, which assumes local thermal equilibrium (LTE) condition. [Figure 7d in Ref 28]

In Fig. 7.37, 2D PIC simulation result is shown for the magnetic field at the time of 24 fs after the laser peak with pulse duration of 40 fs [28]. The laser intensity is 10^{20} W/cm^2 irradiated from the left with focal spot size of 4 μm, and the target is cupper with a thickness 2 μm. The initial surface of the target is at 2.5 μm in the x-axis. The distribution of the magnetic field averaged more than laser period is shown in the figure. It is typical that the axially symmetric magnetic field is generated in both sides of the front and rear of the target if extending the image to three dimensions. The torus (ring-like) magnetic fields are generated on the both sides; however, the vector direction of magnetic field rotations is opposite. The strength of this simulation is about 100 MG in the front side and about 50 MG in the rear side. These values are roughly the value given in (7.9.1).

In addition, strong magnetic field is also generated inside the target, where the return current by the bulk electrons are subject to be induced to flow by the electrostatic field induced by the charge separation due to the escaping of generating hot electrons. At the central region of the target, the high current of hot electrons penetrates into the solid target, consequently high-current by the bulk electrons flows

7.9 Magnetic Field Generation

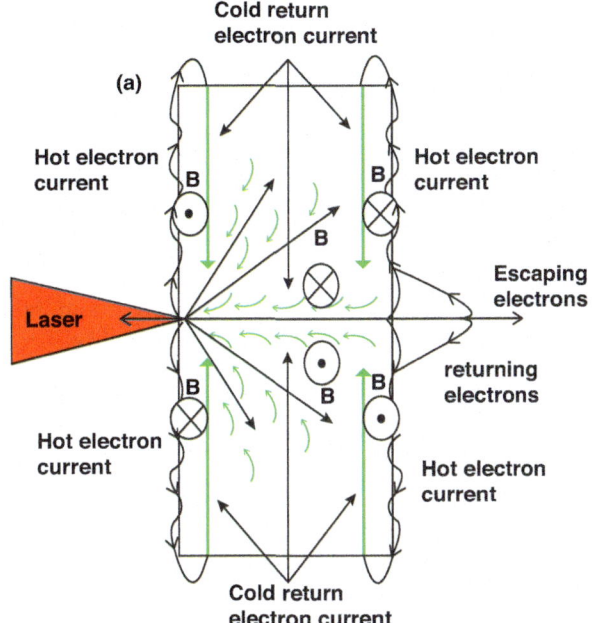

Fig. 7.38 Schematic representation of the non-oscillating B field generation by hot electrons and induced return currents. [Figure 4 in Ref. 29]

like counter stream with the hot electrons. It is well-known that such counter streaming in collisionless plasma induces a plasma instability called as **Weibel instability**. The detail of Weibel instability will be shown later. Simply saying, such counter stream is unstable to small perturbation of the magnetic field in the direction perpendicular to the flow velocities. The kinetic energy of the both flows is converted to the energy of magnetic field until a nonlinear saturation terminates Weibel instability.

It is informative to show a schematic picture showing the physical mechanism of DC magnetic field in laser-matter interaction as seen in Fig. 7.37. It is shown in Fig. 7.38, where rough trajectory of hot electrons and bulk electrons are drawn [29]. The red triangle showing a focused laser generates the hot electrons staring to run randomly in and out of the target. Due to the sheath electric field on the surface in both sides, the hot electrons with most of the electric current can escape about the distance of the Debye length, almost equal to the skin depth. So, the hot electron current flows away from the center to crawl on the surfaces as shown in Fig. 7.38. The hot electrons can escape from the central region because they have large kinetic energy. Then, the generated electrostatic field by charge separation attracts cold electrons to the central region to compensate the missing charge. This is called **return current** and cold electrons flow like green arrows. The current loops are generated near the both surfaces to generate the magnetic fields in the different direction.

Near the laser axis, the hot electrons and return current electrons flow like counter streaming because of collisionless or rarely collisional plasmas. Such plasma state is not stable, and the kinetic energy of the both currents is converted to the magnetic field energy, and of course finally the magnetic energy is also converted to the random motional energy of electrons. This is a thermalize process of statistical mechanics in collisionless plasmas.

7.10 Multi-dimensional Physics in Pre-formed Plasmas

In Chap. 6, we have discussed the physics of relativistic laser propagation in low-density plasmas. This is not simple, and still quantitative formulation based on physics models is still open question. The multi-dimensional effect due to the focused laser propagating in the pre-formed plasma is essential for the physics of relativistic laser propagation and the interaction physics, such as laser absorption, hot electron generation, magnetic field coupling, and so on. The physics of relativistic laser interaction with plasma will be described in detail later. Let us here briefly discuss a typical phenomenon seen in the pre-formed plasmas as multi-dimensional effects, namely, self-focusing, filamentation instability, magnetic field generation, and hot electron acceleration in the self-focused plasma channel.

When a relativistic laser with strength a_0 is focused into the subcritical density pre-formed plasma, its intensity increases toward the focusing point propagating from the low density to the critical density. In general, the power of laser is much higher than the critical power for the self-focusing in (6.5.19), and the laser is subject to self-focus due to the change of the plasma dielectric constants in the laser propagation path. The growth rate of the self-focusing is given as the solution of (6.5.27) by setting the size of the laser beam equal to k_y. The growth rate for large beam size is given in (6.5.29), and it is written for the 1 μm wavelength laser in the form:

$$\frac{1}{\gamma} \equiv \tau_{SF} = 4 \frac{n_{cr}}{n_0 a_0^2} \quad [\text{fs}] \tag{7.10.1}$$

Note that this is a very short time scale compared to the pulse duration for the subcritical density plasma $n_0 \sim n_{cr}$ and relativistic laser $a_0 > 1$. For example, in the case of pre-formed plasma shown with blue line in Fig. 7.2, the main laser intensity is 10^{19} W/cm^2, and pulse duration is 30 fs. Inserting $n_0 \sim n_{cr}$ and $a_0^2 \sim 10$, the growth time of the self-focusing τ_{SF} is $\tau_{SF} = 0.4$ fs. This value is very much shorter than the laser duration 30 fs.

As clear from such evaluation, the most of the experiments of solid targets discussed so far can be regarded that the self-focusing and resultant physics controlled the physical phenomena in the pre-formed plasmas. The self-focusing is

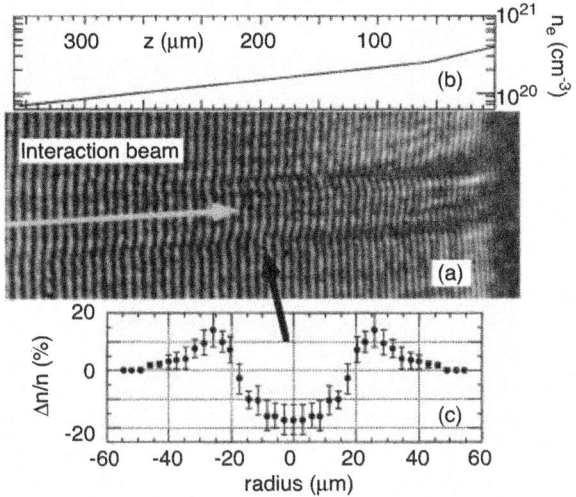

Fig. 7.39 (a) Interferogram showing the relativistic interaction beam channeling; (b) density profile along the propagation axis; (c) radial density profile extracted at a density, 10^{20} cm^{-3}. Error bars in the radial density profile take into account uncertainties on the background density and the fringe pattern localization. The 3.3×10^{18} W/cm^{2}µm^2 interaction beam comes 500 ps after the creation beam on the 30 µm foil and the probe beam 5 ps after the interaction beam. [Figure 1 in Ref. 30]

observed experimentally in [30]. The pre-formed type plasma is produced by irradiating a long pulse laser on CH foil before irradiation of 600 fs laser with $a_0 = 1.5$ intensity. The image by interferometry technique is obtained at 5 ps after the 600 fs laser irradiation. The resultant density profiles along the laser propagation and the cut view of the radial density profile are shown in Fig. 7.39. The 600 fs pulse is focused with focal spot of 4×5 µm^2 by relatively long focus f/6. The measured density scale length is $L \sim 200-300$ µm, which is much longer than the pre-formed plasma by the pedestals.

In Fig. 7.39a, laser beam channel is clearly seen as a single channel and is broken up to small channel near the right edge. The radial electron density profile of the difference from the surrounding density 10^{20} cm^{-3} is plotted from Fig. 7.39a. It is speculated that the ponderomotive force repels the electrons from the laser propagation channel to induce electrostatic field by charge separation. This field works as a driving force for the ions to follow the electrons with a typical velocity given in (7.1.2):

$$V_{PM} \tilde{6} \times 10^8 \ [\text{cm/s}] \tag{7.10.2}$$

This value well coincides with the observed value 5×10^8 cm/s [30].

When the self-focusing proceeds and the self-focused laser intensity increases, the filamentation instability with larger k_y becomes also unstable, and the laser beam tends to be broken up to small filaments. This is already computed related to the experiment shown in Fig. 7.19, and 3D PIC simulation shows that the laser beam is broken up to several beamlets in the pre-formed plasma before approaching the solid target surface [14]. However, the situation is more complicated whether the filamentation instability finally breaks up the relativistic laser to filaments in the low-density pre-formed plasma region or the filaments tend to coalescence due to the additional factor appearing in the laser channel. Let us consider the effect of electron acceleration and resultant magnetic field around the self-focused laser channel.

In the same experiment, static magnetic field of 35 MG is also measured by use of Faraday rotation technique [30]. It is well accepted that the magnetic field is produced along the laser channel by 2D and 3D PIC simulations. In the pre-formed plasma in front of solid targets, it is demonstrated that the absorption is very much enhanced and most of the absorbed energy is used to generate the hot electrons. Except for the JxB heating process, which theoretically concludes the hot electron temperature is given by the ponderomotive scaling in (7.4.4), most of the cases show the hot electron energy is higher than the JxB heating one. There should be another efficient acceleration mechanism in laser-plasma interaction in the pre-formed plasmas. Let us consider briefly the physics expected in the laser channel in the pre-formed subcritical density plasmas.

When a relativistic laser is impinged from the vacuum to the pre-formed plasmas, low-density electrons are pushed toward higher-density region as explained in Chap. 6 by assuming one dimension. In the multi-dimensional case, the laser is finite size, and the charge separation due to the moving of electrons in the laser channel can be compensated by the electrons from the outside of the laser channel. If the charge depletion due to the accelerated electrons with almost the velocity of V_d in (5.3.25) in the channel is compensated by the electrons from the outside, the hot electrons with the velocity V_d, namely,

$$\frac{E}{mc^2} \approx \frac{1}{\sqrt{1 - \widehat{V}_d^2}} = \frac{1}{2}\sqrt{a_0^2 + 4} \to \frac{a_0}{2} \quad (7.10.3)$$

can be generated during the laser pulse. This average energy is almost the same as the ponderomotive scaling in (7.4.4), since both forces have the same origin due to the JxB force.

In addition, the wake field explained in Sect. 6.1 is also produced in the channel. The wake field has longitudinal electric field whose phase velocity is almost the speed of light and induces the acceleration of a fraction of electrons to relativistic energy. The parametric instability discussed in Sect. 6.4 has a large growth rate as shown in (6.4.4) to generate plasma waves propagating forward direction.

Assume, as seen above, that there some physical mechanism to accelerate electrons inside the channel toward the laser incident direction, it is obvious that a strong static magnetic field wrapping the channel is generated by the charge current due to the accelerating electrons. When the single channel has broken up to several channels as in Fig. 7.19, each channel may also have a hot electron flow inside and consequently the static magnetic field surrounding each channel. The Lorentz forces to the neighbor current channels in the same direction of current vectors are attractive force to cause the merger of the channels. So, the coalescence of the broken-up filaments may occur to unify to form a single channel in a certain condition. It is, however, difficult to say which is dominant: the breakup of the laser beam or the coalescence of the beams to form a single beam. It may be difficult to control the phenomena because the physics is highly nonlinear.

References

1. L. Yu et al., Opt. Express **26**, 2625 (2018)
2. T. Mandal et al., Phys. Plasmas **26**, 013103 (2019)
3. L. Chopineau et al., Phys. Rev. X **9**, 011050 (2019)
4. Z. M. Sheng, Chin. Phys. B **24**, 015201 (2015): Phys. Rev. Lett. **88**, 055004 (2002)
5. A. Tarasevitch et al., Phys. Rev. Lett. **98**, 103902 (2007)
6. M.J. Streeter et al., New J. Phys. **13**, 023041 (2011)
7. A.S. Pirozhkov et al., Appl. Phys. Lett. **94**, 241102 (2009)
8. Y. Ping et al., Phys. Rev. Lett. **100**, 085004 (2008)
9. R.J. Gray et al., New J. Phys. **20**, 033021 (2018)
10. A. Yogo et al., Sci. Rep. **7**, 42451 (2017)
11. S.C. Wilks et al., Phys. Rev. Lett. **69**, 1383 (1992)
12. S.C. Wilks, Phys. Fluids B **5**, 2603 (1993); SC. Wilks, W.L. Kruer, IEEE J. Quantum Electron. **33**, 1954 (1997)
13. L. Willingale et al., Phys. Rev. Lett. **102**, 125002 (2009)
14. Y. Ping et al., Phys. Rev. Lett. **109**, 145006 (2012)
15. M.C. Levy et al., Phys. Plasmas **20**, 103101 (2013)
16. M.C. Levy et al., Nat. Commun. **5**, 4149 (2014)
17. J.R. Davis, Plasma Phys. Controlled Fusion **51**, 014006 (2009)
18. T. Nakamura et al., Phys. Plasmas **17**, 113107 (2010)
19. I. Principle et al., Plasma Phys. Controlled Fusion **58**, 034019 (2016)
20. A. Sgattoni et al., Phys. Rev. E **85**, 036405 (2012)
21. M. Passoni et al., Phys. Rev. Accel. Beams **19**, 061301 (2016)
22. J.H. Bin et al., Phys. Rev. Lett. **120**, 074801 (2018)
23. A. Pukov, Z.M. Sheng, J. Meyer-ter-Vehn, Phys. Plasmas **6**, 2847 (1999)
24. D. Khaghani et al., Sci. Rep. **7**, 11366 (2017)
25. M.A. Purvis et al., Nat. Photonics **7**, 796 (2013)
26. M. Dozieres et al., Plasma Phys. Controlled Fusion **61**, 065016 (2019)
27. L.L. Ji et al., Sci. Rep. **6**, 23256 (2016)
28. L.G. Huang, H. Takabe, T.E. Cowan, High Power Laser Sci. Eng. **7**, e22 (2019)
29. B. Albertazzi et al., Phys. Plasmas **22**, 123108 (2015)
30. J. Fuchs et al., Phys. Rev. Lett. **80**, 1658 (1998)

Chapter 8
Chaos due to Relativistic Effect

8.1 Basic Relation of an Electron in Relativistic Laser Field

When plane strong laser with linearly polarized in the y-direction propagates in the x-direction in vacuum, the basic equations for an electron in the plane of x and y under the wave field by the laser are given as follows with inclusion of the force by a longitudinal electric field E_x in the x-direction:

$$\frac{dp_x}{dt} = -ev_y B_z - eE_x \tag{8.1.1}$$

$$\frac{dp_y}{dt} = -eE_y + ev_x B_z \tag{8.1.2}$$

$$mc^2 \frac{d\gamma}{dt} = -ev_y E_y - ev_x E_x \tag{8.1.3}$$

where E_y and B_z are due to the laser field and given with the vector potential in the form:

$$\begin{aligned} E_y &= -\frac{\partial A}{\partial t} \\ B_z &= \frac{\partial A}{\partial z} \end{aligned} \tag{8.1.4}$$

The vector potential of the plane wave propagating in the x-direction is given as a vector in the y-direction $(0, A, 0)$ in the form:

$$A = A_0 \cos(\xi) \tag{8.1.5}$$

$$\xi = \omega(t - x/c) + \xi_0 \qquad (8.1.6)$$

where ξ is the phase of the wave with a constant ξ_0 for a general description. In (8.1.1) and (8.1.3), a constant electric field in the x-direction E_x is assumed as an external field to make the present discussion more general.

According to the relation given in (5.3.13), (8.1.1)–(8.1.3) are normalized and given in the following form:

$$\frac{d\widehat{p}_x}{dt} = -\widehat{E}_x + \widehat{v}_y \frac{\partial a}{\partial \xi} \qquad (8.1.7)$$

$$\frac{d\widehat{p}_y}{dt} = \frac{da}{dt} \qquad (8.1.8)$$

$$\frac{d\gamma}{dt} = \widehat{v}_y \frac{\partial a}{\partial \xi} - \widehat{v}_x \widehat{E}_x \qquad (8.1.9)$$

where a new dimensionless value for E_x is defined:

$$\widehat{E}_x = \frac{eE_x}{m\omega c} \qquad (8.1.10)$$

In what follows, note that the normalized physical quantities are shown without the hut, ^, on the tops of the variables as far as they are clearly identified as normalized quantities. The following relations are used to change the time to that in a moving particle frame:

$$\frac{d\xi}{dt} = 1 - v_x \qquad (8.1.11)$$

$$\frac{d\tau}{dt} = \frac{1}{\gamma} \qquad (8.1.12)$$

Equation (8.1.11) represents the change of phase of the wave (8.1.5) at the position of the electron. It is clear from (8.1.11) that the phase change of the laser field becomes very slow when the electron is accelerated in the x-direction to almost the speed of light. The time τ is the **proper time** of the electron defined by the relativistic dynamics in (5.2.40), and it means the relativistic time in the frame moving with the electron. It is essential that the proper time becomes shorter as the increase of particle energy and the proper time stops in the limit of infinite energy.

The normalized laser vector potential (8.1.5) defined in (5.2.46) is shown as

$$a = a_0 \cos(\xi) \qquad (8.1.13)$$

which should satisfy the propagation relation:

8.1 Basic Relation of an Electron in Relativistic Laser Field

$$\frac{\partial a}{\partial t} + \frac{\partial a}{\partial x} = 0 \qquad (8.1.14)$$

Equation of motions are rewritten in the non-dimensional form as

$$\frac{d}{dt}\left(p_y - a\right) = 0 \qquad (8.1.15)$$

$$\frac{d}{dt}R = (1 - v_x)E_x \qquad (8.1.16)$$

where (8.1.16) is derived by eliminating (8.1.7) from (8.1.9). The new variable R defined below is called a **dephasing rate** and given in the forms [1]:

$$\begin{aligned} R &= \gamma - p_x \\ &= \gamma\frac{d\xi}{dt} \\ &= \frac{d\xi}{d\tau} \end{aligned} \qquad (8.1.17)$$

Note that $R = \alpha$ in (5.3.12) without the external electric field, E_x. With use of the constant α, (8.1.16) can be given in the form:

$$\begin{aligned} R &= \alpha + \int (1 - v_x)E_x dt \\ &= \alpha + \int E_x \frac{R}{\gamma} dt \end{aligned} \qquad (8.1.18)$$

The value of α is determined by the initial condition of an electron. If the longitudinal field is localized, the value of R increases for $E_x > 0$ but decreases for $E_x < 0$ after passing the region where E_x is finite.

For the case without the longitudinal electric field E_x, an electron under the laser field with linear polarization can be solved for a general case with two **constant of motions**, β and α:

$$\gamma - p_x = \alpha \qquad (8.1.19)$$

$$p_y - a = \beta \qquad (8.1.20)$$

In (8.1.19) and (8.1.20), α is a constant given by the initial condition as shown in (5.3.12), and β is also a constant given by the initial condition, although $\beta = 0$ is assumed in the previous discussion as (5.3.8).

Let us assume $\beta = 0$ and use the relation $p_y = a$ in (8.1.19); we easily obtain the following relation:

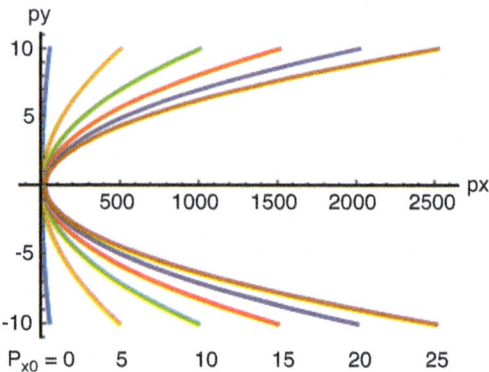

Fig. 8.1 Periodic motion of an electron in the momentum space under a constant laser field with different initial momentum. Laser with $a_0 = 10$ is given for an electrons with initial momentum of 0, 5, 10, 15, 20, and 25. The maximum value of the x-momentum is very sensitive to the initial momentum. The laser is assumed linearly polarized in y-direction. The oscillation frequency of the motion in the y-direction is ω, while 2ω in the x-direction. This is nonlinear oscillation to show chaotic dynamics in a certain condition. In non-relativistic limit, the motion is only in the y-direction

$$p_x = \frac{1}{2\alpha}\left(p_y^2 + 1 - \alpha^2\right) \qquad (8.1.21)$$

The constant α is determined by the initial condition, $p_x = p_{x0}$ and $p_y = p_{y0}$ at $t = 0$. Then, (8.1.21) can be rewritten as

$$p_x = \frac{1}{2\alpha}p_y^2 + p_{x0} - \frac{1}{2\alpha}p_{y0}^2 \qquad (8.1.22)$$

In what follows, consider the property of the solution of (8.1.22) in the case with $p_{y0} = 0$, namely, $\beta = 0$ for $a = 0$ at $t = 0$.

In Fig. 8.1, the relation (8.1.22) is plotted for the case with the laser amplitude of $a_0 = 10$ and the initial x-momentum $p_{x0} = 0, 5, 10, 15, 20, 25$. It is clearly seen that the maximum energy increases dramatically with increase of the initial momentum in the laser propagation direction. The physical reason of the increase of the maximum energy of the oscillation becomes clear by plotting Lorentz factor γ (or p_x) as a function of the phase ξ of the laser field in (8.1.13).

In Fig. 8.2, the relation between (ξ, γ) is plotted for the different p_{x0} same as in Fig. 8.1. The phase of the laser field ξ given in (8.1.6) is given in the normalized form:

$$\xi = t - x \qquad (8.1.23)$$

Since the laser propagates with the speed of light satisfying the relation $x = t$, the laser phase ξ at a particle position x should satisfy the condition that ξ is positive and

8.1 Basic Relation of an Electron in Relativistic Laser Field

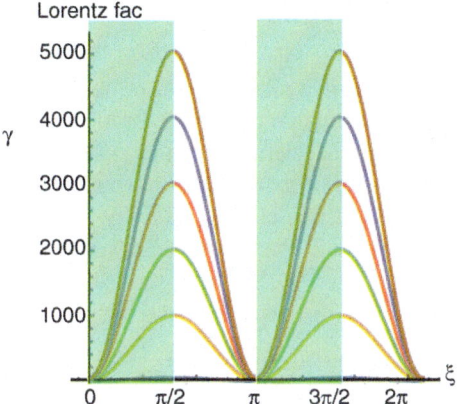

Fig. 8.2 Electron energies as functions of the phase trajectories in laser field for the different initial momentums as in Fig. 8.1. The colored region in the phase is the acceleration phase for electrons. The energy increases because electrons can remain in the acceleration phase with their velocity near the speed of light. The interaction time is longer for the electrons with higher velocity near the speed of light. This is the reason why the maximum energy strongly depends on the initial velocity

increase with time. The plotted trajectories in Fig. 8.2 show electron trajectories stating with different p_{x0} and all stat from $\xi = 0$ point. Each particle is accelerated mainly in the x-direction by **v**×**B** force during the phase $0 < \xi < \pi/2$. After this phase, the direction of the vector **B** becomes opposite, and electrons are decelerated. Since **v**×**B** force oscillates with 2ω frequency, the same process is repeated with every π in ξ.

The reason why the maximum energy is higher for larger initial x-momentum p_{x0} is easily understood as follows. If the x-momentum is large, the speed of the electrons is nearer the speed of light. Since the phase change at the speed of light, electrons running with near the speed of light in the x-direction can stay in the acceleration phase for relatively long time. For this long time, the electrons are accelerated to increase the x-velocity and Lorentz factor by the **v**×**B** force. And further increase of the maximum energy is expected as higher the initial x-momentum and velocity. It is noted that if we can design some electromagnetic field propagating at slightly less velocity of the light, this can be used for the accelerator.

It is very important to know that if there is some external force to change the constant of motion α in (8.1.19), electrons can be accelerated or decelerated. For example, some force increases the p_x due to some external kick in laser field, and then the trajectory can jump from one to the other, or diffuse, in Fig. 8.1 or 8.2 between different curves. If this happens randomly, such process is called **stochastic heating**. Before discussing the stochastic heating, let us see such acceleration though external one action (kicking) in the following sections.

8.2 Laser Direct Acceleration

When an electron is oscillating in a relativistic laser field, it can have a possibility to be accelerated after an abrupt change of a constant of motion by an external force. Since the electron is possibly accelerated by the laser field with new constant of motions, it can be called a kind of **laser direct acceleration**. In Sect. 5.3, we have assumed an adiabatic motion of free electron initially at rest before the laser filed is applied. As shown in Appendix 2, the **adiabatic relation** is satisfied for

$$\alpha = 1$$
$$\beta = 0 \tag{8.2.1}$$

Then, such adiabatic electron has the solution from (8.1.19) to (8.1.21) in the form:

$$p_y = a \tag{8.2.2}$$

$$p_x = \frac{1}{2}p_y^2 = \frac{1}{2}a^2 \tag{8.2.3}$$

$$\gamma = \sqrt{p_x^2 + p_y^2 + 1} = \frac{1}{2}a^2 + 1 \tag{8.2.4}$$

Clearly the motion is periodic, and p_x oscillates with 2ω frequency from a in (8.1.13). The trajectory of the two-dimensional (p_x, p_y) plane is parabolic as shown by (a) in Fig. 8.1. The adiabatic motion of an electron in relativistic laser field has an average energy and average drift velocity in the x-direction of the forms:

$$\langle \gamma \rangle - 1 = \frac{1}{4}a_0^2, \quad \langle v_x \rangle = \frac{a_0^2}{a_0^2 + 4} \tag{8.2.5}$$

It is noted that if the electron is ejected from the laser interaction region non-adiabatically, it has the ejection angle θ from the x-direction as derived from (8.2.2) to (8.2.4):

$$\tan^2\theta = \left(\frac{p_y}{p_x}\right)^2 = \frac{2}{\gamma - 2} \tag{8.2.6}$$

This indicates that for a highly relativistic case, the accelerated electrons are ejected with a narrow angle spread. In addition, the electron energy in (8.2.4) is about a_0 times higher than the ponderomotive scaling energy in (6.3.6).

8.2.1 Acceleration (1)

Let us solve an electron motion with arbitrary constants α and β. How can electrons be accelerated non-adiabatically, if it jumps into a certain phase of the laser field a(t, x) abruptly at the beginning? Then, (8.2.3) can be rewritten as the relation to the two momentums:

$$p_x = \frac{1}{2\alpha}\left(p_y^2 + 1 - \alpha^2\right)$$
$$= p_{x0} + \frac{1}{2\alpha}\left(p_y^2 - p_{y0}^2\right) \quad (8.2.7)$$

where p_{x0} and p_{y0} are the initial values of both momentums. There are two simple cases where the solution has higher maximum energy than the free electron oscillation of (8.2.3).

One is an electron injection with the following conditions:

$$\left(p_{x0}, p_{y0}\right) = (0, 0),$$
$$a = a_0, \quad (8.2.8)$$

Then, the constants are

$$\alpha = 1$$
$$\beta = -a_0 \quad (8.2.9)$$

The electron should be injected at the time of the maximum of a, namely, the timing that the electric and magnetic field changes their sign. Then, (8.2.7) is given as

$$p_y = a + a_0, \quad p_x = \frac{1}{2}(a + a_0)^2 \quad (8.2.10)$$

$$\gamma = \frac{1}{2}(a + a_0)^2 + 1 \quad (8.2.11)$$

So, the electron drifts in the y-direction with average momentum a_0. The maximum energy can reach to

$$\gamma_{max} - 1 = 2a_0^2 \quad (8.2.12)$$

This is four times larger than the adiabatic motion in (8.2.4).

8.2.2 Acceleration (2)

The second simple case is to start the electron motion with large initial momentum in the x-direction. From the general relation (8.2.7), higher maximum energy is obtained if the initial value α is small enough:

$$0 < \alpha \ll 1 \qquad (8.2.13)$$

The condition (8.2.13) can be satisfied with the following initial conditions:

$$p_{y0} = 0, \quad p_{x0} \gg 1$$
$$\alpha = \gamma - p_x \approx \frac{1}{2p_{x0}} \ll 1 \qquad (8.2.14)$$

The particle injection with higher x-momentum into strong laser field is required for this such acceleration. The maximum of the energy is given from (8.2.14) to (8.2.7) that

$$p_x \approx a_0^2 p_{x0}$$
$$\gamma_{max} - 1 \approx a_0^2 \gamma_0 \qquad (8.2.15)$$

where γ_0 is the Lorentz factor of the electron at the injection to laser field. This indicates that the maximum energy of an electron in relativistic laser field can be enhanced in proportion to laser intensity. This is substantial acceleration only when the normalized laser strength is much larger than the unity ($a_0 \gg 1$).

It is important to know that the solutions of the above acceleration cases are oscillating solutions, and in order to obtain the super high-energy electrons with near the maximum energies in (8.2.12) and (8.2.15), the electron has to escape from the plane wave system. Therefore, for studying the statistical distribution of the super high-energy electrons, the boundary condition in space and the finiteness of the laser pulse length is important.

8.2.3 PIC Simulations

Related computer simulation with PIC code has been carried out in Ref. [2]. It is pointed out the strong acceleration as seen above can be obtained for two cases where self-generated magnetic field helps to inject the source of hot electrons into the laser propagating channel with relatively large x-momentum, and the over-threshold ionization provides a fresh electron with relatively large initial momentum in x-direction. PIC simulation has been carried out under the following condition.

8.2 Laser Direct Acceleration

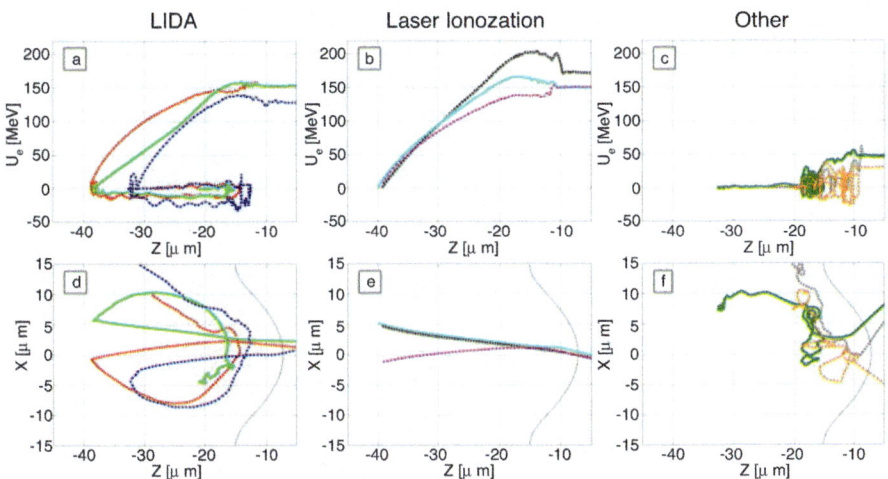

Fig. 8.3 Nine particle tracks which highlight the characteristic features of each injection mechanism. Three LIDA tracks are plotted in red, blue, and green in (**a**) and (**d**); three laser ionization-injection tracks are plotted in black, pink, and cyan in (**b**) and (**e**); three other tracks are plotted in dark green, orange, and gray in (**c**) and (**f**). [Figure 3 in Ref. 2]

A $\lambda = 1$ μm Gaussian laser of 175 fs and about 9 μm intensity width in y-direction is irradiated to the plasma at the intensity of 6×10^{20} W/cm^2, corresponding to $a_0 = 7.2$ ($a_0^2 = 52$). Note that this intensity is the case of $a_0 \gg 1$ as mentioned above. The plasma is a solid target and the pre-formed plasma with its scale length 3 μm. The plasma length with the density less than the critical density is about 20 μm. The aluminum plasma is assumed partially ionized with the charge of +3, and the statistical ionization process via the over-threshold ionization is modeled in the PIC code. Two types of the generation of super-ponderomotive hot electrons are found in [2]. The **super-ponderomotive** hot electron means electrons with the kinetic energy much larger than the ponderomotive scaling temperature defined in (3.16.4), which is about $T_h \sim 5$ MeV for $a_0 = 7.2$.

In Fig. 8.3, typical time evolution of three different kind of acceleration is plotted from Fig. 8.3a–f. For example, the time history of energy and position of three electrons classified as **LIDA (loop-injected direct acceleration)** is shown in Fig. 8.3a, d. It is clear that the electrons come to be accelerated from the outside of the laser beam width (~9 μm) to the strong laser intensity channel and then they go out of the channel to take orbits shown by self-generated magnetic field. In Fig. 8.4, magnetic field in the z-direction obtained in PIC simulation is shown by taking time average to obtain almost steady component. The magnetic field is generated as explained in Sect. 7.9. Almost constant magnetic region spreads in both sides of the laser channel, and its value is about 75 MG. The Larmor radius for the ponderomotive electrons ~5 MeV is about 20 μm. This means an electron escaping from the laser channel with the ponderomotive scaling energy can possibly reenter the laser channel near the vacuum boundary with relatively large initial momentum.

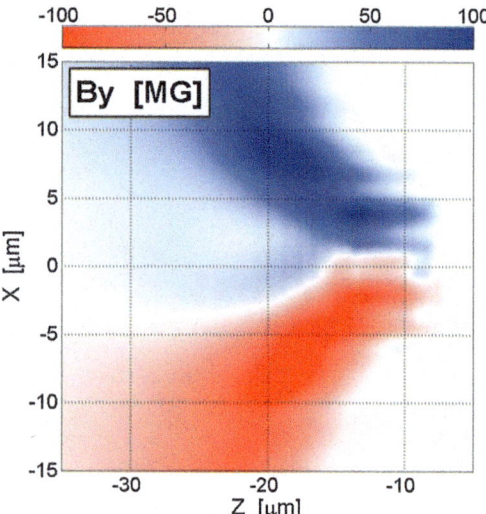

Fig. 8.4 The time and space averaged azimuthal magnetic field at the time when the typical super-hot is roughly halfway through its loop. [Figure 5 in Ref. 2]

If such electron is injected into the laser channel with the x-momentum of ∼5 MeV, it is principally possible for the electron is accelerated to the maximum energy given in (8.2.15), namely,

$$E/mc^2 \tilde{a}_0^2 \times \sqrt{\frac{1}{2} a_0^2 + 1} \approx 270 \qquad (8.2.16)$$

The maximum energy of LIDA electron becomes about 135 MeV. This value can explain the final energy of three electrons in Fig. 8.3a. In Fig. 8.3d, the green and red particle's orbits show strong bending near $x = -40$ μm. Since this is the surface of the boundary between the vacuum and the plasma, electrons are reflected by the sheath field, which helps the loop injection of the electrons.

It is well-known that the above-threshold ionization produces free electrons with almost the same energy as (8.2.5) dominantly in the x-direction as shown in (8.2.6) for a large a_0 lasers. However, it is different from the case of a free electron from the beginning, because the electron interacts with the core of ions via Coulomb force and is ejected with an additional momentum. So, it is expected at the time of just after the ionization that the freed electrons have roughly the initial condition (8.2.14) for strong laser case with $a_0 \gg 1$ like the present simulation. The initial condition is given by the ponderomotive scattering, and the acceleration physics after the initial condition near the vacuum boundary is the same as LIDA as seen in Figs. 8.3b, e. It is noted that for LIDA to act in the system, the laser pulse length should be longer than the traveling time of the looping electrons. Since the interaction region is about 20 μm, the laser pulse should be longer than the time 66 fs (∼20 μm/c).

The other electrons cannot synchronize the laser field as shown in Figs. 8.3c, f. In Fig. 8.5, final electron energy spectrum is plotted with the black solid line, showing almost Boltzmann distribution. The fractions of the electron components generated

Fig. 8.5 The modeled electron energy spectrum integrated over all angles is plotted (long scale) in black. This energy distribution is calculated using every electron crossing a plane 5μm5μm into solid density

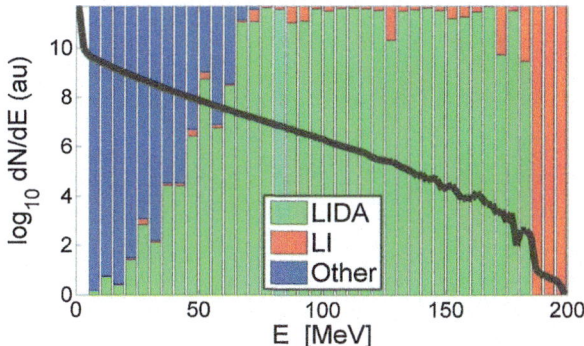

by the three different processes are shown with the color bars in each energy. The highest-energy hot electrons are produced by the initiation of the ionization by strong field, while the most of high-energy hot electrons are generated by the LIDA process.

8.3 Direct Acceleration after Interaction with Longitudinal Field

When the laser propagates in low-density plasmas, the laser interacts with the plasma perturbed by the forward running laser field. The forward running laser may produce density perturbation, and it will possibly create the longitudinal electrostatic field E_x. Let us consider how such electric field locally generated changes the constants of motions of the interacting electrons via non-adiabatic interaction. The change of the dephasing rate R is given in (8.1.18). After the interaction with the longitudinal field, the constant α in (8.1.20) changes to R in (8.1.18). In order to obtain the condition that the electron is further accelerated after the change of R, the dephasing rate R should be as nearer to zero as possible as clear from (8.2.7). It is required that $E_x < 0$. Near the vacuum boundary, this condition is reasonable since the sheath electric field is the direction to push the electrons to the laser propagation direction in general.

Consider that an electron interacts the laser field with the initial condition (8.2.1) and the dephasing rate R changes after the interaction with the localized electric field E_x like

$$R = 1 - \int (-E_x) \frac{R}{\gamma} dt \qquad (8.3.1)$$

Assume that the dephasing rate R reduces from unity to R* (=R of the solution of (8.3.1)) after the finite E_x region. During the interaction, the constant of motion in the y-direction is kept constant so that $\beta = 0$. Then, inserting R* to α in (8.2.7), the maximum x-momentum is given in the form:

$$p_x^{max} \approx \frac{1}{2R_*}(a_0^2 + 1) \tag{8.3.2}$$

Setting that the y-momentum after just after passing the finite E_x region is $p_y = p_y*$, the following relation should be satisfied just at this time:

$$\sqrt{p_x^{*2} + p_y^{*2} + 1} - p_x^* = R_* \tag{8.3.3}$$

Assuming p_x* is much larger than unity, $R*$ has the minimum value for the case with $p_y* = 0$ in the approximate form:

$$R_* \approx \frac{1}{2p_x^*} \tag{8.3.4}$$

Note that the relation (8.3.4) is the same as (8.2.14) and the role of the longitudinal electric field is to jump the initial condition to realize (8.2.14) by the acceleration of electrons in the x-direction with E_x (<0).

In order to see the detail dynamics described above, the basic Eqs. (8.1.1) and (8.1.2) are numerically integrated [1]. The laser amplitude is modeled as a Gaussian pulse in the form:

$$a(x,t) = a_0 \cos\xi \exp\left[-\frac{\xi^2}{2(\omega\tau_L)^2}\right] \tag{8.3.5}$$

The laser parameters are

$$a_0 = 10, \quad \tau_L = 40\text{fs}, \quad \lambda_L = 1\mu\text{m} \tag{8.3.6}$$

In the computation a constant longitudinal electric field is assumed for an interval of 5 μm:

$$\widehat{E}_x = -0.1a_0 \quad 142 < x(\mu\text{m}) < 147 \tag{8.3.7}$$

The potential energy difference over this electric field is quite large to change the dephasing rate $R*$:

$$\Delta\gamma = 16\text{MeV} \tag{8.3.8}$$

This is about the change of $p_x = 31$, namely, the maximum energy (8.2.15) is roughly

8.3 Direct Acceleration after Interaction with Longitudinal Field

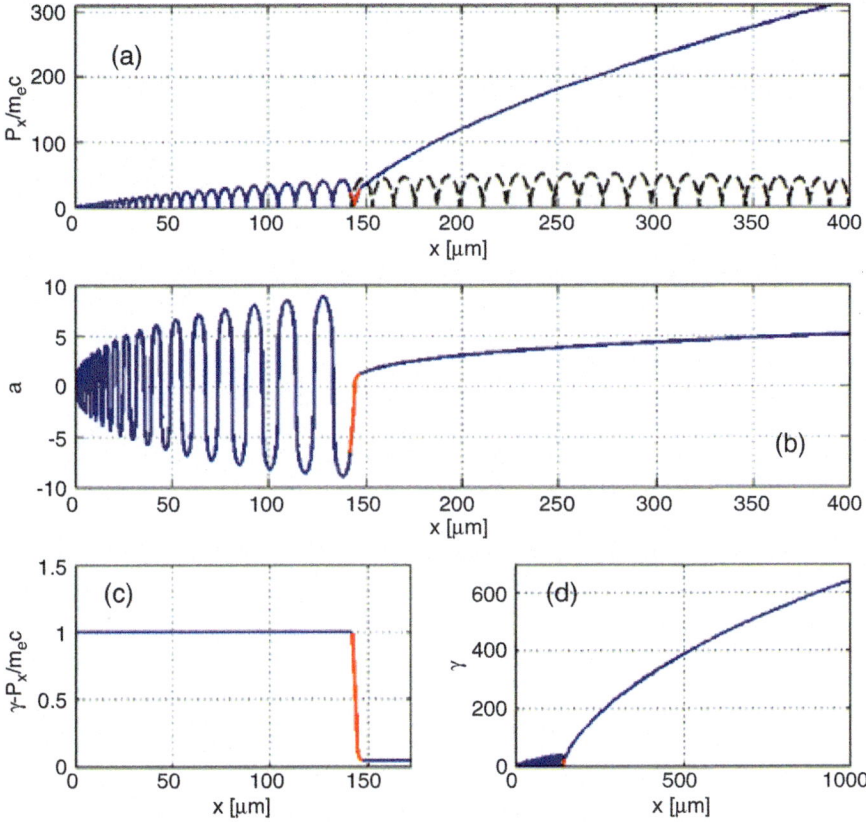

Fig. 8.6 Electron dynamics in a laser field and static field Ex located in the highlighted region. The dashed curve is the axial momentum in the absence of Ex. [Figure 1 in Ref. 1]

$$R* \approx \frac{1}{2p_x^*} \approx 1/62 \tag{8.3.9}$$

$$\gamma^{max} \approx p_x^* a_0^2 = 3100 (\approx 1.5 \text{GeV}) \tag{8.3.10}$$

The numerical result is shown in Fig. 8.6 as a function of the x-coordinate of an electron located at x = 0 and t = 0. An adiabatic motion of the electron is seen before the interaction with E_x. The figure (a) is the normalized x-momentum, showing 2ω oscillation given in (8.2.3), where the black line is the case without E_x and the blue line is the case affected by E_x. The amplitude "a" in figure (b) is equal to the normalized y-momentum and oscillates in the form (8.2.2). After a very short time when the electron is accelerated in x-direction by the longitudinal filed (8.3.7), the motion changes to that of (8.2.7) with $\alpha = R*$.

The change of physical values over the region (8.3.7) is shown in red in Fig. 8.6. During this short time, the dephasing rate R is drastically reduced as seen in Fig. 8.6c. Since the amplitude "a" is proportional to cos(ξ) in (8.3.5), it is clear from Fig. 8.6 that electron acceleration after the interaction (x > 150 mm) is taken place almost in the same phase of the laser:

$$\xi(t, x) = \omega(t - x/c) = \hat{t} - \hat{x} \approx \text{const.} \qquad (8.3.11)$$

This is nothing without the **laser direct acceleration (LDA)**, where electron velocity is nearly equal to the speed of light, and it can continue to increase its energy from (8.1.3) as seen in Fig. 8.6d after the red line timing. In the case of Fig. 8.6, the condition $(-v_y E_y > 0)$ is maintained over the traveling long distance and traveling time t = x/c. How much energy the electron obtains due to such process depends on the size of plasma or particle acceleration distance. The calculation is done up to 1000 μm in Fig. 8.6, which is about 0.3 GeV from Fig. 8.6d and not long enough to be accelerated to the maximum energy theoretically evaluated in (8.3.10).

It is noted that in Fig. 7.26, we have already observed spiky electrostatic field generation in the under-dense plasma produced from the foam material. This is a good example how the electrostatic field in the direction of laser propagation is produced. Of course, the electric field is not only in the favor direction for acceleration to the laser propagation direction. However, some fraction of electrons can be accelerated locally by the electrostatic field and continue to be accelerated by the laser direct acceleration. Such mechanism is also physical candidate to explain the production of super-ponderomotive hot electrons.

When a longitudinal electric field (the electric field in the x-direction) is located in the laser-plasma interaction region, it is suggested that the hot electron energy is higher energy than without the longitudinal field. This is happen dominantly near the timing, $p_x = p_y = 0$ in (8.2.2) and (8.2.3). The addition of the longitudinal field E_x in (8.1.7) localized over the interval Δx changed the momentum in a short time like

$$\Delta p_x = \int_t^{t+\Delta t} E_x dt = \int_x^{x+\Delta x} E_x \frac{\gamma}{p_x} dx \qquad (8.3.12)$$

It is clear that the x-momentum change is biggest for the condition that

$$\begin{aligned} E_x &< 0 \\ p_x &\approx 0 \end{aligned} \qquad (8.3.13)$$

8.3.1 PIC Simulation

The physical mechanism mentioned above is demonstrated in 2D PIC simulation [3]. The laser with $a_0 = 8.5$ propagates in the density channel produced by laser irradiation in the plasma with the electron density of 0.05 the critical density. As seen in Fig. 8.7a, the density channel is produced, and in Fig. 8.7b the longitudinal electric field $-E_x$ is seen to be produced in the density channel region. This electric field is generated to inject the electrons into the channel constantly, because the electrons in the channel drifts to the over-dense direction by the JxB drift motion and the electric field is produced so that the surrounding electrons are injected into the channel region in order to keep the charge neutrality.

A trajectory of a typical accelerating electron is shown with dotted lines in Fig. 8.7. In Fig. 8.7a, the color of the electron trajectory is its Lorentz factor, and it increases along with the x-direction. This indicates that the electron is accelerated by longitudinal electric field in the channel and accelerated further more by the process described above. In Fig. 8.7b the background colors are the time-averaged electric field in the x-direction. The color of the electron trajectory is the value of time-dependent R defined in (8.3.1), and it changes from R = 1 to R < 1 near the time 877 fs. It is observed that at time 830 fs, the electron is almost at rest and starts to be

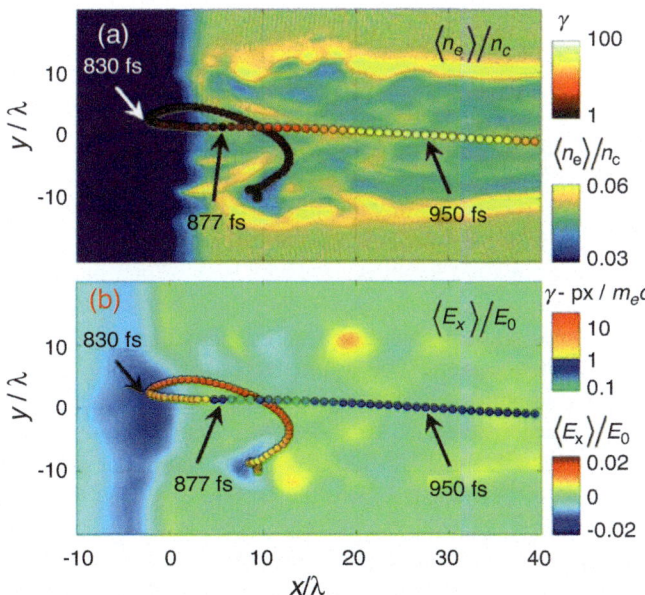

Fig. 8.7 Trajectory of an accelerated electron in a channel. The upper panel shows the trajectory, with the color-coded γ, on top of the time-averaged electron density profile. The lower panel shows the trajectory, with the color-coded dephasing rate R = γ − px/mecR = γ − px/mec, on top of the time-averaged longitudinal electric field. The field and the density are averaged over ten laser periods at 850 fs. [Figure 6 in Ref. 3]

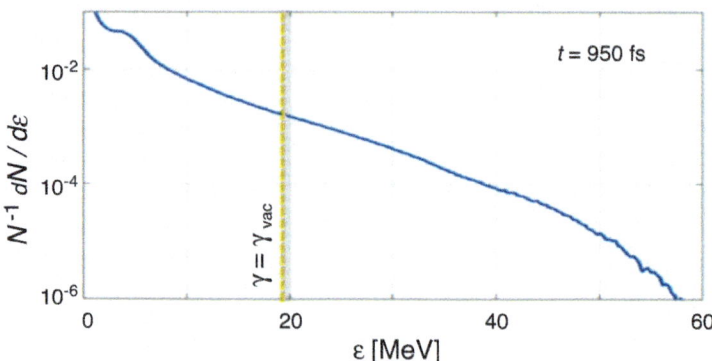

Fig. 8.8 Snapshot of a normalized electron spectrum at $t = 950$ fs from a 2D PIC simulation with $a_0 = 8.5$ and $n_e = 0.05 n_c$. [Figure 5 in Ref. 3]

accelerated to the x-direction by the electrostatic field. Then, due to the laser direct acceleration, the electron starts to be accelerated with a lower value of R* after the time near 877 fs.

At the later time 950 fs, snapshot of electron energy distribution is plotted in Fig. 8.8. Without the effect of the longitudinal electric field, the maximum energy of electrons should be given from (8.2.4) in the form:

$$\gamma_{max} = 1 + \frac{1}{2} a_0^2$$

It is 19 MeV for the present parameter $a_0 = 8.5$. This maximum is plotted in Fig. 8.8 as a reference energy. It is clear that many electrons are accelerated non-adiabatic ways, one of which is acceleration by local longitudinal electric field before electron is accelerated directly by the laser field until the dephasing.

For the case where a solid target is irradiated with a long pulse laser of several picoseconds and strong field of $a_0 = 10$, the abovementioned acceleration may happen several times for selected high-energy electrons. And the hot electron temperature is expected to increase as a function of time. Simulation of 2D and 3D PIC has been done to study the physics of laser-plasma interaction relating to the **fast-ignition** laser fusion in Ref. [4]. Relativistic petawatt laser pulse interacting with over-dense plasma is shown in Fig. 8.9 for the time of 1 ps (a) and 4 ps (b), respectively. The laser is irradiated from the surface $z = 0$ to the solid surface at 80 μm, and the focusing diameter is relatively large 40 μm. The color red shows the energy flux of laser and electrons toward z-direction. The color green outside of the energy flux color shows the electron density. The dashed line indicates the electron density of ten times the critical density ($10 n_c$). It is also found that steady-state magnetic field stronger than 130MG is also observed surrounding the laser beam like Fig. 8.4.

It is seen in Fig. 8.9 [4] that the laser intensity has filamentary structure and the electron orbits may feel a lot of impulses due to the electrostatic field inside

8.3 Direct Acceleration after Interaction with Longitudinal Field 303

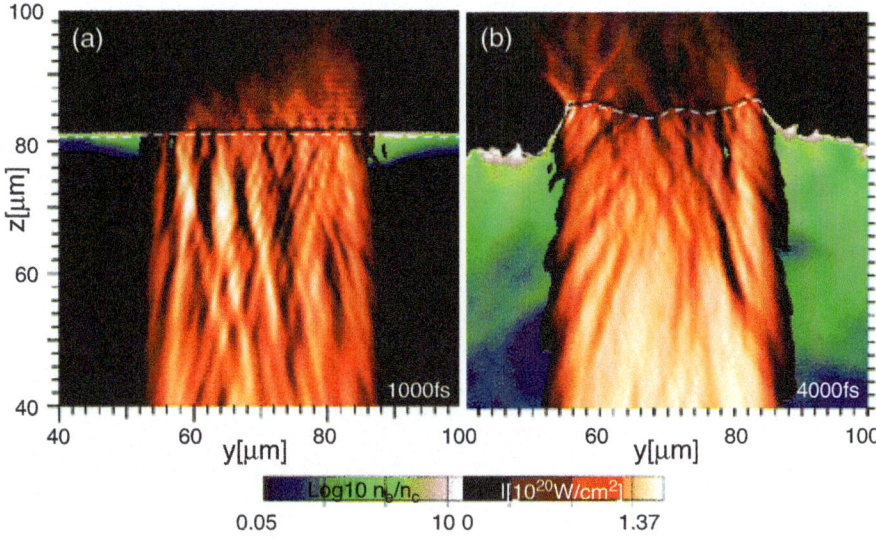

Fig. 8.9 Relativistic petawatt laser pulse interacting with over-dense plasma at 1 ps (**a**) and at 4 ps (**b**); the laser pulse is injected at z = 0, and plasma is initially at z > 80 μm. Energy flux density along z (in red) shows continuously high conversion from the laser into a relativistic electron beam. The dashed line at ne = 10nc shows deformation and motion of the absorption layer. Expansion of under-dense plasma into vacuum (in green) is evident. [Figure 1 in Ref. 4]

relatively wide beam. In addition, plasma is relatively one-dimensional structure so that accelerated electrons in the laser-plasma interaction region are probably confined in the interaction region. As seen in Fig. 8.9b, the low-density plasma region expands in time due to the ablation from the solid target surface. The size of the low-density plasma increases with time, and the laser-plasma interaction region also expanded.

The time evolution of the electron energy spectrum and the density profile are plotted in Fig. 8.10 [4]. The low-density plasma with the density just below the critical density continuously expands, and at t = 5 ps it is about 80 μm. The high-energy tail of the electron distribution has Boltzmann type structure, and effective temperature increases as a function of time. It is noted that the ponderomotive scaling gives the maximum energy of the hot electron $E_{PM} = 7$ MeV for $a_0 = 10$. It is clear that even at 1 ps, the maximum energy is about 50 MeV showing super-ponderomotive hot electrons. It is suggested in [2] that repeating acceleration of LIDA takes place in the interaction region by the help of the strong magnetic field as mentioned above. As shown in (8.2.15), high-energy electrons accelerated multi-times in the interaction region can obtain higher energy with the increase of p_{x0} in the second, third opportunities.

Of course it is not realistic to invoke the physical mechanism of hot electron generation to only one physics, while other mechanisms such as stochastic acceleration to be described later may also contribute to the production of

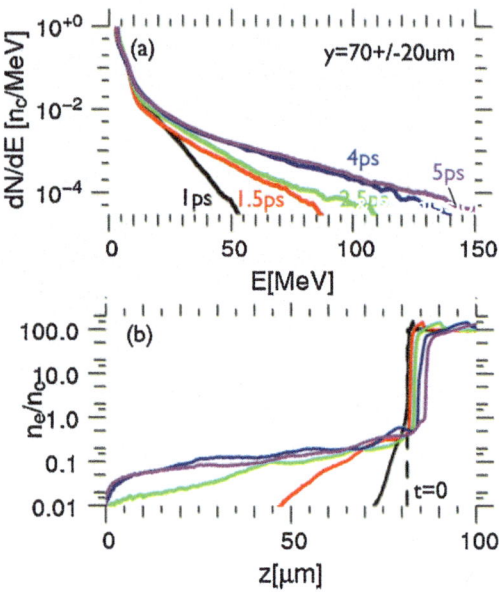

Fig. 8.10 (a) Energy spectra of laser-generated electrons consist of three energy groups, with a high-energy tail that asymptotes at 4 ps; (b) similarly, electron density profiles, averaged across the laser spot, asymptote toward a near plateau at 4 ps; colors in (a) and (b) for the same time steps match. [Figure 3 in Ref. 4]

super-ponderomotive hot electrons. Anyway, this computational result is interesting to know that the hot electron temperature increases as the relativistic laser intensity continues more than hundreds of fs. This means that other physical mechanisms than the ponderomotive acceleration become important when relativistic lasers propagate relatively long low-density plasma of pre-formed plasma or ablation plasma. Then, super-ponderomotive hot electrons are produced in the low-density region.

Since the laser-plasma coupling becomes better in a long pulse lasers, the laser absorption rate is also expected to be higher than the case of shorter pulse. In Fig. 8.11, time history of energy partition to electrons is shown as well as laser absorption fraction. The laser intensity increases in a Gaussian form to the peak intensity at t = 600 fs. The intensity is kept constant after t = 600 fs in the simulation. From 600 fs to 1 ps, the laser absorption rate is about 60%, and mainly the hot electrons are produced by the ponderomotive force as shown in blue line. The super-ponderomotive electrons with energy higher than 7 MeV gradually increase with time, and this red line becomes dominant than the blue component as time proceeds. The absorption rate also increases as the energy flux to the hot electrons increases. After the time 3–5 ps, absorption rate approaches about to 80%. It is also observed that the total energy of the electrons is about 80% of the absorbed laser energy for t > 1 ps. This roughly indicates that after 1 ps, the low-density plasma region is produced so that laser-electron interaction becomes important not only for adiabatic ponderomotive interaction but also non-adiabatic interaction.

8.4 Chaotic Motion due to External Force

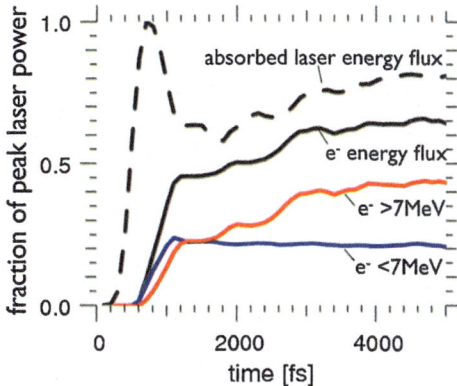

Fig. 8.11 Time history of energy partition in laser-generated electrons, showing sustained absorption of up to 80% (absorbed laser energy flux through z = 0 plane, dashed line) into relativistic electrons (total electron energy flux projected on z, solid black line); also shown are contributions from particles with energies E ≤ 1.5Ep = 7 MeV and > 7 MeV; all values are normalized to peak laser power $P_L \equiv 1.3$ PW. [Figure 4 in Ref. 4]

8.4 Chaotic Motion due to External Force

Let us consider the physics of generation of hot electrons by stochastic process of electron motion in laser field instead of the ponderomotive force we mentioned before. It became clear in Chap. 7 that the laser absorption is high and hot electron temperature is also high in general for the case with pre-formed plasmas. Pre-formed plasma is beneficial to enhance the laser-electron interaction; consequently better coupling is expected. This is the reason for the use of foam layered or structured targets for higher energy conversion of laser to plasmas. We did not study so far the physical mechanism of enhanced absorption in low-density plasma region. In the relativistic laser and plasma interaction, the chaotic heating of electrons becomes most dominant mechanism. Note that this process is in general not seen in the case of non-relativistic laser interaction. Electron orbits in non-relativistic laser are generally stable because it is a simple linear motion, and there is a threshold for the chaotic heating as seen below.

8.4.1 Simple Example [1] (Random Force)

In this section, a brief introduction to the concept of chaos is given for a particle in an external random force given by the equation:

$$\frac{d^2x}{dt^2} = f(t) \tag{8.4.1}$$

This is a simple equation governing Brownian motion, if the external force f(t) is a random force given, for example, by the following form.

$$f(t) = \sum_i \Delta v_i \delta(t - t_i) \tag{8.4.2}$$

For simplicity, one-dimensional motion is considered here, while it is easy to extend (8.4.1) and (8.4.2) to two- and three-dimensional motions. In (8.4.2), Δv_i is by the kicking force at time t_i. Brownian motion of pollen seen on the surface of water is two-dimensional, and it is simply analyzed by assuming two-dimensional random force on the water surface.

Integrating (8.4.1) for a barrow time interval across the nth time of kick, the recurrence relation is obtained as

$$v^{n+1} = v^n + \Delta v|_n \tag{8.4.3}$$

It is well-known that (8.4.3) gives the time evolution of velocity dispersion in the form:

$$\langle v^2(t) \rangle = N(\Delta v)^2 \tag{8.4.3'}$$

where < > is to take the time average and N is the number of kicks to the particle until the time t. Assume that the average of the velocity kicks Δv defined by the relation:

$$(\Delta v)^2 = \langle (\Delta v_i)^2 \rangle \tag{8.4.4}$$

This is related to the diffusion phenomenon. So, inclusion of the random force makes the particle orbit diffusion in the velocity space. The above relations lead to the diffusion equation to the probability function f(v):

$$\frac{\partial}{\partial t} f(v) = D \frac{\partial^2}{\partial v^2} f(v), \quad D = (\Delta v)^2 / \Delta t \tag{8.4.5}$$

where Δt is the average time interval of each two kicks. It is note that the diffusion means the total energy of particles increases in time and this energy is given by the random force in (8.4.2). Let us call such heating as **stochastic heating**.

It is noted that the **classical absorption** (collisional absorption, inverse-Bremsstrahlung absorption) explained in Chap. 2 is also due to the stochastic heating. In the frame moving with an electron under laser field, the Coulomb electric field appears like a delta function to accelerate or decelerate the electron motion, if

8.4 Chaotic Motion due to External Force

there is oscillating motion by laser. This force by the nearest ion appears randomly, and the velocity space of the electron shows the spread with time like diffusion in velocity (energy) space as shown in Chap. 2.

8.4.2 Simple Example [2] (Periodic Force)

It is said that the Poincare pointed out in 1892 that three celestial body motions cannot be integrated and the motions are not predictable. This is the origin of the study of chaos in Hamiltonian system. It is good example to consider the orbit of an asteroid rotating around the sun as shown in Fig. 8.12a [5]. Such a two celestial body problem is solved analytically. According to the solar system formation theory, it is expected there are many asteroids in the solar system. It is not, however, observed in the orbit between Mars and Jupiter as expected from two body motion. It is a good example to know how an integrable asteroid orbit around the sun becomes unstable by the periodic perturbation due to the gravitational force by Jupiter. Here, "unstable" means the total energy of the asteroid becomes positive due to the periodic force by Jupiter to escape from the solar system. In Fig. 8.12b, observation data of the number of asteroids are plotted as a function of the oscillation frequency of asteroid normalized by that of Jupiter. The orbit frequency of asteroids is a simple function of the mean radius of the orbit, and the frequency is a function of the orbit radius. The left is near Mars and the right is near Jupiter in Fig. 8.12b.

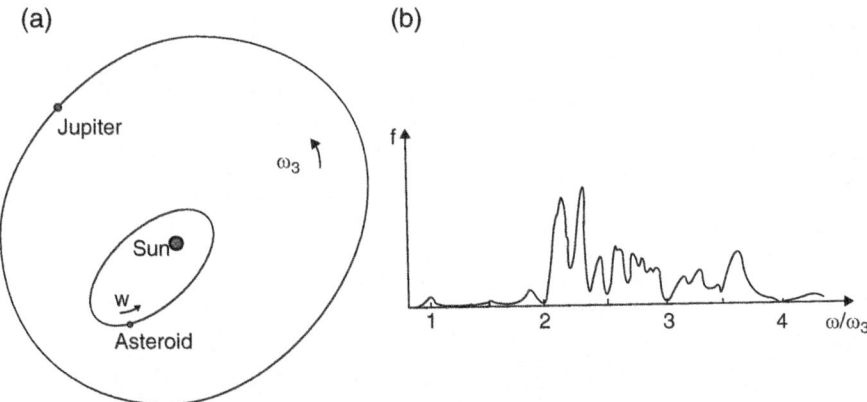

Fig. 8.12 (a) Schematics of an asteroid orbit bounded by the gravity of the sun. The Jupiter gravity affects the Kepler motion of the asteroid as a periodic perturbation force. (b) The observation data of the number of asteroids between Mars and Jupiter. It is inferred that due to the Jupiter perturbation (frequency ω_J) the asteroids rotating with the frequency ω become unstable, and the kinetic energy becomes larger than the binding energy by the sun to escape from the solar system

Fig. 8.13 The particle orbit near the separatrix (saddle) points is unstable to external perturbation. When a particle starts from one separatrix to the center in the figure as red line, an external oscillating force perturbs its orbit dramatically

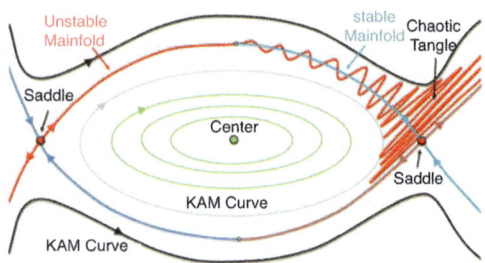

For simplicity, try to model this problem with the following one-dimensional motion, where the binding by the sun is modeled with sine force and the periodic perturbation is modeled by a periodic external force:

$$\frac{d^2 x}{dt^2} = -\sin x + \varepsilon \sin(\Omega t) \qquad (8.4.6)$$

where ε and Ω are constants. When $\varepsilon = 0$, these equations become second order, and such system is called **autonomous** system, and the energy is conserved. However, inclusion of the external force changes the property and permits chaotic solutions. This property depends on the nonlinearity of the potential; the term $\sin(x)$ in (8.4.6). (8.4.6) is a very simple equation, while we can find chaotic motion in general with a finite value of ε. Assume that ε is small enough and the force is perturbation. According to the position in the phase space, the time-integrated velocity changed by the external force is different.

As shown in Fig. 8.13, the particle orbit near the **separatrix (saddle)** points is unstable to any external force [6]. As shown in this phase space diagram, the particle started from the left saddle point (red line) starts oscillating in the potential. On the bottom of the potential, if a small oscillating perturbation is imposed, the particle starts to oscillate as in the figure, and its amplitude is enhanced near the saddle point. Note that this red oscillation is schematic, and the reality is the particle loses or obtains some energy from the external force, and its orbit jumps to another one with different average energy. It is said that the adiabatic condition is not satisfied for particles in the orbit of separatrix and energy transfer easily happens in this case. This means the orbit near the saddle is unstable to any external perturbation as we see below.

In Fig. 8.14a, the Poincare diagram of the orbit by (8.4.6) is shown for the case with $\varepsilon = 0.1$ and $\Omega = 1$. The points in the phase space are those taken as sample points for $t = n2\pi$ ($n = 0, 1, 2,,,$). The particle is initially at the null point ($x = 0$, $dx/dt = 0$) on the separatrix. The separatrix curve is plotted with the red line, and it is the solution of (8.4.6) for $\varepsilon = 0$. The blue dots belt surrounding the separatrix curve indicates that the orbit is not periodic and motion is chaotic, although the orbit is bounded in relatively narrow region in the phase space. It is known that the region of the chaos orbit becomes wider with increase of the amplitude ε and most of the phase

8.4 Chaotic Motion due to External Force

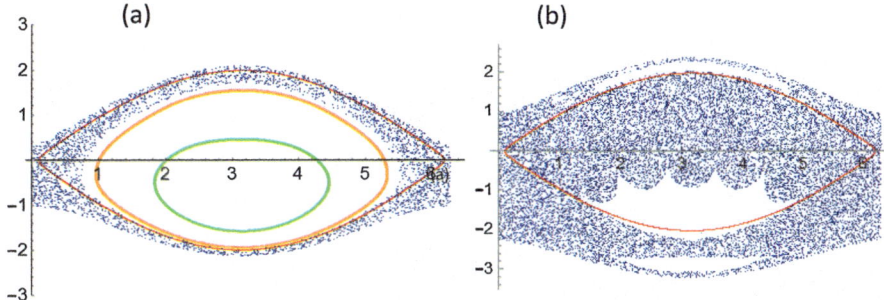

Fig. 8.14 (a) The Poincare map with blue dots are for a particle initially located near the saddle point (x = 0). The parameter in (8.4.6) is ε = 0.1 and Ω = 1. The maps for particles with different initial positions (x = 1 and 2) are plotted with orange and green, and they are stable to such external force. (b) The Poincare map for particle located near the saddle point is plotted for ε = 0.5

space inside and wide region outside of the separatrix both becomes chaos region. This is clear in Fig. 8.14b for ε = 0.5.

From the analogy of the present result to the orbits of asteroids in Fig. 8.12, many bounded asteroids by the solar gravity can be of chaotic motion for a long time, and they can obtain kinetic energy from Jupiter to be free from gravitational bound by the sun. This can explain the observation data in Fig. 8.12b.

8.4.3 Chaos in Propagating Relativistic Wave

It is pointed out in Ref. [7] that some electrons in relativistic laser field can obtain large amount of energy from the laser field. It is the case when laser field perturbs the electron momentum, especially in the direction of the laser electric field in linearly polarized electromagnetic field propagating in vacuum. Let us write equation of motion explicitly for the case where laser is polarized in y-direction and propagates in x-direction. Let us consider the case without the longitudinal electric field in (8.1.1)–(8.1.3). For convenience, use normalized forms of equations and physical quantities according to the definition in (8.1.7)–(8.1.9):

$$\frac{dp_y}{dt} = \frac{da}{dt} \qquad (8.4.7)$$

$$\frac{dp_x}{dt} = -\beta_y \frac{\partial a}{\partial x} \qquad (8.4.8)$$

$$\frac{d\gamma}{dt} = \beta_y \frac{\partial a}{\partial t} \qquad (8.4.9)$$

where β_y are the normalized velocity in the y-direction. The first two equations are the equation of motion, and the last one is equation of energy. Just looking at the equations, it is found that the oscillating motion in y-direction induces the **JxB** force in the x-direction. For the case of non-relativistic laser, the coupling of x and y motions is negligible and no chaotic motion is expected. In addition, it is obvious from (8.4.9) that any motion in x-direction provides energy transfer from laser to electron and vice versa. Only some phase mismatching of electron orbit from the laser electric field results energy transfer.

Let us treat a simple model with random momentum change in y-direction in (8.4.7). In order for energy change by laser field, some unexpected change of β_y or a is required as clear in (8.4.9). Note that the direct energy transfer from laser to electrons is not expected for the case with the random force only in x-direction, say longitudinal waves. In this case, the energy transfer happens indirectly, because the change of v_x changes the phase in the laser filed ξ in (8.1.11).

Consider the case where the random force is given to the y-direction only. It is numerically integrated in [7] for the equations with the following additional term in (8.4.7):

$$\left.\frac{d\widehat{p}_y}{dt}\right|_{random} = \sum_i \Delta\widehat{p}_y \delta(t - t_i) \tag{8.4.10}$$

Integration of (8.4.7) and (8.4.8) with the random force (8.4.10) into (8.4.7) has been done for 5×10^4 particles with different random numbers and different timings [5]. Simulation shows that the randomness of the time interval of the delta function doesn't affect the result and the following time interval and the laser pulse length are chosen:

$$\omega \Delta t = 0.125 \Rightarrow \Delta t = 2 \times 10^{-2} \times 2\pi/\omega \tag{8.4.11}$$

As shown in (8.4.5), this model also results in the diffusion phenomenon, and the diffusion coefficient D is given by:

$$D = \left(\Delta p_y\right)^2 / \Delta t \tag{8.4.12}$$

The simulation has been done for relativistic laser with $a_0 = 3$. The random force in y-direction is defined by giving the diffusion coefficient (8.4.12) as $D/\omega = 0.01$ in the normalized form. Simulation for all electrons are carried out with the initial condition, $p_x = p_y = 0$ at $x = y = 0$ at $t = 0$. It is noted that the given momentum kick in the y-direction is calculated as

$$\Delta p_y = 0.01\omega\Delta t = 1.25 \times 10^{-3} \tag{8.4.13}$$

This is very small random perturbation applied to the electron 50 times per laser cycle.

8.4 Chaotic Motion due to External Force

Fig. 8.15 Energy distribution of electrons accelerated in a laser pulse and an additional transverse stochastic field. [Figure 3 in Ref. 7]

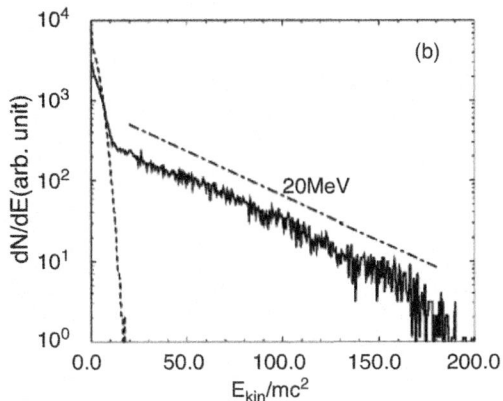

In Fig. 8.15, the energy distribution function of the all test particles is plotted for the time $\omega t = 1800$ (~300 laser cycles and about 1 ps for 1 μm laser), showing a Boltzmann like distribution function. At this time the effective temperature is 20 MeV, which is much larger than that given by the ponderomotive scaling, namely, $T_h = 1.2$ MeV from (7.4.4). The ponderomotive scaling is the electron average kinetic energy given from (8.2.4) for the case of (8.2.1). The stochastic kicks to the y-direction change the values of (8.2.1) at each kick.

In a series of simulation runs, the laser intensity a_0 and the interaction time have been varied [7]. It is found that the effective temperature scales roughly proportional to a_0 (square root of the laser intensity). The effective temperature T_h also increases with the increase of the time of laser particle interaction, approximately like t^α with α between 0.5 and 1. Actually, the hot electron temperature at $\omega t = 900$ is about 12 MeV. In Fig. 8.15, it is seen that the distribution has two Maxwellian. The cold components could not get into the energy transfer regime, and a selected number of the electrons can obtain statistically higher energy.

The time evolution of the x- and y-momentums of an arbitrarily picked electron is shown in Fig. 8.16. It is seen that the trajectory of p_y is slowly spreading from the $p = 0$ point but almost confined around $p_y = 0$ point. At early time, the electron oscillates in the laser field more or less like an unperturbed electron given by (8.2.2) and (8.2.3), and only after sufficient dephasing, it gains energy rapidly in the x-direction. Most of the electrons never reach this phase of strong acceleration and stays in the low-energy pool, seen at the left side of Fig. 8.16 with an effective temperature close to that of pure random walk in momentum space.

Consider the physics controlling the strong acceleration of the electron seen in Fig. 8.16. How can we increase the energy of this electron in the propagating laser field with a constant amplitude? As seen in Fig. 8.16, the peak of the x-momentum gradually increases. Eq. (8.4.9) can be rewritten as

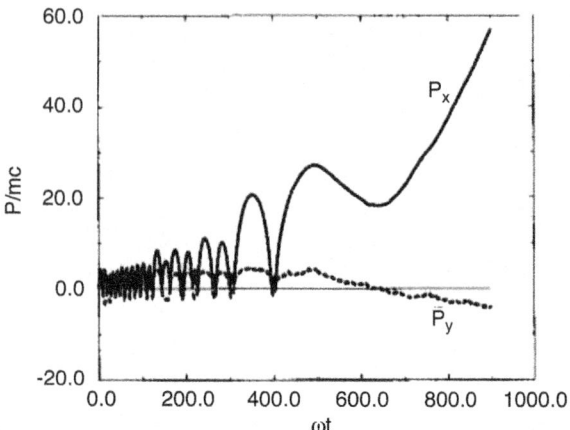

Fig. 8.16 The time evolution of transverse and longitudinal momentum of a typical electron highly accelerated. [Figure 4 in Ref. 7]

$$\frac{d\gamma^2}{dt} = 2p_y \frac{\partial a}{\partial t} \qquad (8.4.14)$$

Inserting (8.2.2) to (8.4.14), the energy of electron cannot increase over long time more than a quarter of the oscillation period. However, we can expect some selected electrons which obtain the energy from laser field continuously, because of only luck in random kicks to keep RHS of (8.4.14) positive.

Consider the mechanism of dramatic electron acceleration in the later time seen in Fig. 8.16. It is like a direct acceleration of the electron by laser field shown in Fig. 8.6. Since the electron is running to the direction of the laser propagation with highly relativistic velocity, the electron feels very slow oscillation of the laser field due to the Doppler shift as shown in Fig. 8.2. In the frame moving with the velocity v_x in the laser propagation direction, the Doppler-shifted laser frequency in this frame is from (5.2.44)

$$\omega* = \gamma(1 - v_x/c)\omega_0 \approx \frac{1}{2\gamma}\omega_0 \qquad (8.4.15)$$

In Fig. 8.16, γ near the final phase at $\omega t = 500$ is about 25, and the electron that interacts in a half phase of the laser field at rest frame is extended to 50 times as follows:

$$\omega*t = \pi \quad \Rightarrow \quad \omega_0 t = 50 \times \pi = 150 \qquad (8.4.16)$$

Since RHS of (8.4.14) is kept positive until the time to $\omega_0 t = 150$, the electrons continue to accelerated. In Fig. 8.16, during the interval of $\omega_0 t = 400 \sim 600$, the p_y is kept positive, and RHS of (8.4.14) is kept positive to directly transfer laser energy to the electron energy. In addition, when the p_y changed the sign from positive to negative, the laser field direction changed to the opposite direction, so fortunately the electron is accelerated dramatically by the DC interaction with the laser field.

8.5 Electron Heating by Laser Field and Induced Plasma Waves

When a relativistic laser is irradiated to low-density plasma such as gas, large-amplitude plasma waves are generated by the Raman scattering as explained in Chap. 6. The induced plasma waves are used to accelerate electrons in the early research stage of the laser acceleration. Then, the forward Raman scattering is expected to be induced predominantly in such parametric instability scheme. From the view point of the present physics, however, not only the induced plasma waves but also the nonlinearity of relativistic laser can also accelerate the electrons; consequently the stochastic acceleration is not avoidable.

Stochastic acceleration in such regime has been studied by 2D3V PIC simulation and corresponding model [8]. The simulation condition assumes a Gaussian laser of pulse length ~ 0.7 ps, focused on low-density plasma [~$n_e = (0.02 \sim 0.07)n_c$] at the intensity 5×10^{19} W/cm^2 ($a_0 = 6$), focused with its spot size of (4–6)λ, where λ the laser wavelength. The simulation zone is varied in the range [$L_x \times L_y$] = [(1000–1500) λ x[(60–200) λ].

In Fig. 8.17, the resultant electron spectra obtained by the PIC simulation are shown about the time when the laser with the pulse width 0.7 ps has already passed the gas plasma. The solid line is the case for a wider focal spot of 6 λ, while the dotted line is the case with 4 λ. As explained later, the electron loss from the interaction region due to the small focusing spot effect affects the profile of the spectra. Both distributions look Maxwellian with a maximum energy cut, while Maxwellian seems due to the dominant particle escape rate at higher energy.

It is shown that the forward Raman scattering (FRS) grows easily to the nonlinear phase for the parameters used in the present simulation. It is reported that the growth factor on the exponent of FRS is G ~ 20 at the end of laser pulse (700 fs).

Fig. 8.17 Electron energy distribution obtained by two-dimensional PIC simulation results at time of 0.7 ps. [Figure 2 in Ref. 8]

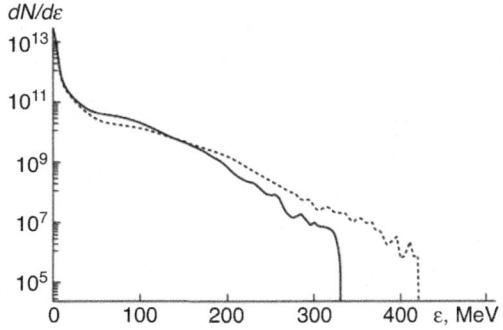

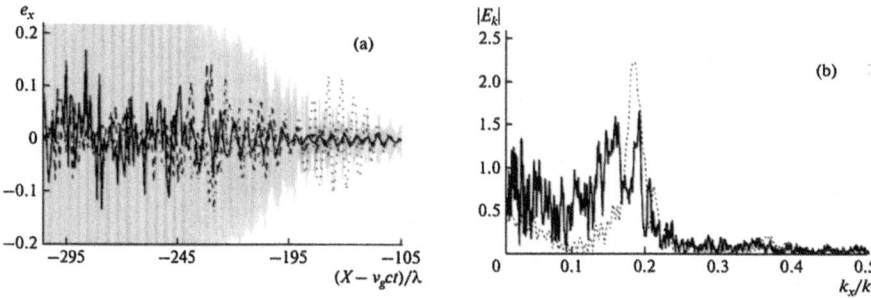

Fig. 8.18 (a) The electric field of laser and plasma waves. (b) Fourier spectra of the electric field Ex by plasma waves

In Fig. 8.18, the electric fields of the laser and plasma waves are shown in (a), and the spectrum of the plasma waves is shown in (b). In Fig. 8.18a, the normalized electric field defined in (8.1.10) with laser frequency ω is plotted for t = 2.6 ps with solid black line and the other three times, while the shorter wavelength and large-amplitude one plotted with gray lines shows the normalized E_y [=a(x,t)] with reduced value as $E_x/10$; note that its peak is a = 6. It is clearly seen that large amplitude of about 0.1 ~ 0.15 is induced and the plasma wave is turbulent. Fourier components of the plasma wave electric field E_x is plotted in Fig. 8.18b with broad spectrum around the peak of FRS matching waves.

8.5.1 Modeling PIC Simulation

The evolution of electron dynamics in the present case can be described by the set of Eqs. (8.1.7)–(8.1.9). Here, however, we have to model the longitudinal electric field in the x-direction due to the turbulent plasma waves. In Ref. [8], the multimode longitudinal waves is modeled as

$$E_x(t,x) = \sum_{j=-N}^{N} E_{0,j}\theta(v_g t - x) \cos\left(\omega_{pe}t - k_{x,j}x + \varphi_{0,j}\right) \quad (8.5.1)$$

The number of the plasma waves is 2 N + 1, and N = 15 is used, and the wavenumber $k_{x,j}$ and amplitude $E_{0,j}$ are roughly fitted to the spectrum in Fig. 8.18b. In (8.5.1), $\varphi_{0,j}$ is random phase and θ is the step function.

In Fig. 8.19, the electron spectrum obtained after integrating (8.1.7)–(8.1.9) is shown by the dotted line for the case with (8.5.1) and no laser field and by solid line for both fields included consistently. The number of the test particles is 2.5×10^4. It is clear that the hot electrons are dominantly produced via electron plasma wave interaction. Then, the chaotic acceleration by the nonlinear property of the relativistic laser interaction accelerates the electrons further more. Compared to the

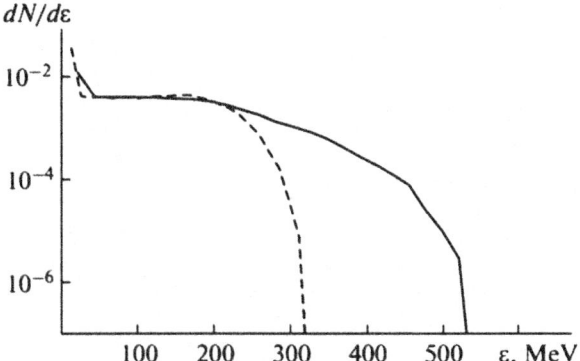

Fig. 8.19 Final electron spectra due to the acceleration only by plasma waves (dotted) and also due to laser field (solid)

spectrum of Fig. 8.17, in addition, it is found that the profiles of the hot electron energy distributions are different. It is pointed out that the difference is mainly due to the escape of electrons in 2D PIC simulation, while it should be noted that the distribution function looks evolving constantly to higher energy region in 1D model calculation. This would be important to compare the electron spectrum including geometry effect.

8.5.2 Quasi-linear Diffusion

In the case only with the turbulent plasma waves, the stochastic acceleration of electrons is well modeled by so-called quasi-linear diffusion model to be well explained in Volume 3. The quasi-linear model has been used in Sect. 2.6 to calculate the classical absorption rate. It is easily extended to the case with many longitudinal plasma waves without the effect by the laser field. In Ref. [8], the formulation is given. There is shown that the interaction of electrons with the plasma waves can be described as a diffusion term by use of the quasi-linear theory. The time change of the electron distribution function by the plasma waves is formulated as

$$\left(\frac{df}{dt}\right) = \frac{\partial}{\partial p_x}\left(D_{QL}\frac{\partial f}{\partial p_x}\right) \quad (8.5.2)$$

$$D_{QL} = \pi \left(\frac{e}{m}\right)^2 \sum_k W_k \delta(\omega - kv_x) \quad (8.5.3)$$

$$W_k = |E_k|^2 \quad (8.5.4)$$

where d is delta function showing that an electron traveling with the phase velocity of the plasma wave obtains or gives energy to each other. In Ref. [8], the quasi-linear diffusion coefficient D_{QL} is obtained as

$$D_{QL} = a_{10}^2 \frac{k_0}{k_{pe}} \tag{8.5.5}$$

In such model, we can solve the diffusion Eq. (8.5.2) for the initial condition of the cold delta function of the electron distribution function. The solution is well-known form:

$$f_e(t) = \frac{n_{e0}}{\sqrt{\pi D_{QL} t}} \exp\left(-\frac{p_x^2}{D_{QL} t}\right) \tag{8.5.6}$$

Inserting the parameters from the PIC simulation, $a_{10} \sim 0.1$, $n_e = 0.03 n_c$, and pulse length ~ 0.7 ps, the effective temperature of (8.5.6), $T_h = 150$ MeV, is obtained. This is not so bad to explain the distribution of the dotted line in Fig. 8.19. Note that only with the plasma wave field, the electrons are accelerated only in the x-direction and the energy is $\varepsilon = (p_x^2 + 1)^{1/2}$. Then, the distribution function is a super-Gaussian to the energy ε, and the hot electron temperature T_h has the following time dependence:

$$f_e \propto \exp\left[-\left(\frac{\varepsilon}{T_h}\right)^N\right], \quad T_h \propto \sqrt{t} \tag{8.5.7}$$

where (8.5.6) is the case for $N = 2$, while Maxwellian is $N = 1$. It is reasonable that the dotted line in Fig. 8.19 looks like a super-Gaussian, although it is not clear $N = 2$ or more.

8.6 Hot Electron Generation

Since the relativistic lasers deposit their energy to the matters only through the collisionless physical process as seen above, it is very important to know how the electrons are obtained random kinetic energy from the relativistic lasers. It requires to study the time evolution of electron energy distribution and to study the mean value of the energy so-called hot electron temperature as function of target and laser conditions. As described so far, the stochastic heating is time-dependent, and only small amount of selected electrons obtained substantial energy from lasers.

In addition, the dominant physics also changes as the plasma density profile is evolved in time. Let us, therefore, study at first the case for relatively short pulse where sharp density profile is kept during the laser pulse irradiated on the solid target surface. Such situation is appropriate for the time scale of ~100 fs laser pulse. It is noted that the physics will change if the short pulse laser interacts with a long-scale low-density plasmas or gas. Then, consider the case for long pulse such as ~ps and multi-ps laser pulses.

8.6.1 Hot Electrons by Femtosecond Lasers

1D and 2D PIC simulations have been carried out to clarify the electron heating process when relativistic laser is irradiated on the solid surface [9]. The effect of the standing wave by the reflected component is also paid attention. The laser is usually assumed to be a plane wave incidenting to the solid surface in the normal direction. In order to resolve the spatial profile in computation, 630 cells per wavelength are used which corresponds to 10 cells per skin depth.

In Fig. 8.20, the electron distribution function is plotted as function of the normalized momentum for early time at t = 50 fs. It is seen that the distributions have step-like structure in the direction of laser propagation at the solid surface regardless of the difference of laser intensity ($a_0 = 3 \sim 24$). The electron energy is normalized by $2a_0mc^2$. This is the ponderomotive potential energy by 2ω frequency as seen in Fig. 6.13. During the short time like ~100 fs in no pre-formed plasma, the laser absorption rate is poor as shown in Fig. 7.8, and a small fraction of energy is converted to the hot electrons with the maximum energy of ~$2a_0mc^2$.

Experimental data of the hot electron energy distribution is reported in [10] for the case of pulse duration of 60 fs and the laser intensity 10^{18}W/cm^2, which corresponds $a_0 = 0.6$ with its wavelength $\lambda = 0.8$ μm. In the experiment, the other long pulse laser is used to form pre-formed plasma. The experimental data is shown in Fig. 8.21 [10]. For the case without the pre-pulse, the lines with minus timing show plateau profile in the energy of 0.5–1 MeV, corresponding to $1 \sim 2a_0mc^2$. This is the property is inferred to be that indicated in Fig. 8.20 by PIC simulation.

For the case when the pre-formed plasma is produced before the relativistic laser is irradiated, the number of the hot electrons increases, and it becomes almost Maxwellian with the increase of electrons with energy less than 1 MeV. It is inferred that the pre-formed plasma helps the stochastic heating of more electrons. According to the theoretical models in low-density plasmas, it seems to produce the super-Gaussian profiles in energy space, while as shown in Fig. 8.17, 2D PIC result, the escape from the boundary side of the laser beam possibly modify such 1D theoretical model result. It is seen by comparing a super-Gaussian of Fig. 8.19 to the Gaussian

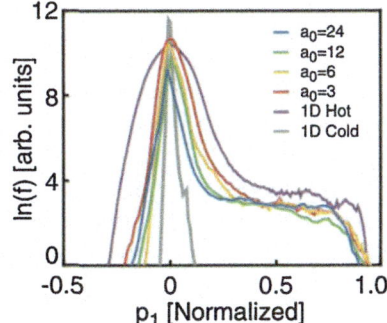

Fig. 8.20 1D and 2D PIC simulation with very fine mesh (630 meshes per laser wavelength) to calculate the vacuum heating type laser solid interaction. The electron distribution functions are plotted for 1D PIC result. [Figure 4a in Ref. 9]

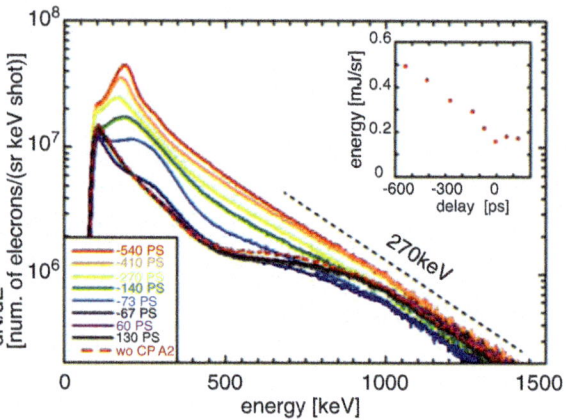

Fig. 8.21 Time evolution of electron energy distribution measured in the experiment with 60 fs and 10^{18} W/cm^2. [Figure 2 in Ref. 10]

of Fig. 8.17. In comparing theory and experiment, electron escape should be modeled in the laser beam region in the low-density plasmas.

8.6.2 Hot Electrons by Picosecond Lasers

There are only a limited number of big laser facilities which can deliver such relativistic lasers with pulse duration longer than ps. It is well-known that the hot electron temperature increases in general with the increase of the pulse duration. For example, the effective temperature increases in proportion to $t^{1/3}$. By use of an electron spectrometer, time-integrated electron energy distribution has been measured.

LFEX laser has been used to measure the hot electron production as a function of pulse length by combining four pulses with 1.5 ps pulse width each [11]. The time-integrated hot electron temperature is reported to increase from 1.5 ps to 3 ps pulses, while it almost saturates around 3 ps and is observed the same for 6 ps as 3 ps pulse. The laser intensity is kept 2.3×10^{18} W/cm^2 for $\lambda = 1$ μm. The time evolution of the hot electron temperature is simulated by PIC code as shown in Fig. 8.22. It is found that the time-integrated hot electron temperature well explains the experimental data.

After the above experiment with LFEX, the precise measurement of time-integrated electron energy spectra has been carried out [12]. Gold cube targets are irradiated by LFEX laser with two different pulse lengths and two different laser intensities. The case of 1.2 ps pulse with laser intensity of 2.5×10^{18} W/cm^2 is compared to the other two cases. It is reported that with increase of laser intensity four times to 1.0×10^{19} W/cm^2 or the pulse duration about four times longer (4 ps), the almost the same energy distribution is obtained in both. It roughly indicates that the spread of the distribution function is a function of $a_0^2 \tau_L$. It is surprising that Maxwell distributions are well fit to both cases and the corresponding 2D PIC

8.6 Hot Electron Generation

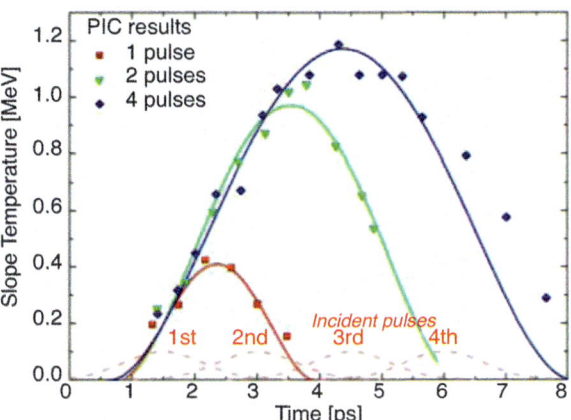

Fig. 8.22 Time evolution of hoe electron temperature calculated with PIC code for three different pulse duration with 2.3×10^{18} W/cm^2. [Figure 2 in Ref. 11]

simulation results well reproduce both distributions. The computational results are shown by black lines.

In the 2D PIC simulation, the same kind of identification of the root of accelerated electrons was also carried out as proposed in [2]. As explained in Sect. 8.2, most of the high-energy electrons are produced by the physical process of LIDA (loop-injected direct acceleration). The 2D PIC simulation is used to identify the contribution of such LIDA in the experiment. It is concluded that for the case with long pulse of multi-picoseconds, not only the single cycle LIDA but also the multiple cycles LIDA contribute substantially to producing the tail of the distribution function [12].

8.6.3 Sheath Potential Effect

As already mentioned in Fig. 7.12, it is pointed out that the hot electron confinement by the sheath potential is important to allow many interactions of the electrons with laser field. Since the pulse length is long enough to expect the long low-density plasma formation shown in Fig. 8.10. So, better coupling and cumulative acceleration are expected in the relatively long low-density plasma region.

Omega-EP laser has been used to study the pre-formed plasma effect on the hot electron energy spectrum of the super-ponderomotive electrons [13]. The experimental data on the hot electron temperature is plotted as a function of the laser pulse length for constant laser intensity of 4×10^{19} W/cm^2 for $\lambda = 1$ μm. It is seen that the hot electron temperature is 1.5 ~ 2.3 MeV for 1 ps laser pulse, depending on the plasma condition, while it increases about 3 ~ 4 MeV at 10 ps pulse. In applying the scaling law to time, it is roughly assumed $T_h \sim \tau^{1/3}$, where τ is the pulse duration.

The importance of the sheath potential to confine the electron and increase the coupling with lasers has been pointed out in [14] with 1D PIC simulation. Regardless of self-generated or externally produced low-density plasmas, the electrons

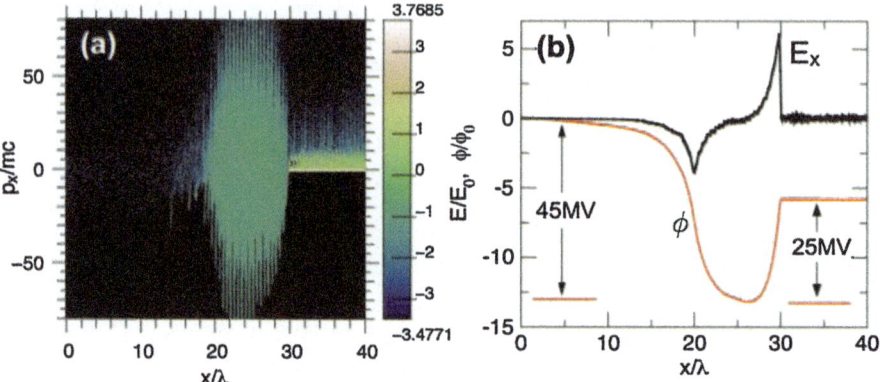

Fig. 8.23 (a) The x-component of the momentum distribution as a function of space. (b) The profiles of the electrostatic field and corresponding electrostatic potential in space

obtain large energy via interaction with laser and easily escape from the interaction region. In the vacuum side, however, high sheath potential almost the same as the hot electron temperature reflects the hot electrons to the interaction region. On the other hand, the strong potential as seen in Fig. 8.23 is produced even at the inward direction. The potential well shown in Fig. 8.23 with red curve is produced to confine the hot electrons. This electrostatic potential becomes important as we have seen in E_x in (8.1.1)–(8.1.3) to change the constant of motion α as the dephasing rate R in (8.3.1).

The effect of the sheath potential to confine the hot electrons in the rage of multi-ps interaction has been studied by focusing on the dephasing rate R in PIC simulation [15]. The laser intensity is 10^{20} W/cm^2 ($a_0 = 8.54$). It is concluded that selected electrons obtain small value of the dephasing rate R in the confined potential to obtain higher energy as shown in (8.3.2). In Fig. 8.24a, it is shown that the plateau low-density profile of electron and ion density are from (a) to (d) (t = 1, 3, 5, 10 ps) formed by the ponderomotive force to produce enough interaction region with laser. The time evolution of the potential is plotted on the right (f). The electron energy distribution is plotted in (e) at each time.

8.6.4 Multi-dimensional Effects

We have discussed so far on the case of one-dimensional geometry. It is important to discuss the multi-dimensional effect of the cutoff density profile and resultant effect to the wave form of the reflected laser component [16]. As already pointed out in Fig. 7.17, the cutoff density profile is modified by the Rayleigh-Taylor-like instability due to the strong ponderomotive force from low-density to the high-density layer. How the laser intensity pattern is modified by the structure and property of the cutoff density layer is shown in Fig. 8.25 [16], where laser intensity is 10^{18}W/cm^2

8.6 Hot Electron Generation

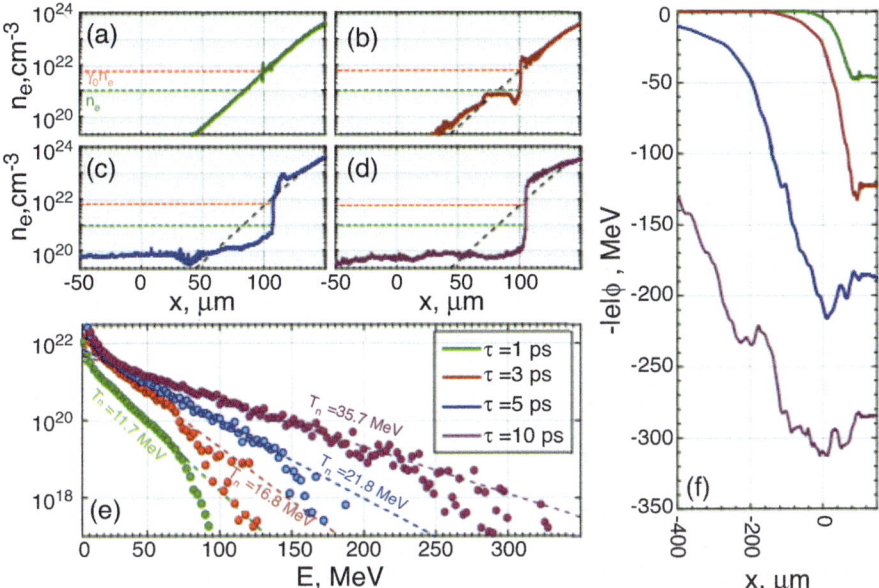

Fig. 8.24 The density modification due to ponderomotive force and the evolution of electrostatic potential. Time evolution of electron energy distribution is also shown

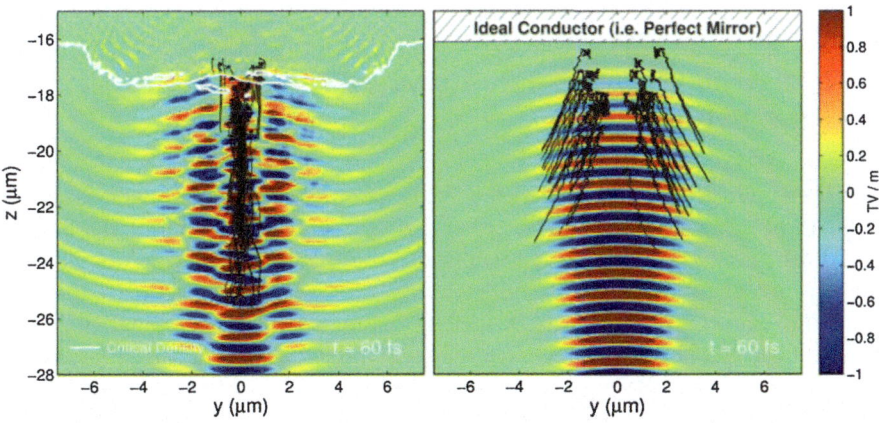

Fig. 8.25 Large deference of the interferometric intensity profile for (**a**) self-consistent plasma and (**b**) fixed density profile

and at the time of 60 fs after the laser front arrived at the critical surface. The right figure showed the electric field profile for the case of the laser reflection by a flat surface with perfect conductivity. It is clear that the laser is reflected in the ideal manner without the density profile of the background.

The left figure is, on the other hand, the same electric field profile observed in 2D PIC simulation for realistic plasma. Since the cutoff density surface is deformed by the ponderomotive force and the conductivity of the over dense region is not perfect, the reflected laser profile is strongly modified compared to the right figure. Typical particle trajectories are also shown with thin black lines in both cases. It is clear that in the ideal case, most of electrons are accelerated by the reflected laser field toward the laser incident direction by following the relation (8.2.4), while in the real plasma, the phase shift and self-generated magnetic field confine the trajectories, and the electron acceleration is boosted.

It is very hard to study quantitatively the physics of coupling and hot electron generation of the structured targets shown in Chap. 7. In order to know a part of multi-dimensional effect, it is useful to introduce the following 2D PIC simulation result. Surface-corrugated solid target is used to simulate the effect of the surface irregularity for the case with laser intensity of 4.65×10^{19} W/cm^2 at $\lambda = 0.53$ μm with linear polarization in the z-direction. The pulse duration is 200 fs.

The target parameter is changed from flat to the sinusoidal surface with the amplitude of $H = \lambda$ and the wavelength of $\Lambda = 2 \lambda$. It is concluded that the absorption efficiency ~5% for the flat target has been increased to ~50% with such surface corrugation. Almost static magnetic field generation is concluded to play an important role in absorption enhancement and stochastic heating of electrons, where the random reflection of the laser also enhances the stochastic of the electron acceleration as seen above.

8.7 Electron Motion in Two Counter-Propagating Relativistic Lasers

There are more general cases where the incidence of relativistic laser induces chaotic motion of electrons in under-dense plasma. As seen in Chap. 7, substantial fraction of incident laser is reflected. It is, therefore, more natural to consider the electron motions in both of incident and reflected relativistic laser field. Of course, depending on the incident condition to a target, there is freedom about the combination of field polarizations, propagating directions, intensity difference, etc. Let us consider the case that the incident and reflected relativistic lasers are counter-propagating in the x- and −x-directions, respectively.

The stochasticity of charged particles in plasma has been studied for a long time. The pioneering work is done by Boris Chirikov in 1959 to study the stability of plasma confinement by open magnetic device, so-called mirror machine [17]. Such chaos appears in many cases in plasma physics, for example, magnetic island in toroidal configurations, ion heating by lower-hybrid waves, wave instability saturated via harmonic generation, etc. In laser-produced plasmas, hot electron generations observed experiment with relativistic lasers are mostly because of the stochastic heating of electrons as see below. It is noted that paying attention to the

8.7 Electron Motion in Two Counter-Propagating Relativistic Lasers

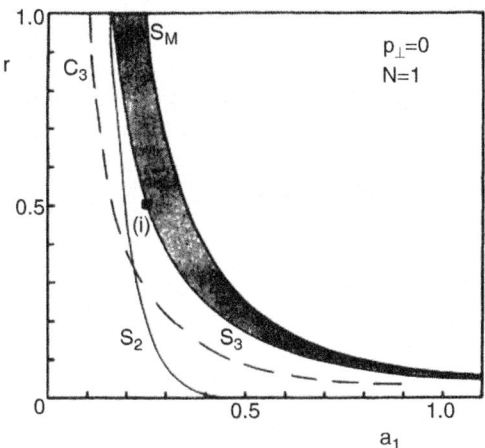

Fig. 8.26 Stochastic threshold obtained analytically based on the perturbation methods. [Figure 4 in Ref. 19]

stochastic nature in relativistic regime, a model experiment to study the physics of stochastic and nonthermal acceleration of cosmic rays has been done [18]. In a series of experiment, non-Maxwellian and power law energy spectra of electrons are observed. It is under research in what condition the power law acceleration probably happens.

The electron motion in the field of counter-propagating two relativistic electromagnetic waves is analytically studied by use of the concept of **Chrikov resonance-overlap criteria** [19]. This is analytical criterion for the onset of chaotic motion in deterministic Hamiltonian system. In Fig. 8.26, the theoretical result for the stochasticity threshold is plotted for two counter-propagating relativistic laser systems [19]. It is pointed out that there are three resonance-overlap points and the criteria and each of the points is shown with three curves, where (a_1, r) are the normalized intensity of incident laser and a fraction of reflected laser, respectively. Note that the threshold is symmetric to (a_1 and ra_1), but not so because of approximation in analysis. However, it is clear that there is a simple relation for the stochasticity threshold in the form:

$$a_1 a_2 \geq 0.10.\tilde{0}5 \qquad (8.7.1)$$

where $a_2 = ra_1$, and the value ½ is evaluated from Fig. 8.26. This relation indicates that roughly saying we have to consider the electrons in plasma suffer such stochastic interaction with the laser field if the laser intensity is more than 10^{18}W/cm^2 for 1 μm laser.

8.7.1 Counter-Propagating Two Laser Systems

When relativistic laser is irradiated on a target surface, reflected laser also interacts with the electrons in low-density plasma. This is also the case when strong reflection is induced by the backward Raman scattering as shown in Sect. 4.9. Since the incident and reflected lasers both interact with electrons, it is clear that an electron orbit becomes non-integrable. As mentioned regarding the chaos in (8.4.6), the basic equation becomes three with the reflected component. Assume in what follows, for simplicity, that two lasers are counter-propagating in the x-direction and both have linear polarization in the y-direction. The vector potential is in the y-direction with the form:

$$a = a_0 \sin(\xi) + a_1 \sin(\zeta) \tag{8.7.2}$$

where both phases are given as

$$\xi = t - x, \quad \zeta = t + x + \zeta_0 \tag{8.7.3}$$

For the case of perfect reflection at the surface $x = 0$, the phase shift $\zeta_0 = \pi$. Assuming $p_y - a = \beta(=0)$, the y-momentum should satisfy the relation:

$$p_y = a_0 \sin(\xi) + a_1 \sin(\zeta) \tag{8.7.4}$$

The equations to the x-momentum and position x becomes

$$\frac{d}{dt} p_x = \frac{1}{\gamma} [a_0 \sin(\xi) + a_1 \sin(\zeta)][a_0 \cos(\xi) - a_1 \cos(\zeta)] \tag{8.7.5}$$

$$\frac{dx}{dt} = \frac{p_x}{\gamma} \tag{8.7.6}$$

Then, the equation to the energy is

$$\frac{d}{dt} \gamma = \frac{1}{\gamma} [a_0 \sin(\xi) + a_1 \sin(\zeta)][a_0 \cos(\xi) + a_1 \cos(\zeta)] \tag{8.7.7}$$

Note that the constant α defined in (8.1.19) is not conserved in this case:

$$\gamma - p_x \neq \alpha \tag{8.7.8}$$

In Ref. [20–22], (8.7.4)–(8.7.6) have been solved numerically. In addition, PIC simulation has also been carried out to see the statistical properties of many electrons in the chaotic regime.

8.7.2 One-Electron Orbit

In Fig. 8.27, numerical result of integrated trajectory is shown for $a_0 = 1$ and $a_1 = 0.75$ (75% reflection), where the maximum normalized time is 500. This corresponds to 260 fs ($=500/2\pi \times 3.3$ fs) for $\lambda = 1$ μm laser. The initial condition of an electron is $x = 0$, $p_y = 0$, and $p_x = 0$. Figure 8.27 is the plot of electron trajectory in the momentum space (p_x, p_y), which should be parabolic for $a_1 = 0$ like (8.1.22). It is clear that the trajectory is chaotic in a certain bounded area in (p_x, p_y) space.

Decreasing the amplitude of the perturbation as $a_0 = 10$ and $a_1 = 1.5$, the **Poincare map** of the time evolution of (p_x, ξ) is plotted in Fig. 8.28. The horizontal axis is the phase ξ defined in (8.7.3). Here, the dots are taken for every multiple of π of the normalized time. Remind that the dots should be on a line given in Fig. 8.2 for

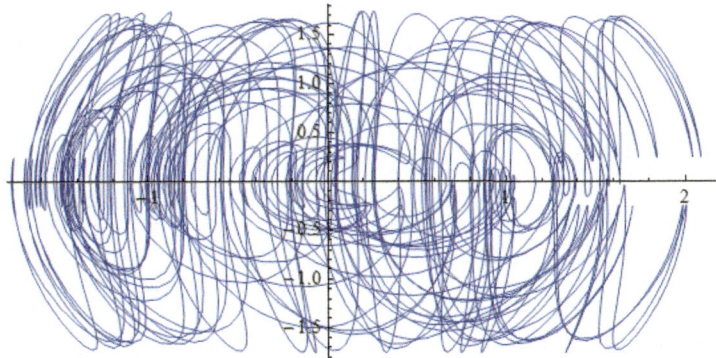

Fig. 8.27 An electron trajectory in the momentum space for the case with $a_0 = 1$ and $a_1 = 0.75$ in the counter-propagating problem

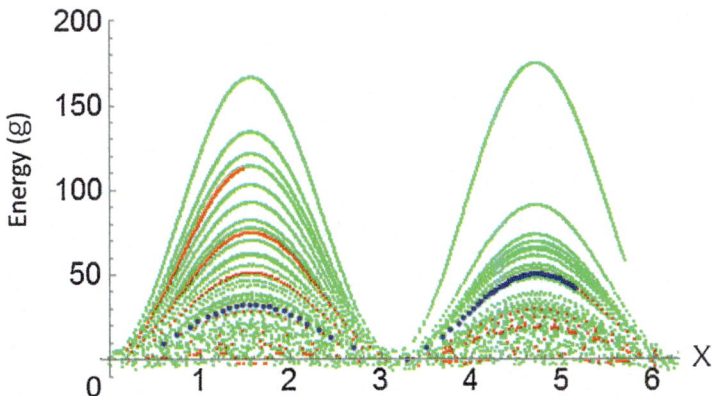

Fig. 8.28 The time evolution of an electron in the counter-propagating system with $a_0 = 10$ and $a_1 = 0.75$. The second wave is weak enough to observe the increase of energy after abrupt changes of α at the phase $\xi = 0$ and π

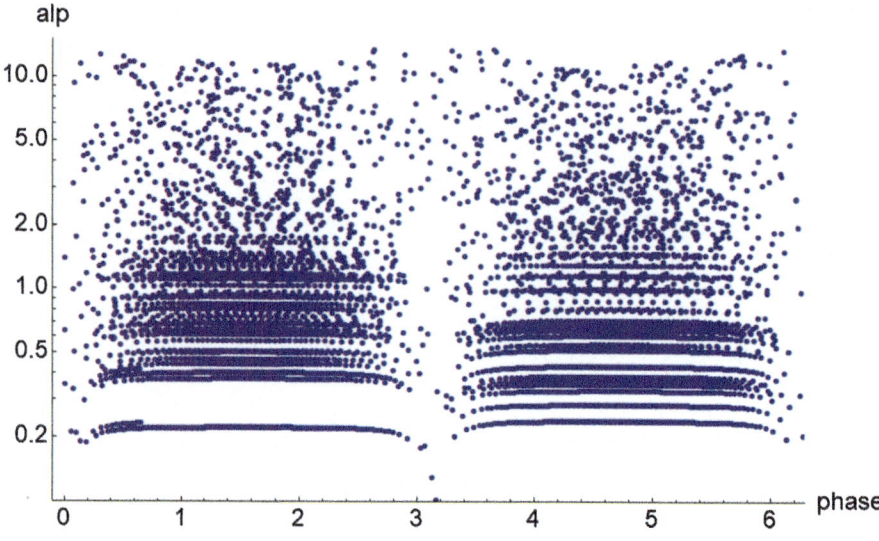

Fig. 8.29 Poincare map of the value of α in the case of Fig. 8.28

the case $a_1 = 0$. The blue dots are taken from the electron trajectory till the normalized time 100 (~52 fs). It is seen that the blue dots start from $(\xi, \gamma) = (0, 1)$ at $t = 0$, take a path in the first increase like the lines in Fig. 8.2, and come to the time limit near the top of the second mountain. The time is too short for the orbit to be chaotic.

Then, the red dots are the points until the time $t = 1000$ (~520 fs). It is seen that the maximum energy increased more than 50 predicted by (8.2.3) and the orbit with the energy γ less than ~30 becomes chaotic. Thanks to the chaotic change of the adiabatic constant α less than unity, the red dot trajectory seems to take trajectory given in (8.1.22) in later time.

The green dots are taken from the trajectory until the time 10,000 (~5.2 ps). The property does not change from the case of the red dots, while the maximum energy approached to $\gamma = 150$. In the green dots, the trajectory below the region with about the maximum $\gamma = 30$ becomes highly chaotic, and they take the relation (8.1.22) with different α values. As we see soon, the value α changes easily at the time when electrons stop moving with $\gamma = 1$.

In order to see which phase in the incident wave the adiabatic constant α changes, Poincare map of α(t) is plotted in Fig. 8.29. In the phase of less acceleration $\alpha > 1$, the value of α is also stochastically changed, while in the phase of acceleration $\alpha < 1$, the electron takes the orbit given only by the incident laser field with a constant value of α. In addition, the value of α becomes discontinuous at the phase $\xi = 0$ or π, where the electron kinetic energy becomes zero, namely, the electron stops in the laboratory frame.

In Fig. 8.29, the electron trajectories are also plotted in (p_x, p_y) plane for $a_0 = 3$ and three values of $a_1 = 0.3, 0.1, 0.05$ with red, green, and blue colors, respectively [20]. It is found that the maximum energy surely increase with the amplitude of a_1 and the mean trajectories are parabolic as given in (8.1.21) with a mean value of α. It

8.7 Electron Motion in Two Counter-Propagating Relativistic Lasers

is clear that the red trajectories are parabolic as average and oscillating due to the field by a_1. Its oscillation period is fast at large p_x and gets slowly near $p_x = 0$.

This result suggests how the perturbation laser changes the motion of electron. When the electron energy is small enough, the perturbation field affects the electron motion so that α changes stochastically at any chance. This is because the electron velocity is low as discussed with a simple model in Fig. 8.13. Once the velocity and energy increase, however, the perturbation field cannot couple the motion of electron moving with almost speed of light in the positive x-direction. As a result, the value of α is changed by the impact of the perturbation field when the electron almost stops at $\xi = n\pi$ (n: integer).

8.7.3 Lyapunov Exponent

Investigation of the stochastic interaction of electrons with two counter-propagating relativistic laser beams was carried out decays ago by Z-M Sheng and Y. Sentoku [21–23] with 1D and 2D PIC codes. Poincare maps for many particles clarified the criteria for stochasticity. The resultant criteria are shown in Fig. 8.30 [22]. The criteria are obtained by tracking the **Lyapunov exponent** (λ) defined by the time evolution of the distance between two particles located very nearby at the initial state. Mathematically, it is defined as

$$\lambda = \lim_{t \to \infty} \frac{1}{t} \sum_{t=0}^{t} \ln \left(\frac{|p(p_0 + \delta p_0) - p(p_0)|}{|\delta p_0|} \right) \quad (8.7.9)$$

This can be rewritten at long time limit as

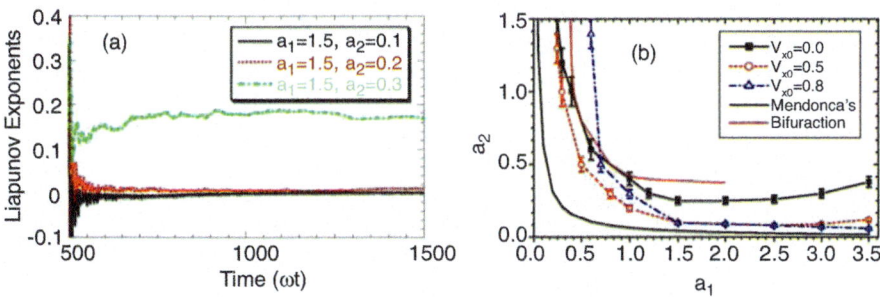

Fig. 8.30 (a) Lyapunov exponents for a test electron moving in counter-propagating laser fields with different incident field amplitudes. (b) Threshold amplitudes for stochastic motion in counter-propagating laser fields obtained numerically for electrons with different initial velocities. Also shown are the thresholds for local stochastic motion by Mendonca. [Figure 3 in Ref. 22]

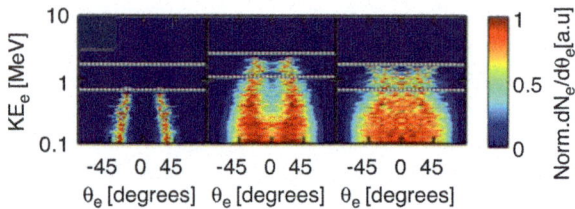

Fig. 8.31 The time evolution of electron distribution in energy and angle space by 2D PIC simulation

$$|\mathbf{p}(\mathbf{p}_0 + \delta\mathbf{p}_0) - \mathbf{p}(\mathbf{p}_0)| = |\delta\mathbf{p}_0| \exp(\lambda t) \quad (8.7.10)$$

So, if the Lyapunov exponent λ is positive, the physical positions of the points initially very near to each other go to very far distant points in the phase space. Such case is defined to be chaos.

In Fig. 8.30a, the time evolution of the Lyapunov exponent is plotted for three different amplitudes a_1 of perturbation beam for a given incident laser $a_0 = 1.5$, where the definition of $(a_1, a_2) = (a_0, a_1)$ in the present text. It is clear that for a_2 larger than 0.3, the distance between two particles is going apart exponentially as green line because of stochastic motion of particles. The criteria diagram shown in Fig. 8.26 is compared to such computational results in Fig. 8.30b. The criteria in Fig. 8.26 is plotted with the solid lines, while the PIC result is shown with marks. The result is approximately expressed as [22]

$$a_0 a_1 > 0.5 \quad (8.7.11)$$

It is clear that Mendonca's theoretical evaluation based on Chrikov-overlap criteria is lower than the PIC simulation result. This is because (8.7.1) is derived by linear approximation applicable to the region $a_2 \ll a_1$ in Fig. 8.30b.

Realistic simulation has been done with 2D PIC code to demonstrate better coupling of laser with the solid plasma by use of corrugated solid target [24]. The time evolution of electron distribution in phase space is plotted for a smooth surface target case. In Fig. 8.31, the time evolution of hot electrons in (γ, θ) is plotted. Using the relation $\tan(\theta) = p_y/p_x$, the parabolic relation in Fig. 8.1 can be speculated. The simulation is done for the condition parameters. The laser peak intensity is 5×10^{19} W/cm^2, the pulse is modeled with an envelope $\sin^2(\pi t/T)$ for $(0 < t < T)$ with T = 200 fs, and the amplitude is $a_0 = 3$. It is reported that the reflectivity is 85% as average. It is seen in Fig. 8.31 that at the left (t ~ 50 fs), electrons are accelerated as (8.2.3), then at the middle (t = 100 fs), the reflected laser starts to make the distribution broad by chaotic coupling, and at the right (t ~ 150 fs), stochasticity makes the distribution broad. As we discussed in Fig. 8.28, the pulse length is not long enough to see enhanced acceleration by changing the value of a, and the maximum electron energy is not large enough compared to the oscillation energy by laser filed.

In order to identify the stochastic heating with 2D PIC simulation, the electron heating with three different laser irradiation cases is studied and compared [Fig. 8 in Ref. 3, in Chap. 7]. The incident angle of laser is 55 degree and $a_0 = 2.5$. In the first case, a laser is irradiated to thin plasma so that the laser passes through the plasma and no reflection. The stochastic heating is not observed. In the second case, two lasers are irradiated from both of front and rear. Then, stochastic heating is observed. In the third case, thick plasma is used to reflect the incident laser. The stochastic heating is also observed.

It is proposed that using such colliding lasers from two directions enhances the hot electron temperature and is beneficial to target normal sheath acceleration of protons [25].

8.7.4 Hot Electron Temperature Scaling

The electron energy distribution is studied for a given density profile. In Fig. 8.32, the electron energy distribution in plasma slab $L = 50\lambda$ and $n_e = 0.01 n_c$ under the two lasers with $a_0 = 3.0$ and $a_1 = 0.5$ is shown for the time of $t = 200\tau$ and 400τ, where λ is the laser wavelength and τ is the oscillation period [21]. Rapid stochastic heating is observed, while the distribution is not Maxwellian but that suggested like in (8.5.7) seems better fit $\sim \exp(-\gamma^\alpha)$, $\alpha = 2$ or 3. In [21], it is found that, at early time well before the hot electron heating becomes saturated, the hot electron temperature and the maximum electron energy scale are proportional to

$$T_h \propto a_0^A a_1^B t^C$$
$$A \approx 2, \quad B \approx 0.5, \quad C \approx 0.5 \sim 1.0 \tag{8.7.12}$$

Note that the case of semi-infinite laser and A is better to be $A \sim 1$ for the case of finite pulse duration [21]. It is pointed out that the scaling powers A and C are obtained the same by the kicking model explained in Sect. 8.4, although it is a single laser case. In addition, the energy dependence of the distribution function in Fig. 8.32 well coincides with the solution of the diffusion model as shown in the next chapter.

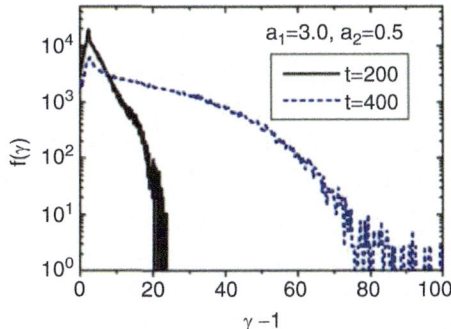

Fig. 8.32 Electron energy distributions from 1D PIC simulations of laser interaction with test electrons in vacuum or a plasma slab at a density $n_e = 01 n_c$ and with a thickness of $L = 550\lambda$. [Figure 7 in Ref. 22]

References

1. P.L. Robinson, A.V. Arefiev, D. Neely, Phys. Rev. Lett. **111**, 065002 (2013)
2. A.G. Krygier et al., Phys. Plasmas **21**, 023112 (2014)
3. A.V. Arefiev et al., Phys. Plasmas **23**, 056704 (2016)
4. A.J. Kemp, L. Divol, Phys. Rev. Lett. **109**, 195005 (2012)
5. H. G. Schuster, *"Deterministic Chaos"* (Physik-Verlag, 1984). Chap. 6. Fig. 105+106
6. W. Schlei et al., J. of Astronaut Sci. **61**, 170 (2014)
7. J. Meyer-ter-Vehn, Z.M. Sheng, Phys. Plasmas **6**, 641 (1999)
8. S.G. Bochkarev et al., Plasma Phys. Reports **40**, 202 (2014)
9. J. May et al., Phys. Rev. E **84**, 025401 (2011)
10. S. Inoue et al., Phys. Rev. Accelerators and Beams **21**, 041302 (2018)
11. A. Yogo et al., Sci. Rep. **7**, 42451 (2017)
12. S. Kojima et al., Commun. Phys. **2**, 99 (2019)
13. J. Peebles et al., New J. Phys. **19**, 023008 (2017)
14. A.J. Kemp, Y. Sentoku, M. Tabak, Phys. Rev. E **79**, 066406 (2009)
15. A. Sorokovikova et al., Phys. Rev. Lett. **116**, 155001 (2016)
16. C. Orban et al., Phys. Plasmas **22**, 023110 (2015)
17. V. Chrikov, Phys. Rep. **52**, 263 (1979)
18. Y. Kuramitsu et al., Phys. Rev. E **83**, 026401 (2011)
19. J.T. Mendonca, Phys. Rev. A **28**, 3592 (1983)
20. S. Rassou, A. Bourdier, M. Drouin, Phys. Plasmas **21**, 083101 (2014)
21. Z.M. Sheng et al., Phys. Rev. Lett. **88**, 055004 (2002)
22. Z.M. Sheng et al., Phys. Rev. E **69**, 016407 (2004)
23. Y. Sentoku et al., Appl. Phys. B Lasers Opt. **74**, 207 (2002)
24. C.E. Kemp et al., Phys. Plasmas **20**, 033104 (2013)
25. J. Ferri, E. Siminos, T. Fulop, Commun. Phys. **2**, 40 (2019)

Chapter 9
Theory of Stochasticity and Chaos of Electrons in Relativistic Lasers

9.1 Vlasov-Fokker-Planck Equations

In the previous chapter, it is shown that a random perturbation to the relativistic electron motions results in the heating of the electrons. This is called **stochastic heating** and one of the most important processes of the hot electron production due to relativistic laser-plasma interaction. Such a random perturbation to the system, in general, is described with additional diffusion term mathematically to the governing equation of statistical distribution of the electrons in the momentum space, $f(t, p_x, p_y)$ in the present case.

As the diffusion model in real space, **Brownian motion** is well-known, and the distribution of any particles under Brownian motion in space is said, in general, to be described with the following **diffusion equation**:

$$\frac{\partial}{\partial t} f(t, x) = D \frac{\partial^2}{\partial x^2} f(t, x) \qquad (9.1.1)$$

For simplicity, one dimension in space is assumed. In (9.1.1), D is the diffusion coefficient and is usually given in the form:

$$D = A \frac{\Delta^2}{\tau} \qquad (9.1.2)$$

where Δ is the **mean free path** and τ is the **collision time**. It is noted that A is a coefficient depending on the situation and usually A = 1/3 or 1/2.

In the present case, the perturbation is expected to contribute to the heating of electrons, and the hot electrons are generated. Then, Brownian motion is random motion in the velocity or momentum space. **Langevin equation** is well-known as a simple model equation to describe such a heating by external random force.

Langevin equation is a master equation to the velocity of any particles in random force and given in the form:

$$\frac{d}{dt}V(t) = -\frac{1}{\tau_c}V(t) + R(t) \qquad (9.1.3)$$

where τ_c is an effective collision time to give **drag force**. In (9.1.3), R(t) is **random force**, and **Markovian process** with Gaussian probability is assumed. Then, the following relation is satisfied:

$$\langle R(t)\rangle = 0, \quad \langle R(t)R(t')\rangle = D_L\delta(t-t') \qquad (9.1.4)$$

where < > stands for taking the average over time.

The precise derivation of **Fokker-Planck equation** from Langevin Eq. (9.1.3) is shown in Vol. 2 relating to the electron transport kinetics. The probability function P(v,t) the same as the distribution function of the particles with velocity v at time t is known to be derived from (9.1.3) as the following equation:

$$\frac{\partial}{\partial t}P(v,t) = \frac{1}{\tau_c}\frac{\partial}{\partial v}[vP(v,t)] + \frac{D_L}{2}\frac{\partial^2}{\partial v^2}P(v,t) \qquad (9.1.5)$$

On RHS of (9.1.5), the first term is the frictional term showing the velocity decrease by the drag force, and the second term is the diffusion term by the random force. The differential equation is a good approximation as long as the random force is small enough as shown below.

9.1.1 Stochastic Diffusion Equation

Let us derive the diffusion-type equation to the numerical model by Myer-ter-Vehn and Sheng [1] shown in Sect. 8.4. Assume that the random force is given only in the y-direction and (8.4.8) has the form:

$$\frac{dp_y}{dt} = \frac{da}{dt} + R(t) \qquad (9.1.6)$$

where R(t) has the same property as R(t) in (9.1.4).

The momentum change at the time "n" is given by integrating (9.1.6) between the short time interval of the kick at $t = t_n$:

$$\Delta p_y{}^n = \Delta^n \qquad (9.1.7)$$

where the kick is $R(t_n) = \Delta^n\delta(t-t_n)$. Then, the same time integration of (8.4.9) gives

9.1 Vlasov-Fokker-Planck Equations

$$\Delta p_x{}^n = -\Delta^n \frac{\cos\xi}{\gamma}\bigg|_n \qquad (9.1.8)$$

Since the force R(t) is random and independent of the phase ξ, it is clear that

$$\begin{aligned}\langle \Delta p_y{}^n \rangle &= \langle \Delta^n \rangle = 0 \\ \langle \Delta p_x{}^n \rangle &= 0\end{aligned} \qquad (9.1.9)$$

However, the dispersion due to the kicks remains in the form:

$$\left\langle \left(\Delta p_y{}^n\right)^2 \right\rangle = \left\langle (\Delta^n)^2 \right\rangle = \Delta^2 \qquad (9.1.10)$$

$$\left\langle (\Delta p_x{}^n)^2 \right\rangle = \left\langle \left(\Delta^n a_0 \frac{\cos\xi}{\gamma}\right)^2 \right\rangle = \frac{a_0^2 \Delta^2}{2\gamma^2} \qquad (9.1.11)$$

where Δ is the mean value of the kick momentums to the y-direction R(t) shown in (9.1.6). In deriving (9.1.11) it is assumed that

$$\langle \cos^2(\xi) \rangle = \frac{1}{2} \qquad (9.1.12)$$

Let us derive the diffusion term in (9.1.5) mathematically with the assumption of Taylor expansion of the time evolution of the distribution function $f(t, p_x, p_y)$. Assuming that the change of the distribution function due to the perturbation is small enough and the change of the distribution by the next kick after the "n" is given in the following Taylor expansion form:

$$f(\mathbf{p}^{n+1}) = f(\mathbf{p}^n) + \Delta\mathbf{p} \cdot \frac{\partial}{\partial\mathbf{p}} f(\mathbf{p}) + \frac{1}{2}\Delta\mathbf{p} : \Delta\mathbf{p} \cdot \frac{\partial}{\partial\mathbf{p}} : \frac{\partial}{\partial\mathbf{p}} f(\mathbf{p}) \qquad (9.1.13)$$

Inserting (9.1.10) and (9.1.11) into (9.1.13), the following diffusion equation is obtained:

$$\frac{\partial}{\partial t} f = \frac{\partial}{\partial p_x}\left(D_x \frac{\partial}{\partial p_x} f\right) + D_y \frac{\partial^2}{\partial p_y^2} f \qquad (9.1.14)$$

$$D_x = \frac{a_0^2 \langle \Delta^2 \rangle}{4\tau_c \gamma^2}, \quad D_y = \frac{\langle \Delta^2 \rangle}{2\tau_c} \qquad (9.1.15)$$

In deriving the diffusion equation, the mean value of the time interval between the subsequent kicks τ_c is introduced.

9.1.2 Vlasov-Fokker-Planck Equation

The **Vlasov-Fokker-Planck (VFP) equation** governing the distribution function f is given in the form:

$$\frac{d}{dt}f(t, p_x, p_y) = \frac{\partial f}{\partial t} + \frac{dp_x}{dt}\frac{\partial f}{\partial p_x} + \frac{dp_y}{dt}\frac{\partial f}{\partial p_y} = \left(\frac{df}{dt}\right)_{random} \quad (9.1.16)$$

where RHS is the contribution by the random force. Therefore, the equation for the distribution of electrons under the relativistic lasers and the stochastic force is obtained:

$$\frac{\partial f}{\partial t} - \frac{a_0^2}{2\gamma}\sin(2\xi)\frac{\partial f}{\partial p_x} + a_0 \sin\xi \frac{\partial f}{\partial p_y} = \frac{\partial}{\partial p_x}\left(D_x \frac{\partial}{\partial p_x}f\right) + D_y \frac{\partial^2}{\partial p_y^2}f \quad (9.1.17)$$

In deriving (9.1.17), (8.4.8) and (8.4.9) are inserted to (9.1.16).

The role of the diffusion on RHS in (9.1.17) is in general small compared to the second and third terms in LHS, while depending on the problem, the diffusion plays an important role to dramatically change of the distribution function. Small diffusion in p_x direction near $p_x = 0$ causes the decrease of the values of α in (8.1.23), and the maximum energy of electrons increases dramatically due to the diffusion. In addition, D_x in (9.1.15) is proportional to $1/\gamma^2$ which is large near $p_x = 0$ point, and this diffusion effect is very efficient to heat the electrons stochastically.

9.2 Stochastic Heating by Laser Filamentation

It is frequently seen that the incident laser becomes unstable to the filamentation instability and the wave phase is not uniform in the y-direction but random in this direction. For example, a snap shot of laser intensity obtained by 2D PIC simulation is plotted in Fig. 9.1 [2]. In such a case, a jump of an electron from a filament to a filament is affected by a rapid change of the phase of the electromagnetic waves. It is shown in Refs. [3, 4] that such a random motion of electrons in relativistic laser field can be modeled by random walk due to the random force induced by rapid phase change.

The basic equations are (8.4.8) and (8.4.9). The laser field is given as $a = a_0 sin(\xi)$, although we assume that the laser phase ξ is not coherent, but randomly changes in the y-direction because of the filamentation. The time derivative is replaced by the derivative by the **proper time** τ defined in (8.1.12). Then, (8.1.12) becomes

9.2 Stochastic Heating by Laser Filamentation

Fig. 9.1 A plot of transverse laser electric field in real space, where the laser must propagate through under-dense plasma before reaching critical density. Notice the combination of both self-focusing and filamentation. [Figure 12 in Ref. 2]

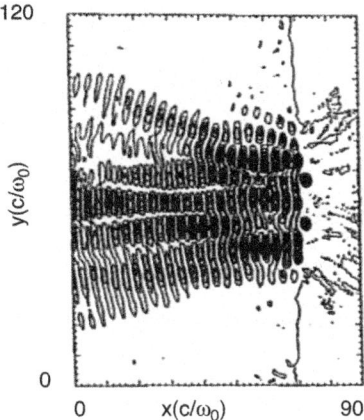

$$\frac{d}{dt}\xi = 1 - v_x = \frac{(\gamma - p_x)}{\gamma} = \frac{\alpha}{\gamma} \quad (9.2.1)$$

This is rewritten with the proper time:

$$\frac{d}{d\tau}\xi = \alpha \quad (9.2.2)$$

Note that the canonical momentum in the y-direction is still conserved.

For the case where the electron moves to the next filament at the time "n," the y-momentum should satisfy the relation at time t from (8.4.8) just after the jump at n:

$$p_y = p_y^n + a_0(\sin\xi - \sin\xi^n) \quad (9.2.3)$$

Inserting a(t,x) to (8.4.9), the following equation is obtained:

$$\frac{d}{d\tau}p_x = -a_0 p_y \cos\xi \quad (9.2.4)$$

Using (9.2.2) and (9.2.3), (9.2.4) is modified like

$$\frac{d}{d\xi}p_x = -\frac{a_0}{\alpha}p_y\cos\xi = -\frac{a_0}{\alpha}\left[p_y^n + a_0(\sin\xi - \sin\xi^n)\right]\cos\xi \quad (9.2.5)$$

Integrating (9.2.5) for a very short time interval across the time of the event "n," (9.2.5) yields

$$p_x = p_x^n - \frac{a_0}{\alpha} p_y^n (\sin\xi - \sin\xi^n) - \frac{a_0^2}{2\alpha}(\sin\xi - \sin\xi^n)^2 \quad (9.2.6)$$

With the same procedure in deriving the recurrence relation (8.4.3), new relations are obtained from (9.2.3) and (9.2.6), respectively:

$$p_y^{n+1} = p_y^n + a_0 \Delta_n^{n+1} \quad (9.2.7)$$

$$p_x^{n+1} = p_x^n - \frac{a_0}{\alpha^n} p_y^n \Delta_n^{n+1} - \frac{a_0^2}{2\alpha^n}\left(\Delta_n^{n+1}\right)^2 \quad (9.2.8)$$

where

$$\Delta_n^{n+1} = \sin\xi^{n+1} - \sin\xi^n \quad (9.2.9)$$

(9.2.9) is a random value with the condition:

$$-2 \leq \Delta_n^{n+1} \leq 2 \quad (9.2.10)$$

Let us assume that the change of the x- and y-momentums in each phase change is small enough and the difference of the distribution function in $\mathbf{p} = (p_x, p_y)$ space defining $f(p_x, p_y, t)$ can be approximated by Taylor expansion same as (8.5.13). Letting for simplicity $\Delta = \Delta_n^{n+1}$, (9.2.8) can be expressed as

$$\Delta p_x = -A p_y \Delta - B \Delta^2 \quad (9.2.11)$$

where A and B represent the coefficients, for simplicity. Taking the average of the random change, the following relations are obtained:

$$\langle \Delta p_x \rangle = -A p_y \langle \Delta \rangle - B \langle \Delta^2 \rangle \quad (9.2.12)$$

$$\langle \Delta p_x^2 \rangle = \left(A p_y\right)^2 \langle \Delta^2 \rangle + 2AB p_y \langle \Delta^3 \rangle + B^2 \langle \Delta^4 \rangle \quad (9.2.13)$$

where it is assumed that the random change satisfies the relation:

$$\langle \Delta \rangle = \langle \Delta^3 \rangle = 0 \quad (9.2.14)$$

In addition, Taylor expansion is possible only for the case satisfying the condition: $\langle \Delta^2 \rangle \gg \langle \Delta^4 \rangle$. Then, the contribution of p_x derivatives in the second and third terms in RHS of (9.1.13) is given as

9.2 Stochastic Heating by Laser Filamentation

$$-B\langle\Delta^2\rangle \frac{\partial}{\partial p_x} f + \frac{(Ap_y)^2}{2} \langle\Delta^2\rangle \frac{\partial^2}{\partial p_x^2} f \qquad (9.2.15)$$

Only the second derivative remains in the expansion to the y-momentum. Finally, for a weak randomness of the laser filaments, the following Vlasov-Fokker-Planck equation is obtained:

$$\frac{\partial f}{\partial \tau} + \left[\frac{a_0^2}{2} \sin(2\xi) + D\right] \frac{\partial f}{\partial p_x} + a_0 \gamma \sin\xi \frac{\partial f}{\partial p_y}$$

$$= D \frac{\partial^2}{\partial p_x^2} \left(\frac{p_y^2}{\alpha} f\right) + D \frac{\partial^2}{\partial p_y^2} (\alpha f) \qquad (9.2.16)$$

$$D = \frac{a_0^2}{2\tau_c} \langle\Delta^2\rangle \qquad (9.2.17)$$

In deriving (9.2.16), the relation (9.2.2) was used. The Eq. (9.2.16) is a VFP equation in the two-dimensional momentum space, and D corresponds to a diffusion coefficient. In (9.2.17), τ_c is the average proper time interval of the change of the phase of the laser field. In reducing the finite difference to (9.2.16), the time dependent value α defined by (8.1.20) is set inside the derivative in order to keep the integrated probability constant in time. In Ref. [3], almost the same diffusion terms as in (9.2.16) were derived, and it was solved numerically only with the diffusion terms. It is noted, however, that, in general, the Vlasov terms are more important for small perturbation assumed in deriving the diffusion term.

Consider how the three terms with the coefficient D in (9.2.16) contribute the time evolution of the distribution function. The first term due to the randomness contributes the advection in the positive p_x direction. As already mentioned in Fig. 8.1, small change of p_x near $p_y = 0$ results an enhanced oscillation amplitude. At the time $p_y = 0$, $sin(2\xi) = 0$ in (9.2.16) and the term D become important to increase the p_x at $p_y = 0$ axis. Even for a small value of D, this diffusion term changes the value of α in (8.1.22), and the oscillation energy of the diffused electrons increases in time.

9.2.1 Numerical Calculation of Test Particles

The random phase change effect on the evolution of the electron distribution was studied by tracking the dynamics of 4000 electrons numerically [4]. This is almost equivalent to solving (9.2.16). It is better to say that (9.2.16) is an approximate equation to solving directly (8.4.8) and (8.4.9). The parameters employed in the calculation were $a_0 = 0.7$ (10^{18}W/cm^2) at $\lambda_L = 790$ nm and pulse duration 100 fs. The laser is focused with the radius 7 μm over the distance of Rayleigh length 200 μm. The magnetic field is also solved, because the magnetic field in the

Fig. 9.2 The momentum (**a**) and energy (**b**) distribution of 4000 electrons accelerated in the stochastic laser field with the phase jump of m = 4

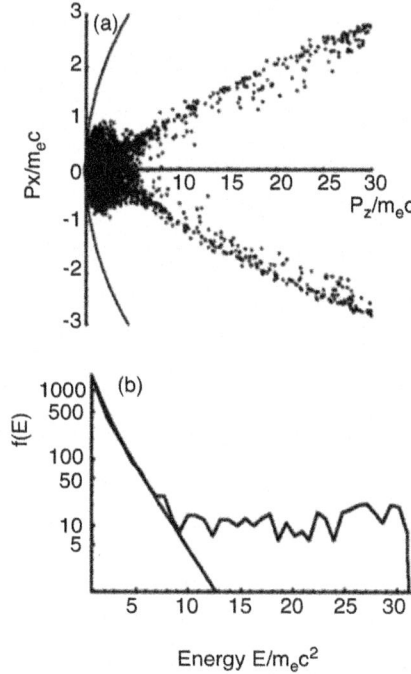

azimuthal direction is important to confine the electrons in the laser-focused region. The magnetic field is calculated by taking account of the current induced by the drift velocity of electrons in (5.3.27).

In Ref. [4], two results for the different number of randomness are shown. The first one is average phase change "m" of two times (m = 2), and the other is four times (m = 4) during the interaction time. It is reported that in the case of m = 2, the distribution evolves as Maxwellian up to the maximum energy of $\gamma = 3.9$, while with increase of m, it becomes $\gamma = 30.9$ for the case of m = 4. It is noted that the maximum energy of the ponderomotive scaling should be $\gamma = 1/2 a_0^2 + 1 = 1.25$ from (8.2.5). Substantial increase of the maximum energy is obtained due to the random change of the laser phase. The distribution of 4000 electrons is shown in Fig. 9.2.

As mentioned about the role of D in the advection term in (9.2.16), the electrons diffuse toward larger p_x near $p_y = 0$ to reduce the effective value of α in (8.1.22). By inserting $p_y = 3$ and $p_x = 30$, we obtain $\alpha = 1/6$ (<<1). The distribution function in Fig. 9.2b has a long tail. It is pointed out that the energy distribution function $f(E)$ can be fitted rather with a **power law** in the form:

$$f(E) \propto (E/E_0)^{-k}, \qquad E_0 = ma_0^2 \qquad (9.2.18)$$

9.2 Stochastic Heating by Laser Filamentation

where k is found to be k = 3. The characteristic energy E_0 is proportional to the average number of the phase change "m" in the interaction region. This m-dependence in (9.2.18) is important to obtain the strong enhancement of the hot electron temperature by the randomness of laser phase.

9.2.2 PIC Simulation of Stochastic Diffusion

As mentioned in Sect. 8.7, counter-propagating two beams makes an electron motion stochastic. In general, such stochastic motion can be approximated with a VFP equation such as (9.2.16), as long as the second beam intensity is much weaker than the main laser intensity. The LHS in (9.2.16) allows the periodic electron motion shown in Fig. 8.1. The RHS with a small diffusion coefficient induces the spread of an electron trajectory along the orbit in Fig. 8.1. This is first demonstrated in [5] (see Fig. 2) with test particle calculations.

By use of 2D PIC simulation, the same problem has been solved for the case where a = 3 laser irradiates the plasma with density $n_e = 10^{-4} n_c$ and the length of 50 μm. The electron density distribution in the momentum space (p_x, p_y) is plotted in Fig. 9.3 [6]. The strength of the counter-propagating laser is (a) $a_1 = 0$ and (b) $a_1 = 0.3$. The snap shots are at t = 648 corresponding to 41.6 fs. In Fig. 9.3b, the diffusion is clearly seen from the solution in Fig. 9.3a, which is the same as one of in Fig. 8.1.

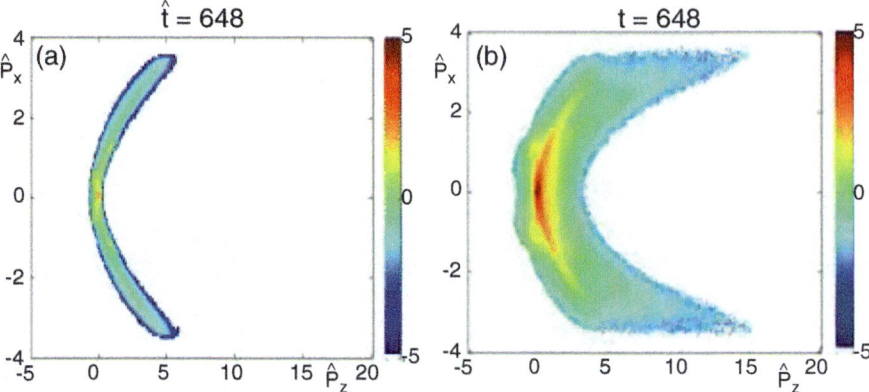

Fig. 9.3 The electron momentum distribution when the main laser with $a_0 = 3$ and the counter-propagating laser $a_1 = 0$ (**a**) and $a_1 = 0.3$ (**b**) at the same time t = 42 fs

9.3 Nonlocal Jump in Energy Space

In deriving Fokker-Planck equation to take into account the effect of stochastic motions from the perturbation force, it is assumed that the momentum change $\Delta \mathbf{p}$ is small enough and Taylor expansion can be used for modeling the stochastic effect in Vlasov equation. In general, it is seen that such a diffusion-type model equation provides exponential function of the electron distribution function. It is well-known in modern statistical physics that **Gibbs-Boltzmann** statistics assumes the correlation of particles is local with a Gaussian probability of momentum change. Consequently it yields Maxwell distribution in the equilibrium state. On the other hand, **Tsallis statistics** [7] based on nonlocal correlation of particles is found to result a non-Maxwell distributions. The distribution is the **Cauchy-Lorentz form** which has a **Lorentzian distribution** as one of the solutions in the equilibrium state. The equilibrium distribution of the Tsallis statistics is shown in Appendix A.4.

9.3.1 Levy's Flights

Such a nonlocal transport is called **Levy flights (jumps)** historically [8]. An example of a test particle trajectory with Levy jumps is shown in Fig. 9.4 in two-dimensional space [8]. The particle in Fig. 9.4 drifts like Brownian motion with small steps, while it jumps in space sometimes over long distances. The transport phenomena of the particle, energy, or any other physical quantities stem from the sum of such jumps. The transport mainly by the nonlocal jumps is called **nonlocal transport** and called in general **anomalous transport (diffusion)**. Such transport has been studied in wide range of physical issues, including chaotic phase diffusion of Josephson junction, turbulent diffusion, molecular spectrum fluctuation, etc. [8]. Relating to the laser plasma, **turbulent diffusion** in implosion dynamics will be discussed in

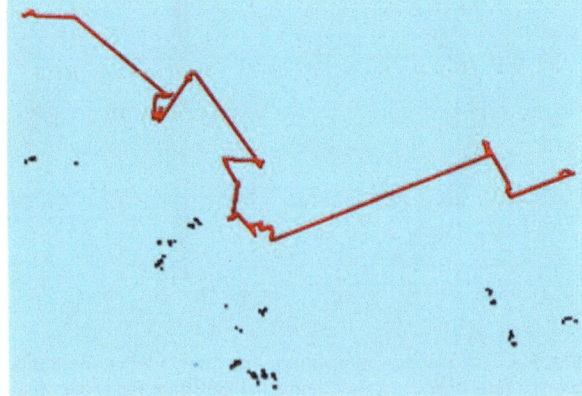

Fig. 9.4 An example of a test particle trajectory due to a combination of Brownian motion and Levy jumps in two-dimensional real space. [Figure 1 in Ref. 8]

9.3 Nonlocal Jump in Energy Space

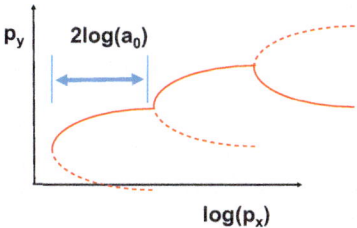

Fig. 9.5 Most lucky-accelerating electron trajectory in log (p_x)-p_y plane. It shows a power law spectrum

Vol. 2. Successful application to the anomalous transport in a large magnetic plasma is shown later by use of the Levy flights.

The Levy flights are composed of **self-similar jumps** as well-known in the **fractal theory** [9, 10]. It should be noted that the importance of the Levy flights was pointed out in [11] in the range of the laser strength $a_0 \sim 10^3$ relating to the better confinement of electron-positron pair particles generated by the vacuum breakdown in multi-colliding laser field. In the present book, the study is limited to the case where the theory can be compared to the experimental data, so that the typical value $a_0 = 1 \sim 10$ is assumed. Before the formulation of the Levy jumps instead of Fokker-Planck type model, it is informative to see how the self-similar jump is possible intuitively in the present relativistic laser electron interaction.

It is already derived that the maximum p_x over a laser oscillation period is give as (8.2.15). For simplicity, consider only the x-momentum in what follows and write p_x as p. If we assume there is a case where after each period the electron jumps into acceleration phase again from time n to the time n + 1. Then, the following similarity relation is obtained:

$$p^{n+1} = a_0^2 p^n \quad \Rightarrow \quad p^n = \left(a_0^2\right)^n p_0 \qquad (9.3.1)$$

Of course, this is for a very lucky electron the phase of laser field of which jumps from $\pi/2$ to π in Fig. 8.2 due to a drift of the electron into the next filament or the laser phase changes abruptly in time. The almost continuous increase of p_x in Fig. 8.16 is that the change of p_y by external kick keeps the electron in an acceleration phase of the incident laser for a long time.

Eq. (9.3.1) indicates that the energy change of a very lucky electron by the external kicks happens at each maximum momentum point. Its energy continues to increase with larger initial value of p in (8.2.15) at each jump. It is easily found that such an electron can move in the *log*(p) space with a constant average velocity and its energy increases after every trajectory of $2\pi/\omega$ period by the factor a_0^2. Such nonlocal jumps (trajectories) are plotted in Fig.9.5 schematically. If we dare to assume that there is no dissipation along such similarity jumps, the following power law energy distribution is obtained:

$$\frac{d}{d \ln(p)} f(p) = \text{const.} \quad \Rightarrow \quad f(p) \propto p^{-1} \tag{9.3.2}$$

It may be possible to obtain the power law spectrum by including such nonlocal jumps in the energy distribution of higher-energy tail.

9.3.2 Integral Form of Random Walk

Return to the original idea of Brown motion, and introduce the **transition probability** density ψ. Then, the time advance of the distribution function $f(p,t)$ by a **Markov (memoryless) stochastic process** is given by the following evolution equation [12]:

$$f(p, t + \Delta t) = \int_{-\infty}^{\infty} \psi(\Delta p, \Delta t) f(p - \Delta p, t) d(\Delta p) \tag{9.3.3}$$

The probability density ψ is defined so that it is normalized to the unity, and $\psi(\Delta p, \Delta t)$ gives the event probability to the jump of the momentum p by Δp.

In general, the transition probability is determined by the property of the scattering particles or the external condition of stochasticity in the environment of Brown motion particles. In the case of classical absorption discussed in Chap. 2, the fixed ion statistical distribution gave the heating rate of the quivering electrons, and the electron distribution function is independent of $\psi(\Delta p, \Delta t)$ in (9.3.3). This is also the same in the present case where electrons are scattered or affected by Levy jumps in the p space because of sequential kicks by or phase changes of perturbation fields. Note that in the case of electron-electron or ion-ion scattering, (9.3.3) cannot be applicable, and the transition probability becomes a function of their distribution functions.

9.3.3 Fokker-Planck Diffusion Model

Under the condition that Δp is small enough compared to the variation of $f(p)$ in (9.3.3), Taylor expansion is possibly used to expand RHS of (9.3.3) in the form:

$$f(p - \Delta p, t - \Delta t) = f(p, t - \Delta t)$$
$$- \Delta p \frac{\partial}{\partial p} f(p, t - \Delta t) + \frac{(\Delta p)^2}{2} \frac{\partial^2}{\partial p^2} f(p, t - \Delta t) + L \cdots \tag{9.3.4}$$

Inserting (9.3.4) into (9.3.3) and assuming that $\Psi(\Delta p)$ is an even function of Δp, the following diffusion equation is approximately obtained:

9.3 Nonlocal Jump in Energy Space

$$\frac{\partial}{\partial t}f(p,t) = D\frac{\partial^2}{\partial p^2}f(p,t), \quad D = \left\langle \frac{(\Delta p)^2}{2\Delta t} \right\rangle \quad (9.3.5)$$

The diffusion Eq. (9.3.5) is the same as (9.1.1) except that (9.3.5) is the diffusion in the momentum space, while (9.1.1) is in the real space. As well-known, the probability density is given by a Gaussian distribution in the **Gaussian and Poissonian statistics**.

Consider that the physical problem is the case where the Taylor expansion is not applicable and the diffusion coefficient is relatively large as suggested in (9.3.1) by Levy's jump in the momentum space. Namely, some electrons obtain large amount of energy due to the external perturbation as suggested in (9.3.1). This is more general. Then, the time evolution of the distribution function is obtained by solving (9.3.3) directly. It is clear that the problem is not so simple, because in general $\Psi(\Delta p, \Delta t)$ is not independent of p and t, and in addition Δt may not be constant. However, it is impossible to taken into account such general freedom in a simple model Eq. (9.3.3) to solve it analytically. Here, the present discussion is focused on the next better model than the Fokker-Planck type model. As seen below, it is a big progress to know how such Levy's jumps and nonlocal transport are important in the stochastic electron acceleration by the relativistic laser field.

9.3.4 Fractional Fokker-Planck Model

In order to solve (9.3.3) directly, Fourier transformation defined as below is used:

$$F(k,t) = \int_{-\infty}^{\infty} f(p,t)e^{-ikp}dp \quad (9.3.6)$$

The probability density is also Fourier transformed as follows:

$$\Psi(k,\Delta t) = \int_{-\infty}^{\infty} \psi(\Delta p, \Delta t)e^{-ik\Delta p}d(\Delta p) \quad (9.3.7)$$

Note that the normalization of the probability corresponds to the following relation:

$$\lim_{k\to 0}\Psi(k) = 1 \quad (9.3.8)$$

Carrying out the Fourier transformation of (9.3.3), we can obtain a new relation by use of the **convolution integral** in Fourier transformation:

$$F(k, t+\Delta t) = \Psi(k,\Delta t)F(k,t) \quad (9.3.9)$$

In order to make the mathematics more clear with a simple example, consider the diffusion equation with a constant diffusion coefficient D:

$$\frac{\partial}{\partial t} f = D \frac{\partial^2}{\partial p^2} f \tag{9.3.10}$$

Taking Fourier transformation of (9.3.10), the equation to each Fourier component for k is derived:

$$\frac{\partial}{\partial t} f_k = -k^2 D f_k \tag{9.3.11}$$

The time integration is easily carried out to yield

$$f_k(t + \Delta t) = e^{-k^2 D \Delta t} f_k(t) \tag{9.3.12}$$

Comparing (9.3.3) to (9.3.12), the probability density in (9.3.3) is found to be given in the form:

$$\begin{aligned}\psi_G(\Delta p, \Delta t) &= \frac{1}{2\pi} \int_{-\infty}^{\infty} e^{-(k^2 D \Delta t - i k \Delta p)} dk \\ &= \frac{1}{2\sqrt{\pi D \Delta t}} \exp\left(-\frac{\Delta p^2}{4 D \Delta t}\right)\end{aligned} \tag{9.3.13}$$

This is a Gaussian probability density. Note that the Gaussian noise is assumed for the random force when deriving the diffusion equation and Fokker-Planck equation. It is clear that the Gaussian probability decays as soon as Δp increases and Levy-type jumps cannot be modeled with such a diffusion equation.

Therefore, the diffusion Eq. (9.3.10) has been extended so that it can model the nonlocal jumps. Such an extension of mathematical model has been proposed, and its equation is called the **fractional Fokker-Planck equation (FFPE)** for modeling Levy flight process proposed in [12]. In this model, the kernel of the probability is modified as

$$-k^2 \Rightarrow -|k|^\alpha \tag{9.3.14}$$

Introduce the **fractional derivative operator** as

$$\frac{\partial^\alpha}{\partial |p|^\alpha} \equiv \frac{1}{2\pi} \int_{-\infty}^{\infty} |k|^\alpha e^{-ikp} dk \tag{9.3.15}$$

The normal diffusion Eq. (9.3.10) is converted to the form:

$$\frac{\partial}{\partial t}f = D\frac{\partial^\alpha}{\partial |p|^\alpha}f \qquad (9.3.16)$$

It is obvious that for the case of $\alpha = 2$, (9.3.16) is the same as the normal diffusion Eq. (9.3.10).

For example, consider the case of $\alpha = 1$. Inserting (9.3.14) to (9.3.12) and time integration for Δt, it is found that the probability density becomes **Lorentzian distribution**:

$$\begin{aligned}\psi_L(\Delta p, \Delta t) &= \frac{1}{2\pi}\int_0^\infty \left\{e^{-k(D\Delta t - i\Delta p)} + e^{-k(D\Delta t + i\Delta p)}\right\}dk \\ &= \frac{2/D\Delta t}{1 + (\Delta p/D\Delta t)^2}\end{aligned} \qquad (9.3.17)$$

Note that (9.3.17) has an asymptotic form proportional to $1/(\Delta p)^2$ for large Δp, namely, in Levy jump region. It is clear that the power law decay is much slower than the exponential decay at a large argument; consequently, the contribution by Levy jumps is modeled in the probability density.

It is useful to know the property of Eq. (9.3.16) for $\alpha = 1$. In the case of $\alpha = 1$, (9.3.16) is the equation of convection to the positive direction in $p > 0$, and the negative direction for $p < 0$. Some physical quantities are transported on the flow with the velocity D, so (9.3.16) is also the equation to describe such **convective transport**.

9.4 Time Evolution of Distribution and Fractal Index α

Let us derive the time evolution of the distribution function by external stochastic force governed by the FFPE in (9.3.16) for a simple condition. In the case of $\alpha = 2$, the solution of the diffusion equation is well-known to be given in the form for the initial distribution of the delta function $f(p,0) = \delta(p)$:

$$f(p, t) = \frac{1}{\sqrt{\pi D * t}}\exp\left(-\frac{p^2}{D * t}\right), \qquad D^* = 4D \qquad (9.4.1)$$

For arbitral α, the solution is given as the inverse-Fourier integral form for FFPE in the form with use of (9.3.15):

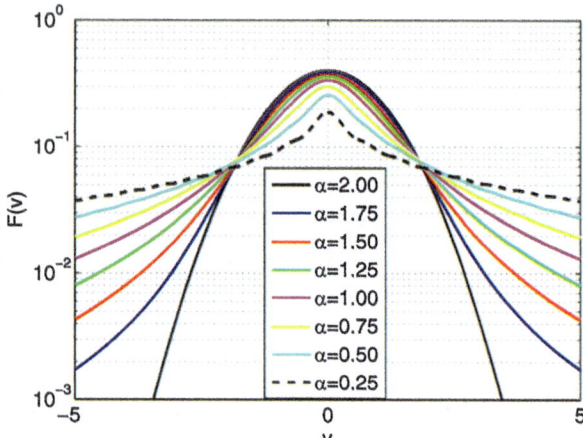

Fig. 9.6 The f^α in (9.4.2) as a function of the momentum p for $\alpha = 2.00$ (black line), $\alpha = 1.75$ (blue line), $\alpha = 1.50$ (red line), $\alpha = 1.25$ (green line), $\alpha = 1.00$ (magenta line), $\alpha = 0.75$ (yellow line), $\alpha = 0.50$ (cyan line), and $\alpha = 0.25$ (black dashed line). [Figure 1 in Ref. 13]

$$f^\alpha(p,t) = \frac{1}{2\pi} \int_{-\infty}^{\infty} \exp(-Dt|k|^\alpha + ikp) dk \quad (9.4.2)$$

This is not integrable analytically except for the case with $\alpha = 1$ or 2. For a given Dt, (9.4.2) is integrated numerically, and the solutions are shown for seven different α-indexes in Fig. 9.6 [13].

The numerical solutions can be approximated in the following form:

$$f^\alpha(p) \propto \frac{1}{[1 + \beta(q-1)p^2]^{1/(q-1)}} \quad (9.4.3)$$

where β is a constant for a given value of α. As indicated in Ref. [13], the precise analytical relation between the **fractal index** α and the **non-extensivity parameter** q is not entirely clear. Two relations have been proposed in [14]:

$$\alpha = \frac{3-q}{q-1} \quad \text{or} \quad \alpha = \frac{1}{q-1} \quad (9.4.4)$$

Note that all solution has power law for large p except for the case of $\alpha = 2$. The power law is given as

$$f^\alpha(p) \propto p^{-2/(q-1)} \quad (9.4.5)$$

For the case of $\alpha = 1$, the distribution function is Lorentzian, and the exact solution of (9.4.2) is easily obtained:

9.4 Time Evolution of Distribution and Fractal Index α

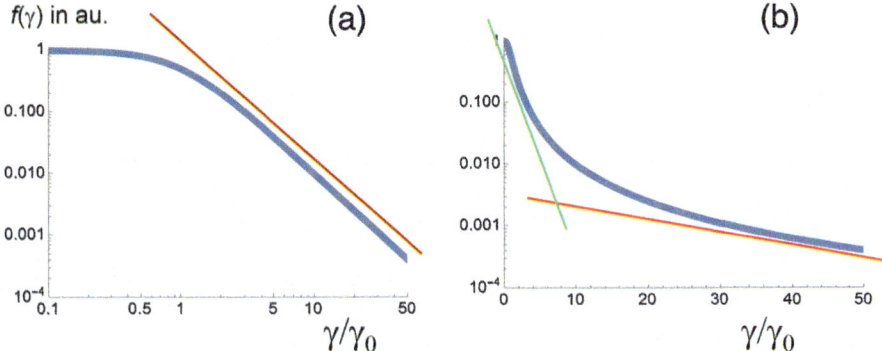

Fig. 9.7 The Lorentzian energy distribution in log-log plot (**a**) and linear-log plot (**b**). Maxwellian distribution is give as straight lines in (**b**), where two Maxwellians (red and green) are plotted with a Lorenian distribution (blue)

$$f^L(p,t) = \frac{1}{\pi Dt} \frac{1}{1 + (p/Dt)^2} \qquad (9.4.6)$$

Note that the solution in the limit q equal to 1 becomes a Gaussian distribution same as the analytical solution of (9.4.2) for α = 2 as plotted by black in Fig. 9.6. In Fig. 9.7, the Lorentzian distribution function (9.4.6) is plotted as a functions of normalized momentum P = p/(Dt). The left figure is log-log plot, while the right is liner-log plot. If the value Dt is larger than the unity, p is approximately equal to the electron energy (p ~ γ). In most of experiments and computer simulations, the liner-log plot is used, because the energy distribution is expected to be Maxwellian, exp. (−Aγ) with a constant A, where with the inverse of A being the hot electron temperature; the distribution becomes linear. It is also frequently shown that the distribution is made of two Maxwellian as plotted by green lines in Fig. 9.7. It seems that it is possible to fit the Lorentzian with two Maxwell distributions in the liner-log plot, while it is the power law in high-energy region as seen in the log-log plot. In taking into account the nonlocal jumps in energy space, the Lorentzian is better than Maxwellian distribution. This is because in such non-equilibrium system, there is no reason that the distribution should be Maxwellian.

9.4.1 Application to Hot Electron Scaling

In order to compare the case with the Gaussian probability, 1D PIC simulation results are shown in Fig. 9.8, where the case (a) is single-laser propagation and (b) is with a counter-propagating second beam being imposed [15]. The simulation condition is that a relativistic laser with a = 3 irradiates the slab plasma with density $n_e = 0.01n_c$ and the length 100 μm. It is clear in Fig. 9.8a that the distribution function does not spread even during a long pulse duration, namely, laser-plasma

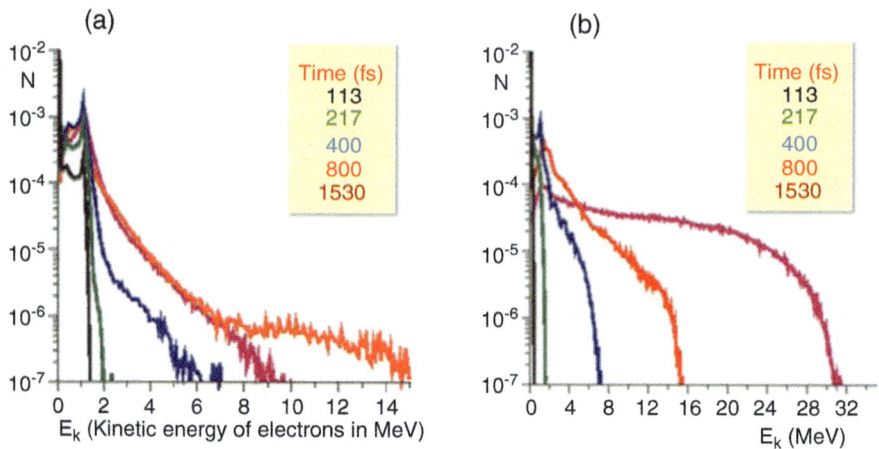

Fig. 9.8 Time evolution of electron distribution by 1D PIC simulation for the cases without the counter-propagating beam (**a**) and with a counter-propagating beam (**b**). Heating rate is dramatically different

coupling is very poor in the case of a single-laser beam. However, when a relatively week counter-propagating beam modeling the reflected laser component from the inner region with the intensity of $a_1 = 0.3$ comes from the counter direction, the electrons are efficiently heated and better coupling is expected. Note that even if the intensity of counter beam is 1/100 of the incident laser, such dramatic change happens for a 1.5 ps pulse.

As we see in Fig. 8.28, the stochasticity due to such a weak and counter-propagating beam can be treated as small perturbation and the diffusion-type approximation is applicable. As the result, it is reasonable to consider that the time evolution of the electron distribution function is approximately given like Gaussian form as shown in (9.4.1). Then, the distribution is proportional to

$$f(E_k) = \frac{1}{\sqrt{\pi Dt}} \exp\left[-E_k^2/(Dt)\right] \quad (9.4.6')$$

where since p>> 1, we have assumed.

$$p \to E_k$$

The average energy of the electrons in plasma is calculated to be

$$E_{av}(t) = \int_0^\infty f(p) p \, dp \propto \sqrt{t} \quad (9.4.6'')$$

It should be noted that the distribution function in Fig. 9.8b is well reproduced by the Gaussian of (9.4.6') in energy space. In such a diffusive stochasticity, it is also suggested that the absorption efficiency decreases in proportion to $t^{-1/2}$ as shown in (9.6.4").

9.4.2 Local and Nonlocal: Gaussian and Lorentzian

Consider that in laser-plasma interaction, filaments of laser shown in Fig. 9.1 induce not only small phase jumps as modeled by Fokker-Planck equation in (9.2.16), but also large phase jumps like Levy flights. Then, it is reasonable to consider that the particle energy is evolved by both stochastic processes, namely, the Gaussian and Lorentzian probabilities in (9.2.16). In Fig. 9.9, 100% Lorentzian is plotted by blue, 90% Gaussian and 10% Lorentzian is by orange, and 1% Lorentzian is plotted by green. In Fig. 9.9, Dt = 50 has been assumed.

In real experiment, it is reasonable to consider that the filamentation and modulation instabilities are induced and electrons are stochastically accelerated in the interaction region. Such an experiment has been reported in Ref. [16], and the authors insisted that the wake field acceleration is taken place in the laser-focused capillary to result the accelerated electrons with the power law spectrum. In the present Levy flight model, it is possible to explain the experimental spectrum with the assumption that the stochastic probability of 90% local Gaussian diffusion and 10% nonlocal Levy's flights.

The experimental data of the electron distribution function is given in Fig. 9.10. It is surprising that the assumption of 10% Levy's jump can reproduce the experimental spectrum. If this is the case, the physics of electron acceleration changes completely. The distribution of Fig. 9.9 is obtained by the stochastic heating only by relativitic laser field. However, the conclusion of Ref. [16] is that the electrons are accelerated by the wake fields generated by the ponderomotive force by the relativistic lasers. Which one is correct is still open question, but identification of which is very important in stochastic particle acceleration physics.

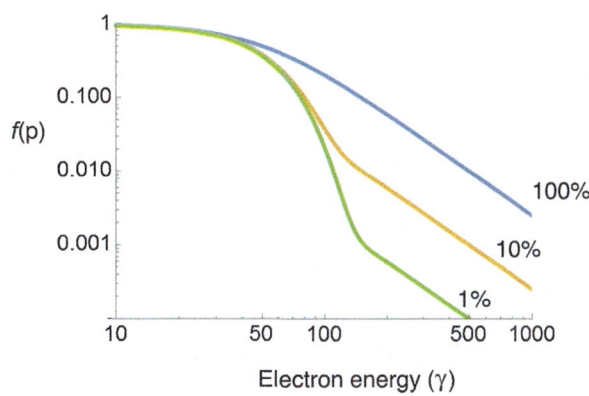

Fig. 9.9 The electron energy distribution obtained with the combination of Gaussian and Lorentzian distributions, where the percent is the fraction of Lorentzian one

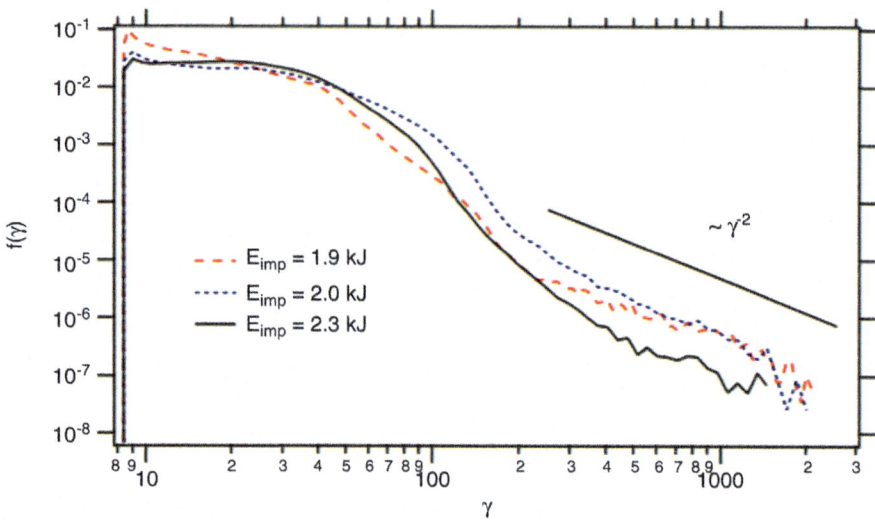

Fig. 9.10 Electron energy distributions from three shots of PW laser experiment at Osaka. It showed a power law at the tail. [Figure 2 in Ref. 16]

Consider the hot electron scaling law to the laser pulse length and the diffusion coefficients for Gaussian case ($\alpha = 2$) and Lorentzian case ($\alpha = 1$). It is clear that the distribution functions of (9.4.1) and (9.4.6) are normalized so that the p-integrals of both distributions are unity. Assume that $p \gg 1$ and the normalized particle energy $\gamma = p$. The time evolution of the normalized energies of (9.4.1) and (9.4.6) is calculated as follows.

The Gaussian case is

$$E^G(t) = \int_0^\infty p f^G(p,t) dp \qquad (9.4.7)$$
$$= E_0^G \sqrt{D^* t}$$

where E_0^G is a numerical constant given as

$$E_0^G = \frac{2}{\sqrt{\pi}} \int_0^\infty \xi e^{-\xi^2} d\xi = \frac{1}{\sqrt{\pi}} \qquad (9.4.8)$$

where the new valuable ξ has been introduced:

$$\xi = p/\sqrt{D * t} \qquad (9.4.9)$$

On the other hand, the Lorentzian case is

9.4 Time Evolution of Distribution and Fractal Index α

$$E^L(t) = \int_0^\infty p f^L(p,t)dp \qquad (9.4.10)$$
$$= E_0^L Dt$$

where

$$E_0^L = \frac{2}{\pi} \int_0^\infty \frac{\xi}{1+\xi^2} d\xi = \frac{1}{\pi} \ln\left(1 + \xi_{max}^2\right) \qquad (9.4.11)$$

In (9.4.11), the non-dimensional new variable is defined as

$$\xi = p/(Dt) \qquad (9.4.12)$$

It is well-known in Tsallis distribution that higher moments of the distribution functions diverge, because of the power law property shown in (9.4.3). In the Lorentzian case, it is reasonable to cut the distribution at the convective velocity $\xi = \xi_{max}$ with a value of order unity. The time evolution of log function in (9.4.11) can be neglected and be regarded almost constant.

It is also important to note that if the maximum ξ is a constant in (9.4.11), the maximum energy of the distribution function evolves with time in the form:

$$\langle \gamma \rangle_{max} \propto D\xi_{max} t$$

9.4.3 Evaluation of Fractional Index α from Experimental Data

The results of the above two cases are intuitively clear. If the diffusion in momentum (energy) space is poor like Gaussian case, electrons cannot couple with laser to transfer the interaction energy into the higher-energy region, namely, poor diffusion. On the other hand, if the diffusion is very efficient to transfer the coupling energy to higher-energy region via Levy flight mechanism, the laser energy is easily absorbed. Such physics of diffusion is simply evaluated by dimensional analysis of (9.3.16). Simple dimensional analysis gives roughly the dependence of the mean momentum (energy γ) in the form:

$$\langle \gamma \rangle \equiv T_h(t) \sim (Dt)^{1/\alpha} \qquad (9.4.13)$$

In many experiments with relativistic lasers, this $\langle \gamma \rangle$ is regarded as the hot electron temperature T_h as seen many times in the book. (9.4.13) is very important to know the coupling process. When the time dependence of T_h is observed experimentally or computationally, we can evaluate the effective diffusion coefficient D and fractal index α.

Assume that the mean energy increase is due to the energy absorption from laser, and use the normalized form of laser energy flux to each electron:

$$\frac{I_L}{n_e} = \frac{\varepsilon_0}{2n_e}|E_L|^2 = \frac{mc^3}{2}a_0^2\left(\frac{\omega}{\omega_{pe}}\right)^2 \qquad (9.4.14)$$

The absorption rate is calculated to be

$$\eta_{ab}(t) = \frac{dT_h/dt}{I_L/n_e} = \frac{1}{2\alpha}\frac{n_e}{n_{cr}}D^{1/\alpha}t^{(1-\alpha)/\alpha} \qquad (9.4.15)$$

In the case of local diffusion $\alpha = 2$, the absorption rate decreases with time, while the absorption rate is kept constant for the case of nonlocal jumps and/or anomalously efficient diffusion for $\alpha = 1$.

In the previous sections, the following rough relation has been derived in Fokker-Planck equations:

$$D \sim \frac{a_0^2}{\tau} \qquad (9.4.16)$$

where τ is the mean time interval between two non-adiabatic jumps. (9.4.16) indicates also that the diffusion coefficient D is proportional to the laser intensity from (9.4.14). Inserting (9.4.16) into (9.4.13) and (9.4.15) with help of (9.4.16), the following important relations are obtained:

$$\begin{aligned} T_h &\propto \sqrt{I_L}t^{1/2} \\ \eta_{ab} &\propto \frac{1}{\sqrt{I_L}t^{1/2}} \end{aligned} \quad \text{for } \alpha = 2 \text{ (Gaussian)} \qquad (9.4.17)$$

and

$$\begin{aligned} T_h &\propto I_L t \\ \eta_{ab} &\propto \text{const.} \end{aligned} \quad \text{for } \alpha = 1 \text{ (Lorentzian)} \qquad (9.4.18)$$

It is very important to compare the both time dependencies (9.4.17) and (9.4.18). The time dependence of the hot electron temperatures from PIC simulations and experiments has been discussed in Sect. 8.8, and it is concluded that in the two laser counter-propagation case, the time dependence of $T_h(t)$ is given in (8.8.12). This time dependence was

9.4 Time Evolution of Distribution and Fractal Index α

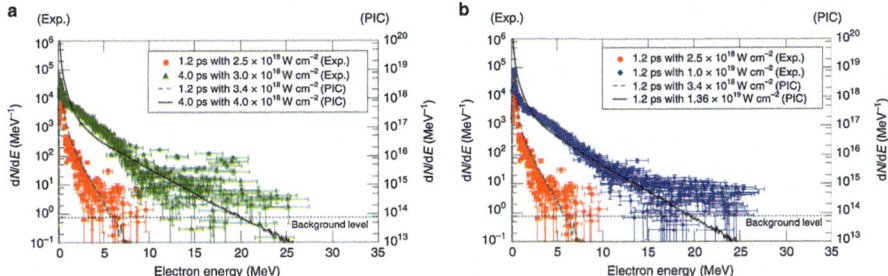

Fig. 9.11 Experimental data of the electron energy distributions in three different shots. Three different shot data **A**, **B**, and **C** are plotted in (**a**) and (**b**) with red, green, and blue, respectively, in order to avoid the overlapping plots of green and blue. The red data is plotted in both. [Figure 2 in Ref. 17]

$$T_h \propto t^C \quad (C = 1/2\tilde{1}) \qquad (9.4.19)$$

The time dependence of the computational results shown in [5] can be well explained by the physical model based on FFPE. It is also important to note that the model simulation with many small kicks in [1] concluded that the effective temperatures scale roughly as (9.4.17) with time and laser intensity.

Finally, compare the theoretical result with experimental data [17], where Levy's jumps are expected in the interaction region, because the pulse lengths are relatively long, and large amplitude of noise fields are inferred to be generated in the plasma due to a variety of instabilities and nonlinear process. The time-integrated electron distribution functions for three cases shown below, A, B, and C are plotted in Fig. 9.11. The corresponding PIC results are also shown there. The laser parameters and measured hot electron temperatures are as follows (T_2 as T_h) [17].

	Intensity (W/cm^2)	Pulse length (ps)	T_h (MeV)
A	2.5×10^{18}	1.2	0.7
B	2.5×10^{18}	4.0	1.7
C	1.0×10^{19}	1.2	1.7

From the above data, the following relation is obtained:

$$T_h = I_L^{0.64} t^{0.73} \qquad (9.4.20)$$

From (9.4.13) and assuming (9.4.16), the inferred fractal index α is obtained as

$$\alpha = \begin{cases} 1.35 & \text{from } t - \text{dependence} \\ 1.56 & \text{from } I_L - \text{dependence} \end{cases} \qquad (9.4.21)$$

The value is an intermediate one between the Gaussian and Lorentzian. Note that as shown below, this index is about 0.8 in highly turbulent Tokamak transport in a stationary state. The fact in (9.4.21) suggests that the turbulence in plasma is not fully developed and picoseconds are not long to observe the dominant contribution of Levy flights in such laser-plasma interactions. It is interesting subject to compare the electron distribution function in [17] with the profile given in Fig. 9.6 for the case with the fractal index in (9.4.21).

In addition, it should be noted that with use of the assumption (9.4.16), the average jumping time τ can be evaluated. This normalized time ($\tau\omega$) is in general believed to be of order unity in the anomalous turbulence in fluid or other turbulent mixing models.

9.5 Model Experiment of Cosmic Ray Physics in the Universe

Since the power law energy spectra are more reasonable for the relativistic electrons produced by lase-plasma interaction in the relativistic intensity regime, consider a possibility to design a model experiment to demonstrate the physics of cosmic ray acceleration in laboratory. Particle acceleration is one of the most important subjects in the study of ultra-short and ultra-intense laser interaction with low-density plasmas as well as many applications of the produced electron and ion beams.

Cosmic rays were found to come from the universe more than 100 years ago. The observation data from many different types of instruments have been accumulated in the log-log space of the observed particle energies as shown in Fig. 9.12. It is amazing that the energy rage in Fig. 9.12 is from 10^9 eV (GeV) to 10^{21} eV and the data show clear dependence of a power law. The blue regions to 10^{15} eV (PeV) are thought to be due to the source of particles generated in our galaxy, while the particles with the energy in the pink region are inferred to be generated out of our galaxy.

The physics of the acceleration mechanisms have been studied intensively for a long time. The most important idea was proposed by E. Fermi [18], and now most of the researchers believe that the so-called the first Fermi acceleration is the most reliable model to explain the huge number of observation data in Fig. 9.12. The first Fermi acceleration is a diffusive acceleration by the accumulation of many small amount of energy gaining through passing of charged particles across the **collisionless shocks**. So, the mathematical model is called **diffusive shock acceleration (DSA)** [19]. It is the most promising model to explain the power law spectrum in the case where particles are accelerated many times around the shocks with non-relativistic velocity. Such non-relativistic but very high Mach number shocks are generated in our galaxy by, for example, supernova explosions. It is said that the cosmic ray with energy less than 10^{15} eV are non-relativistic shock origin.

On the other hand, when the shocks become stronger and the speed of shock waves become relativistic, the Fermi acceleration model faces a difficulty to accelerate the particles. The most simple reason is that the particles randomly traveling in

9.5 Model Experiment of Cosmic Ray Physics in the Universe

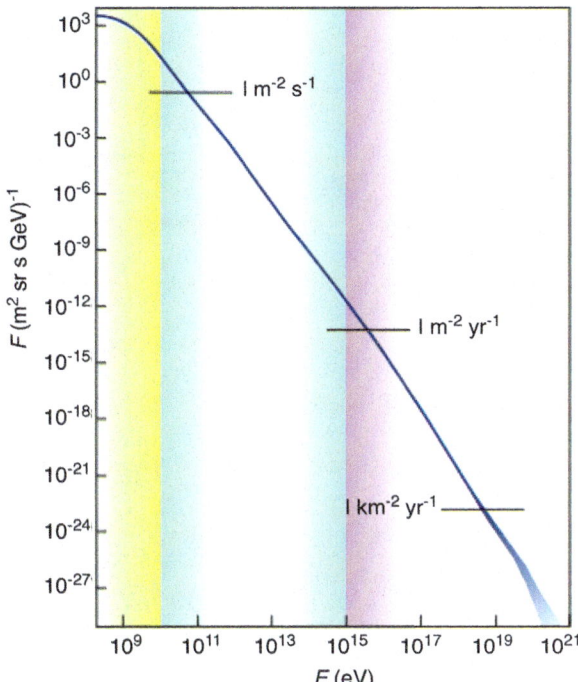

Fig. 9.12 Cosmic ray energy spectrum measured at many observatories worldwide in log-log plot. The particles sources are classified into three regions in the universe. The yellow region is from the solar system, the blue region is from our galaxy, and pink region is from outside of our galaxy

the downstream of the shocks cannot catch up the relativistic shock to keep many bouncing in the both regions before and after the shock front. Alternative models have been proposed, and one of them is directly related to the laser-plasma interaction, namely, so-called wake field acceleration in relativistic plasmas [16, 20].

In the wake field acceleration, large amplitude plasma waves are generated by the shock waves or else. It has been pointed out that the strong plasma waves can be induced by the electromagnetic (EM) waves generated at the relativistic collisionless shocks [21]. What became clear in the study so far in the book is that if the EM waves produced by the relativistic shock are strong enough and its normalized amplitude a_0 is larger than unity, the EM waves can directly accelerate the particles in the universe. If this acceleration is highly stochastic and the energy jumps by Levy's flights are expected for a long time, power law spectrum of particle energy may be reasonable to be produced near the relativistic shock waves.

9.5.1 Relativistic EM Wave Generation by Relativistic Shocks

The data in Fig. 9.12 can be approximated as

Fig. 9.13 Snap shots of density and transverse magnetic field from PIC simulation in the highly relativistic collisionless shock wave propagating from the left to right. The electromagnetic field with relativistic intensity is generated in the compressed down flow region and propagate in front of the shock wave. [Figure 12 in Ref. 22]

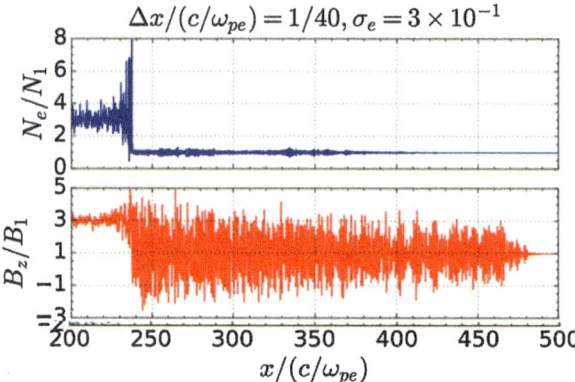

$$f(E) = \frac{dN}{dE} \propto E^{-k} \qquad (9.5.1)$$

The observation data fits well as follows:

$$k = 2.7 \text{ for } 10^{10} \text{eV} < E < 10^{15} \text{eV} \qquad (9.5.2)$$

$$k = 3.4 \text{ for } 10^{15} \text{eV} < E < 10^{19} \text{eV} \qquad (9.5.3)$$

The power k observed on the earth is the particle energy distribution arriving on the detectors after an extremely long distance flight in the galaxy and in the space out of the galaxy. The space is full of turbulence of magnetic field, and the spectra are affected by transport process. So, roughly saying the power law in the cosmic ray generation region is inferred to $k \sim 2$.

Relating to the physics of relativistic collisionless shocks, 2D PIC simulation has been carried out [22]. The Lorentz factor γ_{sh} of the shock wave is assumed to be $\gamma_{sh} = 40$. The cyclotron motion of electrons and positrons in the downstream emits the so-called precursor EM waves. The snap shots of the precursor EM waves are shown in Fig. 9.13, where the density (blue) and the magnetic field of EM waves (red) are shown. In the PIC simulation, relating to the wake field acceleration, the normalized amplitude a_0 of the EM waves are observed, and it is reported that the value reaches about $a_0 = 2.5$ in a standard simulation.

As seen in Fig. 9.13, the EM wave pulse is very long compared to their typical wavelength, and the electrons in front of the shock wave is possibly accelerated stochastically by EM waves as we have seen so far in laser-plasma interaction. Nonlinearity of the relativistic EM waves helps the many Levy's jumps in the energy space to provide the power spectrum of accelerated electrons in the universe.

9.5.2 Model Experiments with Relativistic Lasers

The proof of principle experiment can be designed regarding the direct acceleration of cosmic ray by EM waves in laboratory as a challenging problem of laboratory astrophysics. When a relativistic laser is injected into low-density plasma like gas jet plasma, the experimental data of the electron distribution function can be compared to the theoretical models. If the power law spectra are obtained from the experiment, we can try to identify the best index of α in (9.4.3) and (9.4.4) to identify the acceleration physics. The identification of the index α is a big step to study the anomalous transport in the momentum space in the relativistic fields as has been done for charged particle transport in Tokamak turbulence to be discussed below.

The advantage of the direct acceleration by EM waves is that the energy conversion rate to cosmic rays is expected higher than the acceleration by the wake field generated by the EM waves. Analysis of the details of the experiment data of Figs. 9.10 and 8.22b would be beneficial as the starting point of such research. Finally, it should be noted that the high-energy electrons are accelerated though the jumps in many phases of laser field. That is, it should be careful as mentioned in [16] [Fig. 8.20] that the PIC simulation should be used so as to resolve the fine structure of laser field distribution. Especially, it is almost impossible to study such Levy-type acceleration in the simulation shown in Fig. 9.13, because the precise resolution of the phase of electric and magnetic field in the precursor waves needs a huge number of computational meshes. The better way is to start with a model simulation with EM waves.

In Fig. 9.14, 2D PIC simulation result on the time evolution of the ion and electron energy distributions is plotted [23]. The power law spectra are generated as the envelope of the tails of the distribution functions. The simulation condition is that the Lorentz factor of the propagating velocity of the collisionless shock is $\gamma_{sh} = 15$, the mass ratio is 25, and magnetization parameter is $\sigma = 10^{-5}$. It is observed that the both distributions are showing the power laws with k = 3.0 (ions) and k = 2.5 (electrons) as shown in Fig. 9.14. The time evolution of the maximum energies is also important. It is reported [23] that both ion and electron maximum energies are proportional to

$$\gamma_{max} \propto t^{1/2} \qquad (9.4.17)$$

and Bohm limit in strong magnetic fields is

$$\gamma_{max} \propto t \qquad (9.4.18)$$

Comparing both time dependences to (9.4.13), it is important to know that the cosmic ray generation rate is also in the rage of $1/2 \leq C \leq 1$ in (9.4.13). It is reasonable to consider that the difference of (9.4.17) and (9.4.18) is due to the local or nonlocal properties of acceleration corresponding to the difference of (9.4.17) and (9.4.18), respectively.

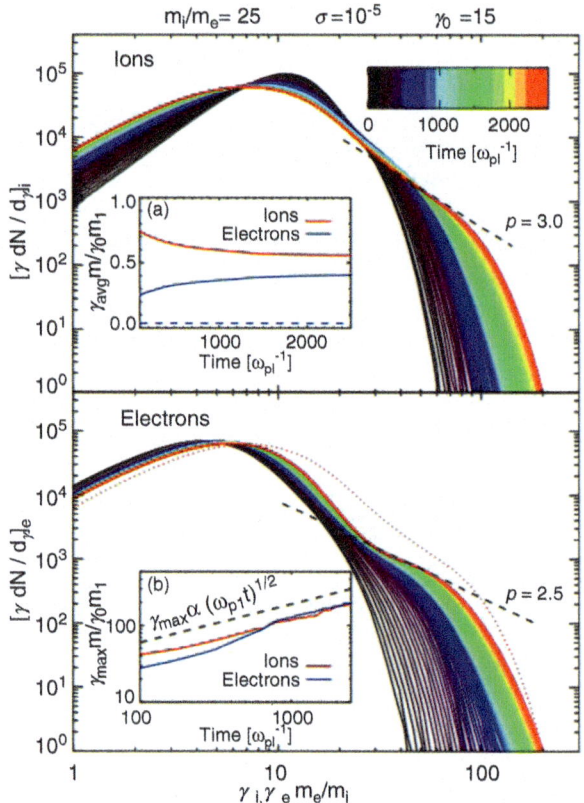

Fig. 9.14 The time evolution of the ion and electron energy distributions around a relativistic collisionless shock wave. The distribution functions progress with power law spectra. [Figure 11 in Ref. 23]

9.6 Fractal Index $\alpha = 0.8$ in Big Tokamak Transports

It is well recognized that the energy confinement of magnetic confinement devices is not classical, but plasma turbulence effect is dominant. This has been studied intensively for a long time as the anomalous diffusion problem. There have been trials to apply the Levy's flight diffusion concept to analyze the anomalous diffusions by magneto-hydrodynamic turbulence (**MHD turbulence**). **Joint European Tokamak (JET)** database is used to identify the fractional index α based on the FFPE model [24]. The databases from JET and ITER-wall experiments were used for obtaining the fractal index α in two-dimensional transport model. The analyzed dataset contains 1256 samples from 868 different plasma shots.

In Fig. 9.15, an experimental profile of the electron pressure versus normalized poloidal axis ρ_p is plotted by black line. The electrons are heated near $\rho_p = 0$ by the external sources. Then, the energy is transported toward the wall near $\rho_p = 1$. This is a simple heat conduction problem for a given heat source, transport toward the wall, and energy loss at $\rho_p = 1$. It is clear how $\alpha = 0.8$ model of FFPE well reproduced the

9.6 Fractal Index α = 0.8 in Big Tokamak Transports

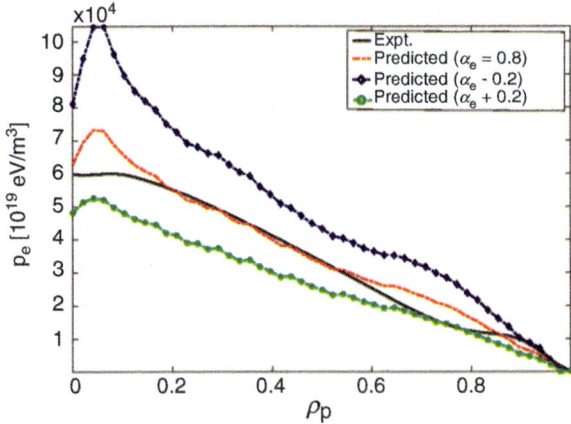

Fig. 9.15 Comparison of the experimental (black solid line) electron pressure profile versus normalized poloidal flux index ρ_p and the predicted profile following the global transport model by FFPE (red dash-dotted line) for the plasma discharges #92071. The predicted pressure profiles with ±0.2 above (blue dotted line with diamond symbols) and below (green dotted line with circle symbols) the computed values of α_e are also shown. This discharge is an ELMy H-mode pulse of hybrid type from ILW with 30 MW total input power (25 MW NBI +6.6 MW ICRH) and regular type I ELMs during. [Figure 4 in Ref. 24]

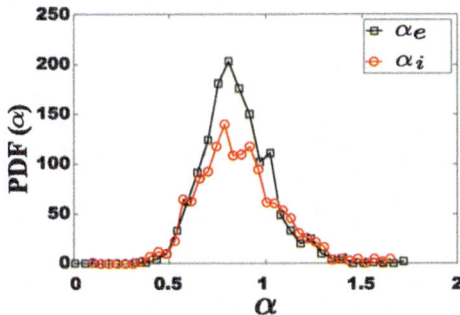

Fig. 9.16 The histogram of the computed α_e (black line with square symbols) and α_i (red line with circle symbols) for the selected JETPEAK dataset. [Figure 3 in Ref. 24]

electron pressure profile, while small change of α + − 0.2 alters the profile significantly.

The fractal indexes for electron and ion transports are plotted for the best fit of each shot in Fig. 9.16. It is clear that α = 0.8 is the best fit and the Levy-type nonlocal transport is induced by MHD turbulence in JET and ITER type large Tokamaks.

9.7 Chaos in Standard Map Model

So far we have assumed that the laser interacts with the electrons in low-density plasma, namely, the plasma effect has been neglected. In the case of very short pulse, the vacuum heating becomes important at the beginning as shown in Sect. 3.10 and in the discussion on Figs. 7.3 and 7.4. In such an ultra-short pulse without the pedestal pulse, the ions are at rest, and their density has the profile with sharp density jump at the solid surface. Then, the most of electrons accelerated toward the vacuum by laser field are affected by the electrostatic field generated by the charge separation as schematically shown in Fig. 6.1.

A theoretical model has been proposed by Bulanov et al. [25] by taking into account of laser force and such an ambipolar field generated by electron motions for the case of the vacuum heating. It is assumed that the ambipolar field is the same force inside and outside of the solid surface. They modeled the physics with the following nonlinear oscillator by external force like the form of (8.4.6) for the case of a relativistic motion. Note that p, x, t, and a_0 are normalized as already shown before:

$$\frac{dp}{dt} = -\varepsilon_p sing(x) + a_0 cost$$
$$\frac{dx}{dt} = \frac{p}{(1+p^2)^{1/2}} \quad (9.7.1)$$

where *sign* (x) is defined as

$$sing(x) = \begin{cases} = 1 & (x > 0) \\ = -1 & (x < 0) \end{cases} \quad (9.7.2)$$

The electric field due to the ambipolar field is given in the normalized form [25]:

$$\varepsilon_p = \frac{e^2 n l}{2\omega\varepsilon_0 mc} \quad (9.7.3)$$

where l is the thickness of surface charge ($=enl$) at the solid surface. Since (9.7.1) can be integrated in every short time (t_n, t_n + 1) after and before crossing the surface x = 0, they are converted to an equation of the **Chrikov standard map** [25, 26]:

$$I_{n+1} = I_n + K sin(\phi_n)$$
$$\phi_{n+1} = \phi_n + I_n \quad (9.7.4)$$
$$(I_n, \phi_n = mod2\pi)$$

where K, I_n and ϕ_n are defined as

9.7 Chaos in Standard Map Model

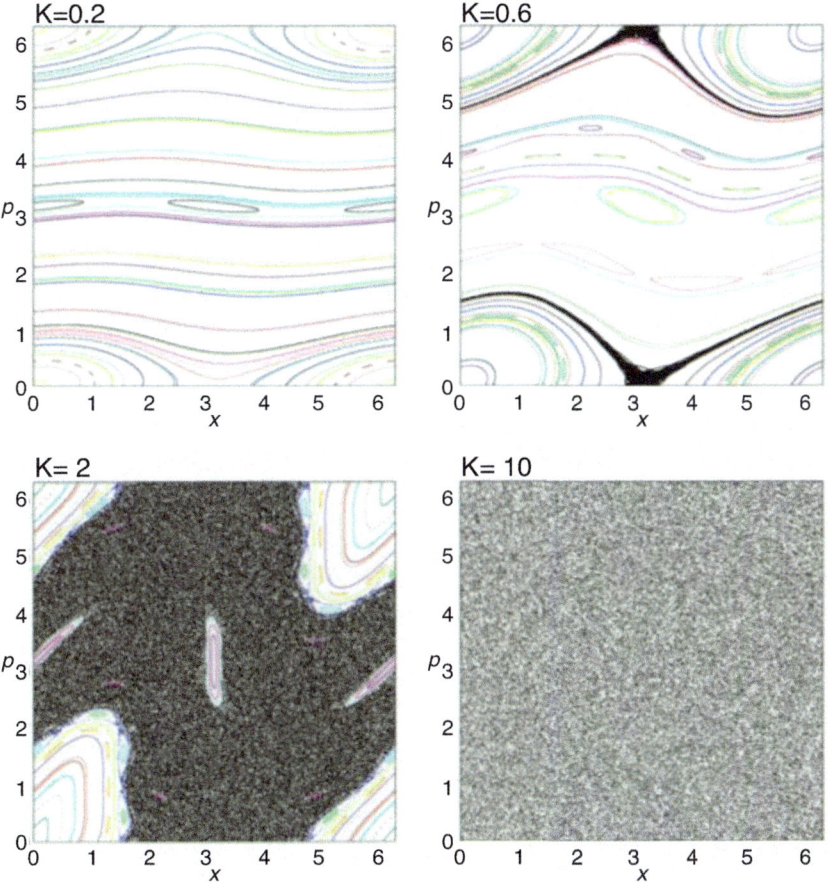

Fig. 9.17 Phase portrait of the standard map for four representative K values. [Figure 1 in Ref. 26]

$$K = 8\frac{a_0}{\varepsilon_p}, \quad I_n = \frac{4}{\varepsilon_p}p(t_n), \quad \phi_n = t_n \tag{9.7.5}$$

It is well-known that the simple Eq. (9.7.4) has the threshold value of K over which the solution becomes chaos. The chaos solution is given [27] for

$$K > 1 \tag{9.7.6}$$

It is clear that for the case $K = 0$, (9.7.4) shows only lines with a given constant in stating with any initial values of (ϕ, I). The (p, x) maps corresponding to (I_n, ϕ_n) in (9.7.4) are shown in Fig. 9.17 for four different K values [26]. Note that (p, x) in the figure is (I, ϕ) in (9.7.4) and the suffix "n" is extended to a very large number; consequently the data are not line but a sum of many dots.

For a small K (K = 0.2), almost all lines are shown as small modification from K = 0 case. It is not seen the connection from a small p to a large p region. This means the solution of (9.7.4) is stable and no increase of p(t) is expected. This is also the same even with the increase of K (K = 0.6). The black region with many dots appears, while they do not make a bridge from small p to large p, so no acceleration is expected. However, for the case when the value K exceeds the critical value (9.7.6), the chaos regions spread, and it becomes possible for the time evolution of p(t) to start from a small value and continuously increase its value with time.

This is due to a sequential jump in each step in (9.7.4). Some solution goes up to the top of the figure (p = 2π) and comes to the bottom to repeat such a routine. Such p(t) shows the selected acceleration by the chaos of the system. It is noted that although (9.7.4) is very simple, the original Eq. (9.7.1) is a nonlinear pendulum equation with an external force by laser field. This problem is similar to that of (8.4.6) and/or the problem of the chaos of asteroids in Fig. 8.12. The basic equations are characterized as the nonlinear oscillation pendulum and an external periodic force. Even the stable orbits of orange and green in Fig. 8.14a have the threshold amplitude of the external force of ε in (8.4.6). The orbits become unstable like the standard map for large external force. In addition, note that the threshold of chaos also depends on the frequency of the external force.

It is also shown [25] that (9.7.6) is approximated with a diffusion equation and its solution has the form:

$$f(t, p) \propto (\chi t)^{-1/3} \exp\left(-\frac{|p|^3}{9\chi t}\right), \quad \chi = \frac{a^2 \varepsilon_p}{2} \quad (9.7.7)$$

As a long time behavior, the distribution function growth as a super-Gaussian (N = 3) in (9.7.7), but $T_h \sim t^{1/3}$.

9.8 Analytical Mechanics of Electron Motions

9.8.1 One Laser Relation

Let us first define the equation of motion of an electron in two laser systems. Consider the relation for the case with only the incident laser. The incident laser is given as

$$a = a_0 \sin(\xi) \quad (9.8.1)$$

where ξ is the phase of the incident laser wave and has the following normalized form:

9.8 Analytical Mechanics of Electron Motions

$$\xi = t - x \tag{9.8.2}$$

The Canonical momentum conservation provides the relation for $\beta = 0$ in (8.1.21):

$$p_y = a = a_0 \sin(\xi) \tag{9.8.3}$$

Then, the basic equations are the following two for the case of incident laser only

$$\frac{d}{dt} p_x = \frac{a_0^2}{\gamma} \sin(\xi) \cos(\xi) = \frac{a_0^2}{2\gamma} \sin(2\xi) \tag{9.8.4}$$

$$\frac{dx}{dt} = \frac{p_x}{\gamma} \tag{9.8.5}$$

As already studied, (9.8.4) and (9.8.5) with (9.8.3) are integrable, because it has a constant of motion. Note that α in this chapter is not the fractal index but α defined in (8.1.20). This corresponds to the equation of radius in the motion of a planet or asteroid around the sun for a given angular momentum. In this case, the second-order differential equation to the radius of the planet is periodic oscillation in a nonlinear potential, and total energy E is the constant of motion.

9.8.2 Adiabatic Constant

In order to study the property of the oscillating solution given in (8.2.2) and (8.2.7), consider the similarity to the harmonic oscillation with Hamiltonian:

$$H = \frac{1}{2} p^2 + \frac{\omega^2}{2} q^2 \tag{9.8.6}$$

It is well-known that the **action variable** J is defined as

$$J = \frac{1}{2\pi} \oint p \, dq \tag{9.8.7}$$

This is an adiabatic constant and used by Bohr to introduce the discretization of electron orbit of a hydrogen atom in the early time of quantum mechanics. It is easy to integrate (9.8.7) to obtain the relation:

$$J = \frac{E}{\omega} \tag{9.8.8}$$

where E is the total energy of the oscillation. It is well-known that E/ω is the adiabatic constant of a simple oscillation as described in Appendix 2. Then, the angular frequency ω satisfies the relation:

$$\frac{\partial E}{\partial J} = \omega \qquad (9.8.9)$$

The **angle variable** θ is defined by

$$\theta = \omega t \qquad (9.8.10)$$

Then, J and θ are new variables after Canonical transformation from (q, p). It is convenient to use (J, ω) in such periodic motion, because J is an adiabatic invariant to be kept constant under small perturbation to the system.

Analogously consider our nonlinear oscillation of electrons by relativistic laser. It is a periodic oscillation, and both the action and angle variables are given in the present two-dimensional case. The action variable in x-motion J_x is given as

$$J_x = \frac{1}{2\pi} \oint p_x dx = \oint \frac{p_x^2}{2\pi\gamma} dt = \frac{1}{2\pi\alpha} \oint p_x^2 d\xi = \frac{1}{\alpha} \langle p_x^2 \rangle \qquad (9.8.11)$$

where the relations (8.1.11) and (8.1.12) are used. In (9.8.11), < > means the average value over the wave phase ξ. The action of the y-direction is also obtained in the form:

$$J_y = \frac{1}{2\pi} \oint p_y dy = \oint \frac{p_y^2}{2\pi\gamma} dt = \frac{1}{2\pi\alpha} \oint p_y^2 d\xi = \frac{1}{\alpha} \langle p_y^2 \rangle \qquad (9.8.12)$$

Setting E the average energy of the electron, the following relation is obtained:

$$E^2 - 1 = \alpha (J_x + J_y) \qquad (9.8.13)$$

The angle variable is given by ξ, and (8.1.11) and (8.1.12) reduce the relation:

$$\xi = \alpha\tau \qquad (9.8.14)$$

Compared to the relation for the harmonic oscillation, the following analogy is obtained:

9.8 Analytical Mechanics of Electron Motions

$$\begin{aligned} J &\leftrightarrow J_x + J_y \\ \omega &\leftrightarrow \alpha \\ \theta &\leftrightarrow \xi \\ E &\leftrightarrow E^2 - 1 \end{aligned} \quad (9.8.15)$$

It is known that in relativistic motion, **super-Hamiltonian** [28] defined by

$$H^s = E^2 - 1 = p^2 \qquad (9.8.16)$$

Note that the super-Hamiltonian is not equivalent to the electron energy. It should be noted that for given amplitude and frequency of laser vector potential in (9.8.1), H^s and $J_x + J_y$ are only functions of the value α. Therefore, α is only one variable to determine the periodic motion of electrons, when a weak perturbation acts on the electrons non-adiabatically over a relatively short time compared to the oscillation period. The change of the oscillating motion of electrons can be studied by focusing on the time evolution of the constant α as seen below.

9.8.3 Numerical Solution for Counter Beam Interaction

In order to see the property of the electron motion when the second laser is propagating in the counter direction with amplitude a_1 and same frequency and wavelength, the corresponding differential equations are solved numerically. The parameters are

$$a_0 = 3, \ a_1 = 0.3$$
$$t_{max} = 260 \, [\text{fs}] \text{ and } 2.6 \, [\text{ps}]$$

In Fig. 9.18, the numerical result is shown, where (a) is (p_x, p_y) diagram, (b) time evolution of the vale $1/\alpha$, (c) time evolution of the electron energy γ, and (d) the phase value divided by 2π as a function of time. Note that the time is shown as a value divided by the laser oscillation period. The maximum time 80 is 260 [fs].

It is important to note that the laser oscillated 80 cycles at a given point, while the phase defined in (9.8.2) has changed only about 6 as seen in Fig. 9.18d. This means the electron runs in the laser wave with about 93% of the speed of light. At the beginning, the electron is at rest and the value of $\alpha = 1$ and $\gamma = 5.5$ by the ponderomotive scaling, while with the increase of $1/\alpha$ in (b), the electron gains more energy as seen in (c), and its trajectory becomes bigger as seen in (a). Note that the mean value of $1/\alpha$ changes only when the energy γ is almost equal to unity. In addition, with the increase of energy, namely, the increase of electron velocity in x-direction, the time change of the phase becomes slow so that the electron can stay in the acceleration phase for a long time.

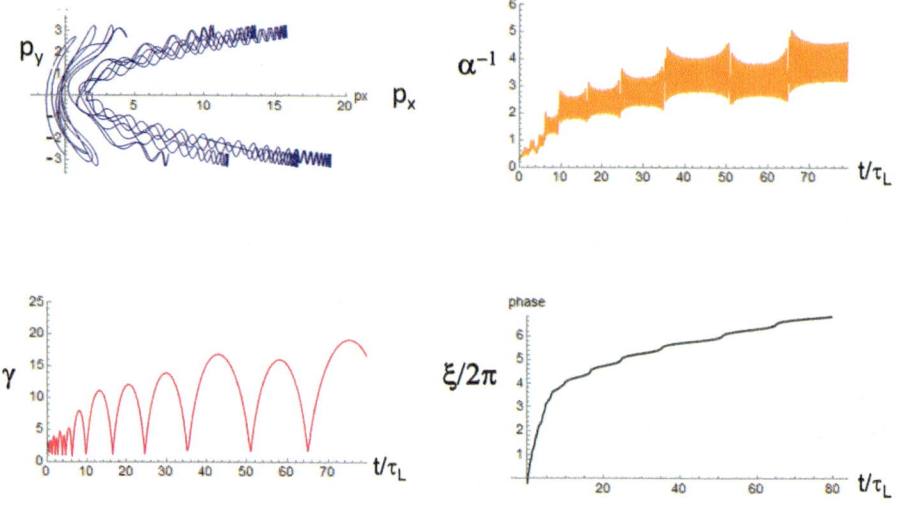

Fig. 9.18 An electron motion in counter-propagating laser beams. The trajectories in the momentum space (**a**), and the time evolutions of $1/\alpha$ value, electron energy γ, and the phase position of the electron. The parameters are $a_0 = 3$, $a_1 = 0.3$ and the maximum time is 260 [fs]

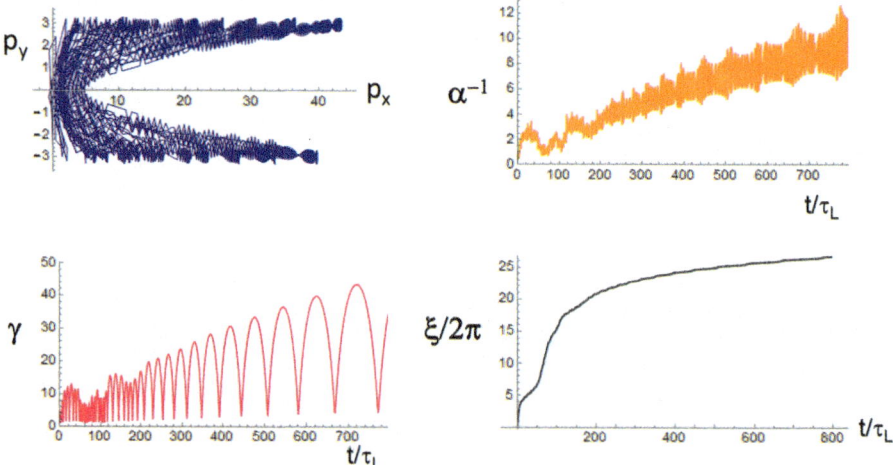

Fig. 9.19 The results of the same condition as in Fig. 9.18, but the maximum time is ten times longer, 2.6 [ps]

In Fig. 9.19, the same figures but for the long time evolution up to ten times (2.6 ps) are plotted. It is important to note that the growth of $1/\alpha$ and the maximum energy γ in Fig. 9.19b, c well coincide. This relation is important to analytically study the perturbation theory as seen later. The increase of the maximum energy looks in the relation of (9.4.19). There are no Levy-type jumps, and the increase of the energy with time is like the diffusion type.

9.8 Analytical Mechanics of Electron Motions

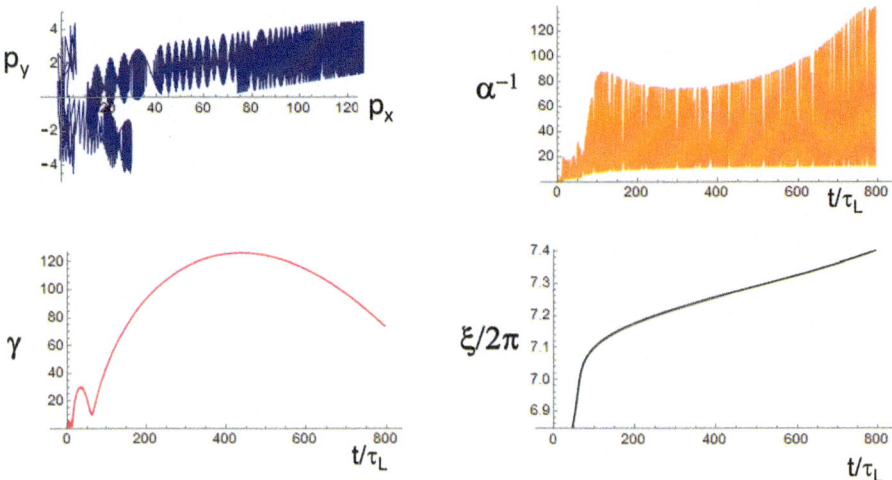

Fig. 9.20 The results for the case with the same condition as in Fig. 9.19 except for $a_1 = 1.5$, five times stronger counter beam

In order to see the change of the property of the energy increase from the diffusion type to the nonlocal Levy jump type, the intensity of the second laser is increased to 25% of the incident laser as

$$a_0 = 3, \quad a_1 = 1.5$$
$$t_{max} = 2.6\,\text{ps}$$

In Fig. 9.20, the resultant figures same as above two figures are plotted. It is surprising to know that the electron is soon kicked to the acceleration phase of the seventh period as seen in (d) and it stays for a long time in phase with laser field to continue to be accelerated. When its energy comes to the maximum of about $\gamma = 120$, the electron moves to the deceleration phase. It is clear from Fig. 9.20c that the electron is in the same phase for most of the time of 2.6 ps and the laser works as a linear accelerator of electron. Although the second beam intensity is only 25% of the main laser, it cannot be like perturbation but changes the dynamics dramatically.

9.8.4 Perturbation Method in Hamilton Equation

Let us consider **Hamilton mechanics** of an electron motion in a relativistic plane laser field $\mathbf{A}_0$ and a perturbation laser $\mathbf{A}_1$ and assume that both vectors are in the y-direction as we have considered so far. The **Lagrangian** of the electron motion in the vector potential field is given in (5.2.13) as the normalized form:

$$L = -\sqrt{1 - v^2} - \mathbf{a} \cdot \mathbf{v} \tag{9.8.17}$$

where $\phi = 0$ is assumed and

$$\mathbf{a} = \mathbf{a}_0 + \mathbf{a}_1 \tag{9.8.18}$$

Canonical momentum $\mathbf{P}^c$ is defined as

$$\mathbf{P}^c = \frac{\partial L}{\partial \mathbf{v}} = \frac{\mathbf{v}}{\sqrt{1 - v^2}} - \mathbf{a} = \gamma \mathbf{v} - \mathbf{a} = \mathbf{p} - \mathbf{a} \tag{9.8.19}$$

Using the Canonical momentum, the Hamiltonian of an electron is given:

$$H(\mathbf{r}, \mathbf{P}^c, t) = \mathbf{P}^c \cdot \mathbf{v} - L(\mathbf{r}, \mathbf{v}, t) \tag{9.8.20}$$

The Hamiltonian can be rewritten as

$$H = \mathbf{p} \cdot \mathbf{v} + \frac{1}{\gamma} = \sqrt{1 + (\mathbf{P}^c + \mathbf{a})^2} = \sqrt{1 + \mathbf{p}^2} = \gamma \tag{9.8.21}$$

It is clear that Hamiltonian represents the total energy of the electron.

Hamilton equation of motion is given as

$$\begin{aligned} \frac{d\mathbf{r}}{dt} &= \frac{\partial H}{\partial \mathbf{P}^c} \\ \frac{d\mathbf{P}^c}{dt} &= -\frac{\partial H}{\partial \mathbf{r}} \end{aligned} \tag{9.8.22}$$

By use of the relation (9.8.22), the time evolution of the Hamiltonian H is derived to be

$$\frac{dH}{dt} = -\frac{\partial L}{\partial t} = \mathbf{v} \cdot \frac{\partial \mathbf{a}}{\partial t} \tag{9.8.23}$$

It is well-known that in the case where Lagrangian does not explicitly depend on the time, the energy is conserved. In the present case, the laser fields $\mathbf{a}$ is a function of time, and the total energy changes with time.

It is easy to derive (8.1.7) and (8.1.8) and (8.1.9) without E_x from (9.8.22) and (9.8.23). Let us assume a weak perturbation case:

9.8 Analytical Mechanics of Electron Motions

$$|\mathbf{a}_1| \ll |\mathbf{a}_0| \tag{9.8.24}$$

Then, Hamiltonian can be expanded with a perturbation:

$$H = H_0 + H_1 \tag{9.8.25}$$

where

$$H_0 = \sqrt{1 + p_{x0}^2 + \left(P_{y0}^c + a_0\right)^2} = \sqrt{1 + p_{x0}^2 + p_{y0}^2} \tag{9.8.26}$$

where the suffix "0" means the solutions without the perturbation and

$$H_1 = \left(P_{y0}^c + a_0\right) a_1 = p_{y0} a_1 \tag{9.8.27}$$

the time evolution of Hamiltonian is

$$\frac{dH}{dt} = \frac{dH_0}{dt} + \frac{dH_1}{dt} \tag{9.8.28}$$

From (9.8.9), we obtain

$$\frac{dH_0}{dt} = v_{y0} \frac{\partial a_0}{\partial t} = \frac{a_0^2}{2} \sin(2\xi_0) \tag{9.8.29}$$

Inserting 0th solution (9.8.3) to (9.8.27), the perturbed Hamiltonian is

$$H_1 = p_{y0} a_1 = a_0 a_1 \sin\xi_0 \sin\xi_1 = \frac{a_0 a_1}{2} [\sin(\xi_0 - \xi_1) + \sin(\xi_0 + \xi_1)] \tag{9.8.30}$$

9.8.5 Adiabatic Approximation

In order to derive the equation to the perturbation in Hamilton equation, consider that the perturbation by the second beam abruptly changes the action variable α in (9.8.15). Our ansatz is that from the analogy of (9.8.8), J changes due to any non-adiabatic force and the energy also change. We assume the energy changes by non-adiabatic force by the second beam though the change of α in (9.8.15).

We assume α changes due to the non-adiabatic interaction with the second beam and also assume that (8.1.20) is satisfied before and after the non-adiabatic interaction (change of the value α):

$$H = \gamma = p_x + \alpha \qquad (9.8.31)$$

In addition, the solution of p_x is also given in the form (8.2.7). Taking time derivative of (9.8.31), we obtain the equation including time derivative of α:

$$\frac{dH}{dt} = \frac{p_y^2}{2\alpha} + \frac{1}{2}\left(p_y^2 + \alpha^2\right)\frac{d}{dt}\left(\frac{1}{\alpha}\right) \qquad (9.8.32)$$

Since the second beam perturbation is assumed to affect only to the time variation of the value α, the first term of RHS in (9.8.32) can be regarded to cancel with H_0 contribution. Then, we obtain the following equation to the value of α:

$$\frac{d}{dt}\left(\frac{1}{\alpha}\right) = \frac{2}{\left(p_y^2 + \alpha^2\right)}\frac{d}{dt}H_1 \qquad (9.8.33)$$

As we see in Fig. 9.18, the value α struggles in the early time, while once $1/\alpha$ increases as seen in Fig. 9.18b, it continues to increase in time. This means an effective energy deposition to electrons continues.

For the case of the counter-propagating laser with same frequency and wavelength studied in [5], more important term in (9.8.30) is the second term. Inserting (9.8.30) to (9.8.32), we obtain the following equation:

$$\frac{d}{dt}\left(\frac{1}{\alpha}\right) = \frac{p_x}{\gamma\left(p_y^2 + \alpha^2\right)} a_1 a_0 \{\cos\left[2x(t) + \Delta\varphi_+\right] - \cos\left[2\omega t + \Delta\varphi_-\right]\} \qquad (9.8.34)$$

where the following relation is used:

$$\frac{dx(t)}{dt} = v_x = \frac{p_x}{\gamma} \qquad (9.8.35)$$

It is clear that the coefficient of (9.8.34) becomes large when the energy γ approaches unity as seen in Fig. 9.19. Then, the electron velocity is also slow enough, so that the first cosine term changes slowly. The change is of course not always the decrease of the value α so as to increase the oscillation energy, while statistical diffusion of α can be expected. It is noted that the oscillation after the abrupt change of α is mainly due to 2ω oscillation by the second term in (9.8.34).

9.8.6 Further Discussion

In the present analysis, there is no threshold for the stochasticity of the electron motion. In the pioneer work by Mendoca [28], the author developed a perturbation theory and studied the criteria for chaos by stating with super-Hamiltonian of an electron. In this case, the resonance condition gives the threshold of the Chirikov resonance-overlap criterion for the onset of chaotic motion in deterministic Hamiltonian systems. The theory was extended by Rax to relating to the relativistic laser electron interaction in the configuration of the counter-propagating two laser problem [29]. In the papers, Hamilton-Jacobi equation is derived to solve the perturbation equation. However, it is difficult to explain these methods without complicated mathematics.

References

1. J. Meyer-ter-Vehn, Z.M. Sheng, Phys. Plasmas **6**, 641 (1999)
2. S.C. Wilks, W.L. Kruer, IEEE J. Quant. Elec **33**, 1953 (1997)
3. T. Nakamura et al., Phys. Plasmas **9**, 1801 (2002)
4. M. Tanimoto et al., Phys. Rev. E **68**, 026401 (2003)
5. Z.M. Sheng et al., Phys. Rev. Lett. **88**, 055004 (2002)
6. S. Rassou, A. Bourdier, M. Drouin, Phys. Plasma **21**, 083101 (2014)
7. C. Tsallis, J. Stat. Phys. **52**, 479 (1988)
8. J. Klafter, M. F. Sclesinger, and G. Zumofen, Physics Today, Feb. p.33 (1996)
9. B. Mandelbrot, *The Fractal Geometry of Nature*, (Freeman, 1982)
10. A.V. Milovanov, J.J. Rasmussen, Phys. Lett. A **378**, 1492 (2014)
11. S.V. Bulanov et al., Aust. J. Plant Physiol. **83**, 905830202 (2017)
12. E. Barkai, R. Metzler, J. Klafter, Phys. Rev. E **61**, 132 (2000)
13. J. Anderson et al., Phys. Plasmas **21**, 122109 (2014)
14. C. Tsallis et al., Phys. Rev. Lett. **75**, 3589 (1995)
15. A. Bourdier, M. Drouin, Laser Part. Beams **27**, 545 (2009)
16. Y. Kuramitsu et al., Phys. Plasmas **18**, 010701 (2011)
17. S. Kojima et al., Commun. Phys. **2**, 99 (2019)
18. E. Fermi, Phys. Rev. 75, 1169 (1949); Astrophys. J. 119, 1. (1954)
19. A. R. Bell, Mon. Not. R. Astro. Soc. 182, 147 (1978); K. M. Schure et al., Space Sci. Rev. 173, 491 (2012)
20. P. Chen et al., Phys. Rev. Lett. **89**, 161101 (2002)
21. M. Hoshino, Astrophys. J. **672**, 940 (2008)
22. M. Iwamoto et al., Astrophys. J. **840**, 52 (2017)
23. L. Sironi, A.Y. Spitkovsky, J. Arons, Astrophys. J. **771**, 54 (2013)
24. S. Moradi et al., Physical Rev. Research **2**, 013027 (2020)
25. S. Bulanov et al., Phys. Plasmas **22**, 063108 (2015)
26. U. Tilnakli, E.P. Borges, Sci. Rep. **6**, 23644 (2016)
27. A.J. Lichtenberg, M.A. Liberman, *Regular and Stochastic Motion* (Springer, 1983). Chapter 3
28. T. Mendoca, Phys. Rev. A, 3592 (1983)
29. J.M. Rax, Phys. Fluids **B-4**, 3962 (1992)

Appendices

Appendix-1: Rutherford Scatterings

It is essential in studying plasma physics to know the Coulomb interaction between charged particles. The fundamental of the problem is well-known as **Rutherford scattering**, and its mathematics is well explained in a textbook on classical mechanics. Historically, it was studied by Rutherford to identify the existence of nuclear charge at the center of atom. He demonstrated this fact from the angle dependence of alpha particle scattering, when the particles are impinged to a thin gold foil. The Coulomb repulsive force scatters the injected alpha particles.

Assuming the Coulomb potential by a central charge is spherically symmetric and given to be U(r), we obtain the equation of motion of a scattered particle in the spherical coordinate:

$$\frac{1}{2}m\left(\dot{r}^2 + r^2\dot{\theta}^2\right) + U(r) = \frac{1}{2}mv_0^2, \tag{A1.1}$$

where m, r, and θ are the particle mass, radial position, and angle coordinate, respectively. The superscript dot means the time derivative. In (A1.1), the v_0 is the initial velocity of the particle being scattering. For simplicity, we assumed that the target charge is fixed, and only the motion of the scattered charge is studied. This is a good approximation in the case when an electron is scattered by an ion in plasma. It is clear that the angular momentum is conserved, namely,

$$r^2\dot{\theta} = bv_0 = \text{const.}, \tag{A1.2}$$

where b is a constant and is called **impact parameter.** Then, the variables are changed as:

Fig. A.1 Schematics of electron Coulomb collision by an ion at rest

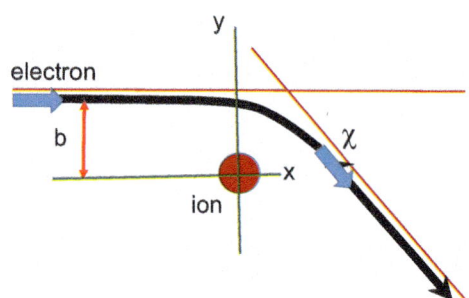

$$\frac{dr}{dt} = \frac{dr}{d\theta}\frac{d\theta}{dt} \qquad (A1.3)$$

$$d\theta = \frac{b}{r^2}\left[1 - \frac{b^2}{r^2} - \frac{U(r)}{E_0}\right]^{-\frac{1}{2}} dr, \qquad (A1.4)$$

where E_0 is the kinetic energy of the particle before scattered. The nearest radius of the scattered particle to the central change r_m is obtained by setting $d\theta = 0$ in (A1.4):

$$r_m = b\left[1 - \frac{U(r_m)}{E_0}\right]^{-\frac{1}{2}} \qquad (A1.5)$$

The scattered angle χ is obtained to be the following relation after integration of (A1.4):

$$\chi(b, E_0) = \pi - 2\int_{r_m}^{\infty} \frac{b}{r^2}\left[1 - \frac{b^2}{r^2} - \frac{U(r)}{E_0}\right]^{-\frac{1}{2}} dr \qquad (A1.6)$$

The picture of such Coulomb scattering of electron by ion is plotted in Fig. A.1.

Let us assume that the charge of the heavy particle is Q and the light one is q, then the potential is given to be:

$$U(r) = \frac{1}{4\pi\varepsilon_0}\frac{qQ}{r} \qquad (A1.7)$$

The nearest distance is calculated to be:

$$r_m = \frac{b^2}{-b_0 + (b_0^2 + b^2)^{\frac{1}{2}}}, \qquad (A1.8)$$

where

Appendices

$$b_0 = \frac{1}{8\pi\varepsilon_0} \frac{qQ}{E_0} \tag{A1.9}$$

This impact parameter corresponds to the initial value b for the case of 90 degree scattering.

Integrating (A2.6) we obtain the relation:

$$\chi(b, E_0) = 2 \sin^{-1}\left\{\frac{b_0}{(b_0^2 + b^2)^{1/2}}\right\} \tag{A1.10}$$

$$\tan\left(\frac{\chi}{2}\right) = \frac{b_0}{b} \tag{A1.11}$$

The differentiation of (A2.10) gives:

$$\left|\frac{db}{d\chi}\right| = \frac{b^2}{2b_0 \cos^2(\chi/2)} \tag{A1.12}$$

We introduce the **differential cross section** σ introduced by Rutherford:

$$\sigma(\chi)d\Omega = 2\pi b db \tag{A1.13}$$

$$d\Omega = 2\pi \sin\chi d\chi \tag{A1.14}$$

These relations lead the following famous formula of Rutherford scattering:

$$\sigma(\chi) = \frac{b_0^2}{4\sin^4(\chi/2)} = \frac{b_0^2}{(1 - \cos\chi)^2} \tag{A1.15}$$

The resultant cross section depends only on charge through b_0^2 and doesn't depend on the signs of the charges.

Finally it should be noted that the Rutherford formula of (A1.15) is exactly reproduced by the scattering cross section derived in quantum mechanical scheme.

Appendix-2: Adiabatic and Nonadiabatic

Periodic Motion

Consider the situation where an electron is in a periodic motion due to external electric field or magnetic field. A familiar example is cyclotron motion rotating along an external magnetic field. The equation of motion is given in the form:

$$\frac{d^2x}{dt^2} = -\omega_c^2 x, \tag{A2.1}$$

where it is assumed that the electron motion is in (x,y) plane, and magnetic field is in the z-direction. In (A2.1), ω_c is the electron cyclotron frequency given as:

$$\omega_c = \frac{eB_0}{m} \tag{A2.2}$$

The adiabatic motion is defined by an approximate solution of (A2.1) when the magnetic strength B_0 changes very slowly in time. If the following adiabatic condition is satisfied, it is possible to find a constant of motion in time:

$$\frac{1}{\omega_c}\frac{d\omega_c}{dt} \ll \omega_c \quad \Leftrightarrow \quad \left|\frac{1}{B_0}\frac{dB_0}{dt}\right| \ll \omega_c \tag{A2.3}$$

The constant is called adiabatic constant and given in the form in the present case:

$$J = \frac{K}{\omega_c} \tag{A2.4}$$

K is the time averaged kinetic energy over the cyclotron motion. This means with the increase of magnetic strength, it is possible to accelerate electrons in cyclotron motion. This concept is used to accelerate electrons by a compact device and is called **Betatron accelerator**. Note that keeping J constant in a magnetic bottle such as **mirror machine**, it is easy to find the motion of electrons in external magnetic field varying its strength in the z-direction. It is useful to note that the property that the cyclotron frequency is independent of the energy K, the idea of **cyclotron accelerator** was proposed by Lawrence in the 1930s in the early stage of nuclear physics research.

The conservation of J is common to any case where the oscillation or periodic motion is given in (A2.1). For the case that a pendulum of the string length l changes in time, using the relation of frequency of the pendulum:

$$\omega_p - \sqrt{g/l(t)} \tag{A2.5}$$

Periodic motion is not necessarily the harmonic oscillation given by (A2.1). The readers are familiar of the adiabatic cooling or heating in thermodynamics. This also comes from the conservation relation of J in three-dimensional gas particle collision with the wall in a box. If the compression velocity is much slower than the average velocity of gas particles, thermal velocity, J is conserved in three-directional motion. Using this mechanical view, it is easy to derive the term of the pressure work (PdV) in the first law of the thermodynamics:

Appendices

$$d\varepsilon = -Pdv + \Delta Q \qquad (A2.6)$$

Derivation of Adiabatic Constants

In the case with time variation in ω_c, the solution can be assumed in the form:

$$x = A(t)\, exp\left(i \int \omega_c dt\right) \qquad (A2.7)$$

Then, (A2.1) reduces the following equation to A(t):

$$\frac{d^2 A}{dt^2} + i\left(2\omega_c \frac{dA}{dt} + \frac{d\omega_c}{dt} A\right) = 0 \qquad (A2.8)$$

Since (A2.3) is satisfied, it looks that the all term left in (A2.8) are much smaller than the fundamental oscillation terms. However, try to find the solution of (A2.8) after neglecting the smallest term, the first one of the second derivative of A. Then, (A2.8) can be easily solved to obtain:

$$A(t) = \frac{A_0}{\sqrt{\omega_c}}, \qquad (A2.9)$$

where A_0 is a constant and the solution is found to be:

$$x = \frac{A_0}{\sqrt{\omega_c}} \exp\left(i \int \omega_c dt\right) \qquad (A2.10)$$

This is the solution of adiabatic motion.

It is useful to relate (A2.10) to the **Wentzel–Kramers–Brillouin (WKB)** approximation in solving Schrodinger equation. In the case of stationary solution of Schrodinger equation shown in Fig. 2.1, WKB method is an approximate way to solve the tunneling effect. If the time derivative is replaced with space derivative, (A2.1) corresponds to a normalized Schrodinger equation. If the wavenumber of the wave function is large enough compared to the variation of the potential structure, (A2.10) can be the solution to this problem. Show the mathematics of this relation. Replace t and x in (A2.1) with space z and wave function of $\psi(z)$:

$$t \to z$$
$$x(t) \to \psi(z) \quad (A2.11)$$
$$\omega_c^2 \to \frac{2m}{\hbar^2}[E - U(z)]$$

Then, depending upon the local sign of E−U, the following two solutions to be considered differently:

$$\frac{2m}{\hbar^2}[E - U(z)] = \begin{cases} k(z)^2 & (E - U > 0) \\ -K(z)^2 & (E - U < 0) \end{cases}, \quad (A2.12)$$

where E−U < 0 is the tunneling region in Fig. 2.1, also corresponding to the U_2 region in Fig. 2.2.

Following the mathematics which derived (A2.10), the following relation is obtained for stationary Schrodinger equations:

$$\psi(z) \propto \begin{cases} \frac{1}{\sqrt{k}} \exp\left(i \int k \, dz\right) \\ \exp\left(-\int K \, dz\right) \end{cases} \quad (A2.13)$$

In the tunneling region, the wave function is exponentially decay as shown in (A2.13). This integral on the exponent increases varies fast to zero with the increase of strength of laser electric field. The factors on the exponent in (2.1.4) and (2.1.15) are obtained after integration of K(z).

Nonadiabatic Case

Nonadiabatic case is just the opposite one of (A2.3). If the magnetic field changes in time faster than the cyclotron oscillation period, J in (A2.4) doesn't conserve, and the second derivative in (A2.8) becomes dominant term. In the case where additional force works externally in (A2.1) for a short time τ satisfying the relation $\omega_c\tau << 1$, it can also regarded nonadiabatic force to periodic motion. It can be easily imaged in the case where a humper hits a pendulum. Even with the same frequency in (A2.1), the adiabatic constant J in (A2.3) changes due to the energy by the hummer. In the case of WKB approximation, it is clear that WKB approximation breaks at the interface of the potential in Fig. 2.2.

Appendix-3: PIC Simulation

It is well-known that from the begging of the invention of computers in the 1950s, plasma physics has been a pioneering field to advance the computation for studying complex phenomena in science. Plasma physics is many body interaction systems requiring high-speed computer for a variety of problems to check theory and to analyze and design experiments. **Particle-in-cell (PIC)** simulation is a direct modeling to solve charged particle motions in electric and magnetic fields. The early works before the 1980s are reviewed by a pioneer of PIC simulation method, John Dawson, in Ref. [1], where the numerical algorism is also well described.

Particles in Computer

The PIC, a computer simulation for the first principle, is to solve Maxwell Eqs. (1.3.1), (1.3.2), (1.3.3) and (1.3.4) by coupling with many of particle motions for electrons and ions governed by (1.3.8) and time integration of velocity to obtain particle positions:

$$\frac{d\mathbf{r}}{dt} = \mathbf{v} = \frac{\mathbf{p}}{m\gamma} \qquad (A3.1)$$

Such simulation method is called particle-in-cell (PIC) simulation, and the readers see many results using PIC simulation in this book. Especially, in Chap. 7 of relativistic laser and solid interaction, many PIC code results are used for analyzing most of experimental data.

It is of course difficult to solve (1.3.8) and (A3.1) for all particles in real system with huge number of electrons and ions. Since the compotation time is so large with real mass ratio m_i/m_e, it is usual to assume a reduced mass ratio, say $10 \sim 100$. In addition, a limited number of particles are used so that simulation has been done to clarify the core physics. Due to the rapid progress of computer capability as shown in Fig. 1.14, recently, PIC simulation is more realistic as the first principle simulation. However, one has to be careful about the PIC simulation result, because the accuracy depends on the numerical modeling and number of meshes and particles. Here a brief of the PIC simulation algorism and numerical conditions are explained so that even nonspecialist in PIC code can see the appropriateness of computational condition and resultant analysis.

At first, consider the time integration of (1.3.8) and (A3.1). For simplicity, equations are solved for a particle with charge q and mass m. In computer, the time is discretized with the time interval Δt. The time integration of (1.3.8) and (A3.1) is given at a given position **r** to be:

$$\frac{\mathbf{p}^{t+\Delta t/2} - \mathbf{p}^{t-\Delta t/2}}{\Delta t} = q\left[\mathbf{E}^t + \left(\frac{\mathbf{p}}{m\gamma}\right)^t \times \mathbf{B}^t\right] \quad (A3.2)$$

$$\frac{\mathbf{r}^{t+\Delta t} - \mathbf{r}^t}{\Delta t} = \left(\frac{\mathbf{p}}{m\gamma}\right)^{t+\Delta t/2} \quad (A3.3)$$

Note that the physical quantities on RHSs in (A3.2) and (A3.3) are located at the center of time integration in order to keep higher accuracy.

Maxwell Equations in Computer

Calculating the electric current density **j** as the sum of all particle currents as a function of space at time t + Δt by the use of (A3.2) and (A3.1), Maxwell Eqs. (1.3.1) and (1.3.2) have to be also time integrated as:

$$\frac{\mathbf{B}^{t+\Delta t} - \mathbf{B}^t}{\Delta t} = -\nabla \times \mathbf{E}^{t+\Delta t/2} \quad (A3.4)$$

$$\varepsilon_0 \frac{\mathbf{E}^{t+\Delta t} - \mathbf{E}^t}{\Delta t} = \frac{1}{\mu_0} \nabla \times \mathbf{B}^{t+\Delta t/2} - \mathbf{j}^{t+\Delta t/2} \quad (A3.5)$$

Inserting given **E** and **B** at time t in (A3.2), it is well-known that with this simple numerical method, the particle energy cannot conserve even without **E** field, because of the numerical error. Numerical algorithm called Buneman-Boris method [2] is widely used to avoid this numerical difficulty and keep the magnetic field gives only rotation of particle in momentum space. In Fig. A.2a, the grid and particles in 2D PIC are schematically shown. The **E** and **B** are defined at each grid point, while the

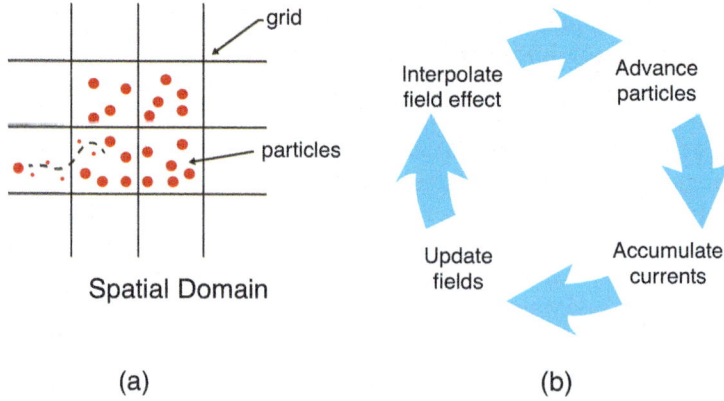

Fig. A.2 A brief of PIC simulation scheme and algorithm. PIC simulation grid and particles in 2D PIC case are schematically shown (**a**). The calculation flow in each time step for PIC code is shown (**b**)

position and momentum of the particles are given at each position. In solving (A3.2) and (A3.3), the E and B at each point of particle are interpolated with the values of the grid points nearby, while (A3.4) and (A3.5) are solved by summing up the nearby particle contribution near each grid. At each time step, the flow of calculation is given as shown in Fig. A.2b.

To describe how to model the giant particles in PIC code is out of the present scope, and the detail numerical methods are explained, i.e., in books [2]. It is noted that the PIC scheme describes fundamentally collisionless plasma phenomena.

Surprising Progress of Computing

One of the pioneering research with PIC code for laser plasma is carried out to study resonant absorption [3] as discussed in Chap. 3. The computational research of laser-plasma interaction with PIC code started to publish synchronized with ICF research paper by Nuckolls et al. [7, Chap. 1] in 1972. The PIC simulation has been done with the code developed to study electron Weibel instability in two-dimensional system. Due to the restriction of computer speed as shown in Fig. 1.14, the space grid in x and y was 50×50, and 10^5 giant particles were used for the calculation over 2000 times Δt [4]. Even under such restriction of computer speed, the physics are well obtained and compared to the corresponding theoretical results.

With the help of progress of computer capability, grand challenge of computing has also been done for plasma PIC simulation. It is reported that VPIC, a first-principles 3D electromagnetic charge-conserving relativistic kinetic particle-in-cell code, was used for Peta (10^{15}) flops computing at LANL to study laser-plasma interaction physics [5]. Note that the used numbers are 325 billion particles on $8256 \times 512 \times 512$ mesh, namely, about 0.4 trillion particles and 2 billion cells, 200 particles per cell.

With such a progress of computer speed, PIC code is widely used as tool to compare experimental results, especially in the case of ultra-short pulse relativistic laser-plasma interaction. Since the pulse duration is very short less than 1 ps, the full-time simulation is now possible with PIC code even in two or three dimensions. Contemporary PIC code algorithm for laser plasma with relativistic and ultra-short pulse is reviewed, for example, in [6, 7]. Reference [6] mainly focuses on the plasma-based particle acceleration, while [7] focuses on the extension to quantum electrodynamics (QED) in relativistic laser-plasma interaction. The Coulomb collision is also important in laser-plasma interaction in cold-dense plasma such as solid density. Recently, most of PIC code includes the collision term based on Coulomb binary collision [7]. Field ionization effect is also installed to PIC code as described in [7]. Now researchers try to include the atomic process such as ionization and recombination process in PIC code [8]. In addition, ultra-intense laser and electron interaction are very sensitive to field profile running near the speed of light, since the electrons highly accelerated by the field are also running near the speed of light. The interaction time in the same phase among waves and particles is essential in the formation of very high-energy electrons in the tail of electron distribution. So, ultra-high-order Maxwell equation solver has been developed by keeping extreme scalability for electromagnetic PIC simulations of plasma, for example, [9].

In laser fusion and HED physics studies, solid matter should be slowly compressed over the time of nanoseconds. In such non-relativistic laser plasma case described in Chaps. 2, 3 and 4, PIC simulation is still difficult to simulate the whole physics scenario over the long pulse time. However, PIC code is very useful to simulate core dynamics and obtain some theoretical model for physics elements in long pulse plasma. From the beginning such as resonant absorption study, PIC code has been used mainly for such purpose and committed a lot to clarify the core physics in new regime. The author, however, is afraid that through writing this book, he usually finds many papers where a certain experiment was done, and PIC simulation was also done to obtain the coincidence with the experimental data after fine-tuning of the PIC condition. This is recent tendency after a tremendous increase of computer speed and the ultra-short relativistic laser can be simulated for a whole time of laser pulse as readers see examples in Chap. 7. The author hopes young people doesn't forget that computer simulation is a tool to find the core physics in complex plasma they are challenging.

There are several open use PIC codes. For example, EPOCH code in [7] is widely used internationally not only for relativistic plasma physics but also nonlinear quantum electrodynamics (QED) physics such as positron productions, radiation damping, and so on expected to happen in ultra-intensity irradiation over 10^{20}W/cm^2.

References

1. J. Dawson, Phys. Plasmas **2**, 2189 (1995); Phys. Fluids **5**, 445 (1962)
2. R.W. Hockeny, J.W. Eastwood, *Computer Simulation Using Particles* (IOP publisher, 1988), p. 111; C.K. Birdsall, A.B. Langdon, *Plasma physics via computer simulation*, (McGraw hill, 1981/CSC Press, 2004)
3. J.P. Freidberg et al., Phys. Rev. Lett. **28**, 795 (1972)
4. R.L. Morse, C.W. Nielson, Phys. Fluids **12**, 2418 (1969)
5. K.J. Bowers et al., J. Physics Con. Ser. **180**, 012055 (2009). https://github.com/lanl/vpic
6. A. Pukov, in *Accelerator School in CERN* (2016). https://e-publishing.cern.ch/index.php/CYR/article/view/220
7. T.D. Arber et al., Plasma Phys. Control. Fusion **57**, 113001 (2015)
8. R. Royle et al., Phys. Rev. E **95**, 063203 (2017)
9. H. Vincenti, J.-L. Vay, Comput. Phys. Commun. **228**, 22 (2018)

Appendix 4: Tsallis Statistics at Maximum Entropy

In the Gibbs-Boltzmann statistics, the thermodynamically equilibrium state is determined by taking the maximum of the particles in a system. This is also given as the steady-state solution of Langevin Eq. (9.1.3) for the case the random force is a Gaussian noise. Let us derive the steady- state solution of the Langevin equation with Levy's jumps random force. In order to obtain generalized Einstein relation, particle mass and the frictional viscosity and the diffusion coefficient are explicitly included. Then, the equation of motion of a particle with a viscosity and external random force R(t) is given by Langevin equation:

$$\frac{dp}{dt} = -\frac{\nu}{m}p + R(t), \qquad (A.4.1)$$

where m and ν are a particle mass and the viscosity coefficient. The time average of the random force <R(t)> should be vanished.

It is well-known that starting with (A.4.1) and assuming R(t) is random function with the probability of Gaussian profile, (9.1.5) leads to the following equation to the distribution function f(p,t):

$$\frac{\partial}{\partial t}f = \frac{\nu}{m}\frac{\partial}{\partial p}(pf) + D\frac{\partial^2}{\partial p^2}f \qquad (A.4.2)$$

Let us extend Fokker-Planck Eq. (A.4.2) to FFPE as:

$$\frac{\partial}{\partial t}f = \frac{\nu}{m}\frac{\partial}{\partial p}(pf) + D\frac{\partial^\alpha}{\partial |p|^\alpha}f \qquad (A.4.3)$$

The property of this equation has been studied, and it is shown that (9.4.8) has meaningful only for the case [Ref. 11, Chap. 9]:

$$0 \leq \alpha \leq 2 \qquad (A.4.4)$$

Since the fractional operator is defined by Fourier transformation as (9.3.15), it is easy to modify (A.4.3) to the following equation to the Fourier component:

$$\frac{\partial}{\partial t}F^\alpha = -\frac{\nu}{m}k\frac{\partial}{\partial k}F^\alpha - D|k|^\alpha F^\alpha \qquad (A.4.5)$$

It is not straightforward to solve this partial differential equation analytically. The best way is to solve (A.4.5) numerically with finite difference for each k-component and to sum up all k-solutions at each time step.

The property of (A.4.5) is inferred from the properties of the two terms on RHS in (A.4.5). The first term is the friction mathematically indicating that F propagates to the negative direction in k space. Namely, the profile condenses near k = 0; consequently the profile becomes smooth in real p space. The second term is the decay term exponentially, and the decay time is short for large k, but its time is of course depends on the fractional index α.

It is easy to obtain the solution of (A.4.5) for the time derivative equal to zero (steady state). The steady-state distribution is given by integrating (A.4.5) in k-space:

$$F_{ss}^{\alpha}(k) = A\exp\left(-\frac{mD}{\nu\alpha}|k|^{\alpha}\right) \tag{A.4.6}$$

Carry out the inverse-Fourier transformation of (A.4.6):

$$f_{ss}^{\alpha}(p) = \frac{A}{2\pi}\int_{-\infty}^{\infty}\exp\left(-\frac{mD}{\nu\alpha}|k|^{\alpha} + ikp\right)dk \tag{A.4.7}$$

It is surprising that (A.4.7) is the same integration as (9.4.2), and the solution is given in the same form as (9.4.3). Then, the following relation is obtained [1]:

$$D = \frac{2^{\alpha-1}}{\Gamma(\alpha+1)}\left(\frac{T}{m}\right)^{\alpha/2}\nu, \tag{A.4.8}$$

where Γ is the Gamma function. This is called the **generalized Einstein relation**. In (A.4.8), the effective temperature T is a given parameter, and the β in (9.4.3) is also a function of T. In case of Maxwell distribution for $\alpha = 2$, the relation $\beta = 1/T$ is obtained.

Reference

1. E. Barkai, Phys. Rev. E **68**, 055104(R) (2003)

Index

A
Ablation pressure, 18
Absorption rate, 52
Adiabatic, 16, 369, 375
Adiabatic constant, 292, 363
Airy function, 85
Anomalous transport, 340
Antenna, 48
Autonomous system, 308

B
Barrier-suppression ionization, 37
Beam trapping, 225
Bessel function, 97, 190
Betatron accelerator, 376
Bremsstrahlung, 79
Broadband effect on instability, 161
Brownian motion, 331
Brunel heating, 124, 242

C
Canonical momentum, 170
Cauchy-Lorentz form, 340
Caviton, 135
Channel formation, 282
Chaos, 20, 309
Chirped pulse amplification (CPA), 8
Chrikov resonance overlap criteria, 323
Circular polarization, 188
Classical absorption, 17, 69, 72, 79, 306
Coherence, 44
Collective phenomenon, 104

Collective Thomson scattering (CTS), 156
Collisional absorption, 16
Collision frequency, 50
Collisionless absorption, 104
Collisionless shock, 354
Collision time, 50, 61
Compton scattering, 191
Conservation law, 260
Constant of motion, 180, 289
Contrast ratio, 245
Convective transport, 345
Cosmic ray, 354
Coulomb collision, 53
Coulomb logarithm, 58
Coupled oscillator model, 149
Coupling parameter, 69, 100
Cut-off density, 13, 66
Cut-off frequency, 66
Cyclotron accelerator, 376

D
de Broglie length, 60
Debye-Huckel equation, 56
Debye length, 15, 56
Debye shielding, 56
Debye sphere, 57
Debye wave number, 57
Decay-type instability, 154
Density bunching current, 213
Density profile modification, 144
Dephasing rate, 289
Dielectric constant, 67
Diffusion model of absorption, 73

D

Diffusive shock acceleration (DSA), 354
Dispersion relation, 66
Dispersive wave, 143
Driver model, 108
Drude model, 51
Dynamic pressure, 141

E

Eddington limit, 50
Eight-figure motion, 46, 186
Electrical resistivity, 51, 61
Electromagnetic wave, 43
Electron classical radius, 49
Electron conductivity, 50
Electron plasma frequency, 18, 52
Electron plasma wave, 104
Electrostatic fluctuation, 150
Electrostatic wave, 43
Energy relaxation time, 64
Envelope soliton, 135
Equation of state (EOS), 89

F

Fast-ignition, 302
Femtosecond laser, 317
Fermi-Dirac distribution, 95
Fermi energy, 29
Fermi temperature, 11, 89
Field ionization, 30
Filamentation instability, 136, 226, 259
Fokker-Planck equation, 22, 332, 342
Forced oscillation, 113
Form-layered target, 263
Fourier-Laplace component, 42
Fourier-Laplace transformation, 142
Fractal Fokker-Planck equation (FFPE), 344
Fractal index, 346
Fractal theory, 341
Frequency shift, 226
Fusion energy research, 6

G

Gaussian and Lorentzian, 349
Gaussian and Poissonian statistics, 343
Gekko XII, 7
Generalized coordinate, 170
Gibbs-Boltzmann, 340
Ginzburg curve, 109

H

Hamiltonian equation, 367
Helmholtz equation, 84
High-energy-density physics (HEDP), 17
Higher harmonic generation (HHG), 194, 209, 236
Hole boring, 225, 252
Hot-electron, 123, 316, 347

I

Ideal plasma, 59
Impact parameter, 373
Inertial confinement fusion (ICF), 5
Inverse Bremsstrahlung, 17, 80
Inverse-Compton scattering, 191
Inverse Landau damping, 68
Ion-acoustic wave, 139
Ionosphere, 67
Ion sound wave, 141
Ion sphere radius, 89
ITER, 12

J

$J \times B$ force, 231

K

Keldysh parameter, 36
Klein-Gordon equation, 57

L

Laboratory astrophysics, 6
Lagrange coordinate, 124
Landau cut, 59, 99
Landau damping, 68, 107, 143
Langevin equation, 331
Large amplitude plasma wave, 118
Larmor emission, 49
Laser direct acceleration (LDA), 292, 300
Laser implosion, 5
Laser intensity, 14
Laser strength parameter, 178
Laser synchrotron source (LSS), 194
LASNEX, 82
Levy's flight, 340
Linearized equation, 105
Linear mode conversion, 115
Linear polarization, 187

Index   387

Local thermodynamic equilibrium (LTE), 11
Longitudinal wave, 42
Loop-injected direct acceleration (LIDA), 295
Lorentz gauge, 169
Lorentzian distribution, 345
Lorentz transformation, 172
Lyapunov exponent, 327

M
Mach number, 141
Magnetic field generation, 279
Magnetic fusion, 12
Maiman, T.H., 4
Matching condition, 146
Maxwell equation, 13, 40
Maxwell stress tensor, 168
Mean free path, 17, 331
Micro-tube plasma lenses, 277
Mirror machine, 376
Modulation instability, 136
Momentum density, 168
Motional electric field, 176
Multi-dimensional effect, 320
Multiphoton absorption, 32, 95
Multiphoton ionization, 32
Multiple Compton scattering, 199

N
Nano-pillar array target, 275
National Ignition Facility (NIF), 7
Nonadiabatic, 375
Nonlinear classical absorption, 96
Nonlinear Compton scattering (NCS), 199
Nonlinear inverse Bremsstrahlung absorption, 96
Nonlinear QED, 200
Nonlinear radiation scattering, 188
Nonlinear saturation, 160
Nonlinear Schrodinger equation, 133
Nonlinear Thomson scattering, 199
Nonlocal transport, 340
Non-LTE, 11
Normalized laser strength, 15

O
Oil shock, 6
Over-threshold ionization, 34

P
Pair distribution function, 96
Parametric instability, 146, 220

Pedestal, 239
Pendulum resonance, 112
Perturbation method, 367
Photon force, 203
PIC simulation, 23, 294, 301, 379
Plasma dielectric constant, 68
Plasma frequency, 18, 52
Plasma mirror, 82
Plasma skin depth, 138, 223
Plasma wake field, 205
Ponderomotive (PM) force, 19, 131, 181, 217
Ponderomotive potential, 132, 218
Ponderomotive scaling, 234, 266
Power law, 338
Poynting vector, 13
p-polarized laser, 84
Principal integral, 112
Proper time, 176, 288
Pump depletion, 112

Q
Quasi-linear diffusion, 78, 315
Quotidian EOS (QEOS), 91

R
Random force, 305
Random walk, 342
Ranking-Hugoniot relation, 261
Rayleigh-Taylor instability, 257
Reactive media, 52
Re-circulation, 248
Relativistic beaming, 175, 197
Relativistic Doppler shift, 177, 197
Relativistic laser, 20
Relativistic mass correction, 206
Relativistic oscillating mirror (ROM), 235
Relativistic skin depth, 229
Relativistic transparency, 207
Relativistic vacuum heating, 237, 242
Resistive media, 52
Resonance absorption, 107
Return current, 243
Run-away electron, 62
Rutherford scattering, 54, 373

S
Self-focusing, 137, 221
Self-similar jump, 341
Separatrix (saddle), 308
Sheath potential effect, 319
Singular point, 111

Skin depth, 129, 229
Soliton, 135, 143
Sound wave, 141
Special relativity, 167
Spitzer's collision frequency, 98
s-polarized laser, 84
SRS saturation physics, 157
Static structure factor, 96
Stimulated Brillouin scattering (SBS), 153
Stimulated Raman scattering (SRS), 153, 220
Stochastic diffusion, 339
Stochastic heating, 244, 291, 306, 334
Stochasticity and chaos, 22, 331
Strength parameter, 46
Structured target, 273
Supercomputer, 24, 381
Super-ponderomotive, 272, 295

T
Target normal sheath acceleration, 266
Thomson scattering, 48, 189
Threshold, 38, 152
Time-dependent Schrodinger equation, 32
Tokamak transport, 358
Transverse wave, 42
Tsallis statistics, 340, 383
Tunneling effect, 29, 85
Tunneling ionization, 35

Turbulent diffusion, 340
Turning point, 85

U
Ultra-short pulse, 81
Uncertainty principle, 33

V
Vacuum heating, 123
Van der Waals force, 53
v×B force, 212, 291
Vlasov-Fokker-Planck equation, 334
Volkov solution, 35, 200

W
Warm dense matter (WDM), 2, 81
Wave-breaking, 119, 216
Wavelength, 43
Wavenumber, 42
White dwarf, 11
WKB approximation, 16, 35, 377
Working function, 29

Y
Yukawa potential, 57

The manufacturer's authorised representative in the EU is Springer Nature Customer Service Centre GmbH, Europaplatz 3, 69115 Heidelberg, Germany. If you have any concerns regarding our products, please contact ProductSafety@springernature.com

Printed and bound by CPI Group (UK) Ltd, Croydon, CR0 4YY

27/03/2026

02079485-0001